Matrix Diagonal Stability in
Systems and Computation

Eugenius Kaszkurewicz
Amit Bhaya

Matrix Diagonal Stability in Systems and Computation

Birkhäuser
Boston • Basel • Berlin

Eugenius Kaszkurewicz
Amit Bhaya
Department of Electrical Engineering
Federal University of Rio de Janeiro
COPPE/UFRJ
Rio de Janeiro 21945-970, Brazil

Library of Congress Cataloging-in-Publication Data
Kaszkurewicz, Eugenius.
    Matrix diagonal stability in systems and computation / Eugenius
Kaszkurewicz, Amit Bhaya.
      p. cm.
    Includes bibliographical references and index.
    ISBN 0-8176-4088-6 (hardcover:alk. paper)
    1. Differentiable dynamical systems.   2. Matrices.   3. Stability.
    I. Bhaya, Amit. II. Title.
    QA614.8.K378   1999
    515'.352—dc21   99-23774
                                                     CIP

AMS Subject Classifications: 93C, 93D, 15A, 34D

ISBN 0-8176-4088-6
ISBN 3-7643-4088-6   SPIN 10682901

Typeset by the authors in LaTeX.
Printed and bound by Maple-Vail Book Manufacturing Group, York, PA.
Printed in the United States of America.

9 8 7 6 5 4 3 2 1

This book is dedicated to our families.

Contents

Preface

This monograph presents a collection of results, observations, and examples related to dynamical systems described by linear and nonlinear ordinary differential and difference equations. In particular, dynamical systems that are susceptible to analysis by the Liapunov approach are considered. The naive observation that certain "diagonal-type" Liapunov functions are ubiquitous in the literature attracted the attention of the authors and led to some natural questions. Why does this happen so often? What are the special virtues of these functions in this context? Do they occur so frequently merely because they belong to the simplest class of Liapunov functions and are thus more convenient, or are there any more specific reasons?

This monograph constitutes the authors' synthesis of the work on this subject that has been jointly developed by them, among others, producing and compiling results, properties, and examples for many years, aiming to answer these questions and also to formalize some of the folklore or "culture" that has grown around diagonal stability and diagonal-type Liapunov functions.

A natural answer to these questions would be that the use of diagonal-type Liapunov functions is frequent because of their simplicity within the class of all possible Liapunov functions. This monograph shows that, although this obvious interpretation is often adequate, there are many instances in which the Liapunov approach is best taken advantage of using diagonal-type Liapunov functions. In fact, they yield necessary and sufficient stability conditions for some classes of nonlinear dynamical systems. In other words, in many cases a diagonal-type function represents "the farthest one can go" with the Liapunov approach and has the added virtue of simplicity.

Strongly related to these diagonal-type Liapunov functions are several classes of matrices that, in most cases, describe the interconnection structure of associated dynamical systems. Chapter 2 is devoted to a discussion and presentation of the basic results for these classes of matrices; the most important being the so-called class of diagonally stable matrices. It is shown at different places in the book that in many cases the necessary and sufficient conditions for stability of nonlinear systems are, surprisingly, necessary and sufficient conditions for diagonal stability of a certain matrix associated to the nonlinear system.

There is also a strong correlation between these classes of matrices and the robustness of the dynamical systems that they are associated with. The results in Chapter 3 and the examples presented, in Chapters 4, 5, and 6, that range from neural networks and computation to passive circuits and mathematical ecology confirm this correlation.

Thus, loosely speaking, the unifying theme of this monograph is the presence and role of "diagonal stability" and the associated diagonal-type Liapunov functions in various stability aspects of certain, fairly widespread, classes of dynamical systems. Of course, there is no claim to completeness, either in terms of theoretical results on matrix diagonal and D-stability or in terms of applications; however, there is an extensive bibliography of over three hundred items. Reviewers and colleagues have often pointed out that the terms diagonal and D-stability are somewhat imprecise. However, a search made in any one of the science and technology databases, such as INSPEC or ISI's Web of Science, reveals hundreds of papers that use these terms. Thus the authors feel that it is justifiable to bow to tradition and continue to use these "traditional" terms in preference to the alternatives proposed in the literature (e.g., Volterra–Liapunov stability, Arrow–McManus stability, D^+L-**stab**, $\mathcal{D}^+$-**stab**, etc.).

Many applications of diagonal and D-stability are commented on in the sections entitled Notes and References that end every chapter and can be regarded as pointers to most of what is not touched upon in the book. These Notes also indicate and discuss the sources consulted; some others that are closely related, but have not been embedded in a deeper discussion in the text, are also included. It should also be pointed out that some long or uninformative proofs have been omitted in the interests of readability. Once again, the notes at the end of each chapter indicate where these may be found.

It is the authors' belief that the target public at which this monograph is aimed consists of graduate students and researchers in the fields of control, stability of dynamical systems, and convergence of algorithms, and that they would benefit from the contents of this book. Readers with interdisciplinary interests will also benefit from the wide range of topics from different disciplines that are included in the examples treated. The results presented are not all new, although almost all have been derived within the last two decades. The novelty of this book resides mainly in the unifying perspective provided by the matrix stability formulation of the results—in particular, matrix diagonal stability and its variants. Evidence that there is still something to be gained from this vantage point is given by the fact that interesting new results continue to crop up, as has been the case recently, for example, with neural networks and variable structure systems.

Familiarity with linear algebra and matrix theory, as well as difference and differential equations, is the mathematical background expected from the reader of this book. A prior knowledge of systems and control theory, including Liapunov stability theory, is also expected, at least in sufficient measure to provide motivation for the problems studied. Fortunately, there are many excellent books that cover this background—more details are given in the Notes and References to Chapter 1. For the reader who has a mathematical background but lacks a control background, the authors recommend a quick look at the introductory Section 1.1, which gives an overview of the book, with an attempt to explain control-related jargon to the uninitiated reader, as well as the appendices to Chapters 2 and 3, which review Liapunov and Stein equations and stability theory, respectively.

A word on notation is in order here: the Halmos symbol ■ indicates, as usual, the end of a proof of some mathematical assertion (lemma, proposition, theorem, etc.), whereas an "empty Halmos" □ indicates the end of the statement of some mathematical assertion that will not be proved explicitly in the book, either because its proof is implicit in the preceding discussion or because it will take the reader too far afield. In the latter case, "chapter and verse" citation of a reference where the proof may be found is given. Numbering of items as well as other notation is standard enough not to merit special mention here.

The authors would like to acknowledge the persons, agencies and institutions that directly and indirectly contributed to the realization of this monograph. Professor Šiljak for his interaction with both authors, principally the first author during a sabbatical at Santa Clara University; Professor Liu Hsu for having initiated, together with EK, the early efforts in this area; Professors Biswa Nath Datta and Carlos S. Kubrusly for having enthusiastically encouraged the project to write this book from the earliest days; Professors Stephen Boyd and Shankar Bhattacharyya as well as the anonymous reviewers for having made many useful remarks on draft versions; the Brazilian Ministry of Education, the government agencies CAPES, CNPq, FINEP, and, in particular, the PRONEX program of the Ministry of Science and Technology for having supported the research of both authors over the years; the Graduate School of Engineering (COPPE) of the Federal University of Rio de Janeiro (UFRJ) for having provided the necessary infrastructure and support for the authors' research over the years in which this book was written, Mara Prata for having transformed the Linear B in which parts of the manuscript were written to a

LATEX version and Professor Ramon Romankevicius Costa for help with the figures. AB would like to thank his parents in India for a debt that, as he realizes each day, is too great even to begin naming, to his sisters for unflagging moral support, and to his family in Rio for having stoically borne irritability, absences, and all the other charming habits of a first-time author; special thanks to Felipe who never lost faith in "zibuk" and to a chaotic little attractor called Asmi for thoroughly enjoyable delays, and to Lucia and Barbara for keeping the pressure up!

Rio de Janeiro, Brazil   Eugenius Kaszkurewicz
October 18, 1999   Amit Bhaya

1
Diagonally Stable Structures in Systems and Computation

This introductory chapter is devoted to examples that originate from different applications and that illustrate the way in which some special classes of dynamical systems dovetail with the concepts of matrix diagonal stability and the associated diagonal-type Liapunov functions.

1.1 Introduction

The class of diagonal matrices has many pleasant properties and, conversely, if a class of matrices is required to have many such properties, then, roughly speaking, it must be the diagonal class.

The objective of this book is to present classes of nonlinear dynamical systems that possess a so-called "diagonally stable structure" that is privileged in an analogous manner. Furthermore, it is also shown, by means of various examples, that this class of dynamical systems occurs in many applications in circuits and systems, in computation using asynchronous iterative methods, in control, and so on.

Terminology used throughout the book is now introduced. The term *dynamical system* refers to a set of difference or differential equations that determine the evolution of a vector in $\mathbb{R}^n$ that is also referred to as the *state vector* or simply the *state*. The evolution occurs in the *state-space* $\mathbb{R}^n$ in *discrete-time* in the case of a difference equation and in *continuous-time* in the case of a differential equation, and the path that describes this evolution in the state-space is referred to as a *(state) trajectory* or *solution*.

The term *linear dynamical system* refers to

$$x_i^+ = \sum_{j=1}^{n} a_{ij} x_j, \quad i = 1, \ldots, n \tag{1.1}$$

where the *state vector* $x = (x_1, \ldots, x_n)^T \in \mathbb{R}^n$, the *system matrix* $A = (a_{ij}) \in \mathbb{R}^{n \times n}$, and the scalar x_i^+ denotes dx_i/dt in the continuous-time case, or $x_i(t+1)$ in the discrete-time case.

A linear dynamical system with an exogenous variable, referred to as a *control* or *input vector*, is written as follows.

$$x_i^+ = \sum_{j=1}^{n} a_{ij} x_j + \sum_{k=1}^{m} b_{ik} u_k, \quad i = 1, \ldots, n \tag{1.2}$$

The adjective control used for the vector u arises from the fact that it can be chosen so as to control the evolution in time of the state vector $x(t)$, i.e., the trajectory of the system. In fact, in *linear state feedback control*, the input vector u is chosen as a linear function of the state, $u = Kx$, so that the controlled system becomes $x^+ = (A + BK)x$, which is again in the form of system (1.1), although the system matrix has changed from A to $A + BK$. The matrix $A + BK$ is referred to as the *closed-loop* system matrix. Such a choice of a *feedback matrix* K is known as *state feedback stabilization* when stability of the system is the desired property.

To explain the term *diagonally stable*, consider the matrix equations:

$$A^T P + PA = -Q \tag{1.3}$$
$$A^T PA - P = -Q \tag{1.4}$$

The equation (1.3) (respectively, (1.4)) is referred to as the Liapunov (respectively Stein) equation in A. If there exist positive definite matrices P and Q satisfying the equation (1.3) (respectively (1.4)), then the matrix A is said to be *continuous-time* or *Hurwitz stable* (respectively *discrete-time* or *Schur stable*). Another well known characterization of stability is in terms of the eigenvalues of the matrix A: If all eigenvalues of A lie in the open left half complex plane (i.e., all have negative real parts), then A is Hurwitz stable; if all eigenvalues lie within the open unit disk in the complex plane, then A is Schur stable. The quadratic form $x^T Px$ is referred to as the *quadratic Liapunov function* associated to system (1.1) and traditionally denoted $V(x)$. The stability being referred to is asymptotic stability: Namely, if the initial condition is nonzero, the resulting trajectory of (1.1) goes to the zero solution asymptotically. If the positive definite solution P is, in addition, *diagonal*, then, in the continuous-time case, the matrix A is referred to as *Hurwitz diagonally stable* and the quadratic form $V(x) = x^T Px$ as the associated diagonal quadratic Liapunov function.

Such diagonally stable matrices and the associated diagonal quadratic Liapunov functions crop up in the stability analysis of a wide variety of systems: variable structure systems, digital filters, Lotka–Volterra systems in mathematical ecology—to name a few examples. This is because, in these examples, the underlying mathematical models can all be considered to be special cases of the following set of n nonlinear differential or difference equations:

$$x_i^+ = \sum_{j=1}^{n} a_{ij} \Phi_{ij}(x,t), \quad i = 1, \ldots, n \qquad (1.5)$$

where for all i and j, $\Phi_{ij} : \mathbb{R}^n \times \mathbb{T} \to \mathbb{R}^n$ satisfy certain so-called sector conditions, and, as before, x_i^+ denotes dx_i/dt in the continuous-time case, when $\mathbb{T} = \mathbb{R}^+$; or $x_i(t+1)$ in the discrete-time case, when $\mathbb{T} = \mathbb{N}$.

As first observed by Persidskii [Per69], for a special case of (1.5), such models admit a class of *diagonal-type Liapunov functions*. These are functions that are, in the simplest case, quadratic Liapunov functions of the the type $x^T P x$, where x is a vector and P a diagonal matrix, with all diagonal entries positive, referred to briefly as a positive diagonal matrix. More generally, these are functions that are conducive to the use of matrix diagonal stability in proving that they decrease along the trajectories of the dynamical system in question. A historical note appropriate here is that the first use of a diagonal-type Liapunov function seems to have been made by Volterra in his classic studies of fish populations in the Adriatic [Vol31].

In the discrete-time case, for the class of diagonal quadratic Liapunov functions, a certain converse result holds for a class of linear time-varying systems whose trajectories are the same as those of (1.5); namely, that if the whole class of linear time-varying systems is to be stable, then the matrix $|A| := (|a_{ij}|)$ must admit a diagonal solution to its Stein equation, i.e., must be diagonally stable. Equivalently, the class of systems in question admits a simultaneous diagonal quadratic Liapunov function.

In several applications, a variation of the concept of diagonal stability also arises—this is the concept of additive diagonal stability. A matrix A is called *additively diagonally stable* if $A + D$ is diagonally stable for any nonpositive diagonal matrix D.

Informally, a dynamical system that can be written, perhaps after a change of coordinates, in the form (1.5) with the matrix A diagonally stable, is said to possess a *diagonally stable structure*. There are also several variants of equation (1.5), presented in Chapter 3, that are also susceptible to analysis by diagonal-type Liapunov functions, using the concept of additive diagonal stability, thereby broadening considerably the class of diagonally stable structures.

Two closely related terms that are frequently used in the book to qualify stability are *robust* and *absolute*. Both terms apply to a specified class of dynamical systems, denoted $\mathcal{C}$. Stability of a given dynamical system is said to be *robust* if all its neighbors in $\mathcal{C}$, defined in some suitable sense, are stable. The dynamical system is then said to possess the property of *robustness*. A criterion is called an *absolute stability condition* if it ensures that *all* members of the class $\mathcal{C}$ are stable. In this sense, absolute stability is an extreme form of robust stability in which it is required that all members of a class be stable, rather than just some neighborhood. Suppose, for example, that a dynamical system S is defined for each value of a parameter p and that $\mathcal{C} := \{S(p) : p \in \mathcal{P}\}$, for some open set of parameters $\mathcal{P}$. Then a given dynamical system $S(p_0) \in \mathcal{C}$ is said to be *robustly stable* if there is some neighborhood of p_0, N_{p_0}, contained in $\mathcal{P}$, such that $S(p)$ is stable for all $p \in N_{p_0}$. A stability condition is said to be robust if it guarantees the stability of a given dynamical system, usually called the *nominal system*, as well as that of some set of its neighbors. An absolute stability condition guarantees the stability of a whole class of systems. Of course, if the whole class turns out to be exactly the set of neighbors or vice versa, then the two definitions coincide. An interesting historical observation is that the term absolute stability originated in the Russian literature and still tends to be preferred in this literature. The term robust originated around 1972 in the Western control literature, borrowed from the statistics literature (see the preface to [Dor87]), in which it tends to be preferred to the term absolute. Prior to 1972, the literature used terms such as stability of *uncertain* systems, *sensitivity* or *roughness* of stability, etc. An important extension of the discrete-time version of (1.5) includes time-varying delays on the right-hand side and can be written as follows.

$$x_i(k+1) = \sum_{j=1}^{n} A_{ij}(x_1(k), \ldots, x_n(k))x_j(d_{ij}(k)), \qquad (1.6)$$

for $i = 1, 2, \cdots, n$, $k = 0, 1, 2, \cdots$, and where $d_{ij}(k) \in \{k, k-1, k-2, \cdots, k-d\}$, for all k, for some integer $d > 0$ and for $i, j = 1, \cdots, n$.

Diagonal-type Liapunov functions also allow simple proofs of the asymptotic stability of neural circuits and asynchronous iterative computations and systems with delays since the dynamical equations describing these systems can be put in the form (1.6), with appropriate definitions of the matrices A_{ij}. However, for these systems, *quadratic* diagonal Liapunov functions do not always suffice to show stability and more general diagonal-type functions must usually be resorted to.

Diagonally stable structures arise in another context where the question of robustness of stability can be reduced to that of the invariance of asymptotic stability of a certain matrix under multiplication by certain subclasses of the class of positive diagonal matrices. In the continuous-time case, if the stability of a given matrix A is maintained under premultiplication by a diagonal matrix with positive diagonal entries, then A is said to be *Hurwitz D-stable* and this type of matrix stability is called Hurwitz D-stability. It was first studied in a price stability problem in the continuous-time case in economics. Since then applications have surfaced in a variety of fields, but the problem of effective characterization of D-stability has so far resisted solution. More details on this as well as on the connections between diagonal and D-stability can be found in Chapter 2.

The other sections of this chapter present a series of examples that serve to illustrate these introductory remarks, and also to motivate the reader to get interested in the book. Some remarks on the examples in this chapter and others elsewhere in the book are worth making. Several of these examples have been much studied in the literature and the results shown here are not necessarily new, but are put in a perspective where certain structural features are taken advantage of in order to get an understanding of the solution to the problem posed in each example. To this end, simple examples have been chosen. However, if readers encounter some difficulty in understanding a particular example completely, they should observe that an effort has been made to provide cross references to the chapters in which a more complete explanation of the ideas involved in the examples is given.

1.2 Robust Stability of a Mechanical System

Consider a simple mechanical device that consists of a disc fixed to a massless elastic shaft, with stiffness coefficient k, at the center. Assume that friction is internal to the shaft. Let the damping coefficient be denoted f, and let y_1 and y_2 represent the deflections of the disk, regarded as a particle of mass m located at the center of the shaft. Then, in a coordinate system rotating at the angular velocity of the shaft, the linearized equations of motion of the device are as follows [Zie68].

$$
\begin{aligned}
m\ddot{y}_1 + f\dot{y}_1 - 2m\omega\dot{y}_2 + (k - m\omega)y_1 &= 0, \\
m\ddot{y}_2 + f\dot{y}_2 + 2m\omega\dot{y}_1 + (k - m\omega)y_2 &= 0.
\end{aligned}
\tag{1.7}
$$

Assume that the parameters are normalized so that $k/m = f/m = 1$ and let $x := (y_1, \dot{y}_1, y_2, \dot{y}_2)^T \in \mathbb{R}^4$. The dynamics of this mechanical system is written in the state-space form as follows.

$$
\frac{dx}{dt} = A(\omega)x
\tag{1.8}
$$

where $x \in \mathbb{R}^4$, the angular velocity ω is considered to be a perturbation parameter with respect to the nominal value $\omega = 0$, and the matrix $A(\omega)$ has the form below:

$$
A(\omega) = \begin{bmatrix}
0 & 1 & 0 & 0 \\
\omega^2 - 1 & -1 & 0 & 2\omega \\
0 & 0 & 0 & 1 \\
0 & -2\omega & \omega^2 - 1 & -1
\end{bmatrix}
\tag{1.9}
$$

This system was first analyzed from a robust stability viewpoint in the book [Šil78] to illustrate the large scale systems technique and later considered to display the merits of the robustness analysis techniques of [Xin87] and [TV90]. In the references cited above, different approaches were utilized to derive different ranges for the parameter variations in this specific example.

The objective here is to use this example to show that the identification of a particular structure readily provides a simple diagonal Liapunov function that guarantees stability of the dynamical system within a range of variation for the parameter which coincides with the range given by the necessary and sufficient Hurwitz stability criterion; consequently this is the best one can do with Liapunov functions. Notice that a linear system is the simplest instance of model (1.5), and thus the example treated in this section can be regarded as an application of the Persidskii-type Liapunov function referred to in Section 1.1.

Previously obtained estimates of the stability domain in the parameter space are as follows: $\{\omega : \omega \in [0,1)\}$ in [TV90]; $\{\omega(t) : \omega(t) \in [0,0.05)\}$ in [Šil78, Chapter 1], noting however that the upper bound can be improved to 0.13, using an optimized vector Liapunov function technique (valid for time-varying ω), and $\{\omega : \omega \in [0,0.08768)\}$ in [Xin87].

Once the special acyclic structure of the matrix (1.9) is identified, it is known (see Section 2.2) that there exists a diagonal quadratic Liapunov function of the form

$$V(x) = x^T P x \qquad \text{with } P = \text{diag}(p_1, p_2, p_3, p_4).$$

In this case a solution for matrix P is diag $(1 - \omega^2, 1, 1 - \omega^2, 1)$ and clearly to ensure P positive definite, necessarily $\omega^2 < 1$. With this Liapunov function, along the trajectories of (1.8), one has $dV/dt = x^T Q x$ with $Q = -\text{diag}(0, 2, 0, 2)$. Consequently dV/dt is a negative semidefinite function with $dV/dt = 0$ on the set $L := \{x : x_2 = x_4 = 0\}$.

From the system equations, it is clear that the largest invariant set contained in L is the singleton containing the origin (since $x_1(t) = x_3(t) = 0$, for all t). LaSalle's theorem states that when dV/dt is negative semidefinite, the trajectories of the system converge to the invariant set [LL61]. Invoking this theorem, it follows that the trajectory $x(t) = 0, \forall t \geq 0$ of $dx/dt = A(\omega)x$ is globally asymptotically stable under the condition $\omega^2 < 1$.

It is easily verified that $\omega^2 < 1$ is exactly the necessary and sufficient Hurwitz stability condition for the characteristic polynomial of $A(\omega)$. Consequently, a simple quadratic diagonal-type Liapunov function leads to a sufficient condition for the stability of this system, and since it turns out to be necessary as well, it is not conservative. It should also be pointed out that the diagonal-type Liapunov function proposed is actually an energy-type Liapunov function and this becomes clear when it is written out in full as

$$V(y_1, \dot{y}_1, y_2, \dot{y}_2) = (1 - \omega^2)y_1^2 + \dot{y}_1^2 + (1 - \omega^2)y_2^2 + \dot{y}_2^2$$

showing the energy terms explicitly.

This example displays a property (see Proposition 2.3.2) which essentially says that the conditions for diagonal semistability of a given matrix A in Schwarz form are equivalent to the Hurwitz stability conditions of this matrix and also to the existence of an associated parameter-dependent diagonal Liapunov function that provides necessary and sufficient stability conditions.

Although the matrix $A(\omega)$ in (1.9) is not in Schwarz form, it has acyclic-3 structure (see Section 2.2) for which the same property holds, under certain circumstances.

Notice that this Liapunov function is not *simultaneous* [BY89, KB93b]; that is, for each value of the parameter ω in the given range, a different $V(x)$ ensures the asymptotic stability of the system. Such a Liapunov function is referred to as *parameter dependent* and, in general, leads to less conservative stability conditions than a simultaneous Liapunov function (see Section 6.4 for additional examples), since it does not ensure stability when, for example, the parameters are time-varying.

It should however be pointed out that, at the price of some additional computation that has been avoided in this introductory example, it is also possible to use a single simultaneous block diagonal Liapunov function,

$$V(x) = x^T P x, \text{ where } P = \begin{bmatrix} 0.380 & 0.124 & 0 & 0 \\ 0.124 & 0.842 & 0 & 0 \\ 0 & 0 & 0.380 & 0.124 \\ 0 & 0 & 0.124 & 0.842 \end{bmatrix},$$

that guarantees stability for the parameter range $|\omega| < 0.908$, even when the parameter ω is allowed to be a function of time [ABFK98, BK99].

This simple example serves to illustrate the point that the identification of a certain structure ensures a stability property under a nonconservative condition. This does not usually happen with methods that are applicable in more general circumstances. In this example, other methods that are applicable in the case of time-varying parameters are not set up to take advantage of either the structural information about the matrix $A(\omega)$ or, for instance, the information that the parameter is invariant, in order to get a less conservative stability condition.

The field of robust stability and control is vast and burgeoning. The collections [Dor87, DY90] give an overview of the various approaches up to the 90s, while [BGFB94] has become the classic reference on the linear matrix inequality approach, which is also relevant to much of this book, in the particular case of diagonal stability. The parametric approach to robust control is surveyed in [Šil89] and treated in the book [BCK95]. It should be mentioned that there are many other approaches to robust control (for example, [DDB95, ZDG96]) that are not closely related to the approach considered in this book.

1.3   The Lotka–Volterra Ecosystem Model

The mathematical model for two competing species derived by Lotka and Volterra and widely used in mathematical population biology is arrived at by considering a simple form of interaction between a predator species x_1 and its prey x_2. The prey population is assumed to constitute the total (exclusive) food supply for the predators at any given moment. The total food consumed by the predators in one unit of time is taken to be proportional to the number of predator–prey encounters, which, in turn, is assumed to be proportional to $x_1 x_2$. Hence the per capita food supply for the predators at time t is proportional to $x_2(t)$. If social phenomena are ignored, then the growth equation for the predator can be written as:

$$\dot{x}_1 = (a_{12} x_2 - c_1) x_1,$$

where a_{12} and c_1 are positive constants. In an analogous way, if the prey species is assumed to have a constant per capita food supply available that is sufficient to sustain its growth in the absence of predators, then the differential equation that describes its dynamics is of the form:

$$\dot{x}_2 = (c_2 - a_{21} x_1) x_2,$$

where a_{21} and c_2 are positive constants.

With this second order system background, consider the generalized multispecies Lotka–Volterra model, given by the equations:

$$\dot{x}_i(t) = c_i x_i(t) + \sum_{j=1}^{n} a_{ij} x_i(t) x_j(t), \qquad (1.10)$$

for $i = 1, \cdots, n$. An important difference between the second order model and the model (1.10) is that, in the former, there is no nonlinear term of the type $a_{11} x_1^2$, while such a term appears in the latter, since it is not assumed that $a_{ii} = 0$. This nonzero self-interaction term is a result of the requirement that each population has a logistic growth in the absence of cross interactions (i.e., when all the a_{ij}s are zero). The logistic growth is in turn usually justified by reference to the effect of crowding and the limitation of resources in the environment. Crowding refers to the increase in the interactions or collisions between pairs of individuals in the competition for resources (including space). Although the assumption that $a_{ii} = 0$ is quite common in the literature [GMM71], this assumption is not made in the analysis that follows.

The model (1.10) was introduced by Lotka and Volterra in the early 1900s and has been studied extensively ever since. Variations and extensions of this model occur in such diverse problems as the kinetics of autocatalytic chemical reactions, the management of fish populations, the propagation of genetic traits, the spread of epidemics, and pest control. In most applications one wants to know the long term behavior. In other words, the pertinent questions are of the type: Does each x_i eventually settle down to a limiting value? If so, is the value independent of the initial conditions?

This model (1.10) is not explicitly in the form (1.5). However, an exponential change of variable (which, in a somewhat different notation goes back to Volterra himself) suffices to bring it into this form. Assuming that no species disappears $(x_i(t) \neq 0, i = 1, \cdots, n)$, one can write:

$$(\dot{x}_i(t)/x_i(t)) = c_i + \sum_{j=1}^{n} a_{ij} x_j(t), \quad \text{or, equivalently,}$$

$$d(\ln x_i(t))/dt = c_i + \sum_{j=1}^{n} a_{ij} \exp(\ln x_j(t)). \tag{1.11}$$

Defining $y_i(t) := \ln x_i(t)$, equation (1.11) can be rewritten as:

$$\dot{y}_i(t) = c_i + \sum_{j=1}^{n} a_{ij} f(y_j(t)).$$

Assuming the existence of a positive equilibrium y^*, i.e., a vector $y^* = (y_1^*, \cdots, y_n^*)^T$, with $y_i > 0$, for all i, that satisfies $c_i + \sum_{j=1}^{n} a_{ij} f(y_j^*) = 0$, one can write the incremental model:

$$\dot{z}_i(t) = \sum_{j=1}^{n} a_{ij} g_j(z_j(t)), \tag{1.12}$$

where for all i, $z_i(t) := y_i(t) - y_i^*; z^T(t) = (z_1(t), \cdots, z_n(t))$, $g_i : \xi \mapsto \exp(\xi + \ln x_i^*) - x_i^*$; therefore $g_i(0) = 0$ and $g_i(\xi) \cdot \xi > 0$ for $\xi \neq 0$. Consequently system (1.12) is a particular case of the Persidskii model, (1.5) with the associated diagonal-type Liapunov function (see Theorem 3.2.3) being given by:

$$V(z) = \sum_{i=1}^{n} p_i \int_0^{z_i} g_i(\tau) d\tau. \tag{1.13}$$

The usual Volterra function used in, for example, [Goh76, Goh77, Goh78, Šil78, TAT78, SL83], is given by:

$$V(x) = \sum_{i=1}^{n} p_i [x_i - x_i^* - x_i^* \ln(x_i/x_i^*)].$$

Note that $V(x)$ is also a diagonal-type Liapunov function that appears to be quite different from $V(z)$ in (1.13). In fact, a little algebra, using the relation $z_i = \ln(x_i/x_i^*)$, shows that the two functions are identical.

The integral-of-nonlinearity kind of Liapunov function in (1.13) was first used by Persidskii [Per69] and is referred to as a Persidskii-type Liapunov function. Computing $\dot{V}(z)$ along the trajectories of (1.12) one has:

$$\dot{V}(z(t)) = g(z(t))^T (A^T D + DA)g(z(t)),$$

where $g(z)^T = (g_1(z_1), \cdots, g_n(z_n))$ and $D = \frac{1}{2}\mathrm{diag}\,(p_1, \cdots, p_n)$ and provided that the matrix A (known as the *community matrix*) is diagonally stable, $\dot{V}(x)$ is negative definite, and the asymptotic stability of the equilibrium of system (1.10) follows. The positive diagonal matrix is also known as a *Volterra multiplier* in honor of Volterra's discovery of its role in stability (see [Red85a, Red85b] and the references therein). The observation that system (1.12) is exactly the Persidskii model, which in turn is a particular case of (1.5), although implicit in the papers cited above, seems to have first been made explicit in [KB93b].

The above discussion may be summarized by formulating the result that diagonal stability of the community matrix $A = (a_{ij})$ implies the asymptotic stability of model (1.10). It should be noticed, however, that the Liapunov function (1.13) and the associated diagonal stability condition do not guarantee asymptotic stability with more general, time-varying perturbations acting on this biological system. In order to guarantee robustness under these circumstances, a more restrictive condition on the community matrix A is required, namely that a *comparison* matrix constructed using the moduli of the entries of matrix A is to be diagonally stable (see Corollary 3.2.11). This is the price that must be paid to guarantee robustness, since this condition ensures the existence of a common or simultaneous Liapunov function for the whole class of perturbed systems. This fact also provides a way of interpreting a result [Šil78, Chapter 5] that requires a comparison matrix to be diagonally stable in order to ensure exponential connective stability of the equilibrium of (1.10).

Many aspects of mathematical biology, including population biology, are treated in [Mur93]. Dynamical system models of ecosystems are studied in [May74, SL83] and in [Kua93] for ecosystems with delays. Much additional background and a completely different approach to Lotka–Volterra types of models, from the point of view of statistical physics, can be found in [GMM71]. Finally, it should be pointed out that diagonal stability and related diagonal-type Liapunov functions are also useful in systems of the type

$$\dot{x}_i = N(x)f_i(x_i)\left(c_i + \sum_{j=1}^n a_{ij}g_j(x_j)\right)$$

as well as to other generalizations of (1.10), involving time delay and diffusion (see, for example, [RW84, RZ81, WB78, Kua93, CC96]). A very clear exposition of the role of diagonal stability (called VL-stability) in the global stability of n-species communities is given in [HS98] which also contains many additional applications to evolutionary games and population dynamics.

1.4   Convergence of Asynchronous Computations

In this section, a simple example is given of the use of a diagonal-type Lia-punov function in the convergence analysis of a parallel iterative method to solve the linear equation $Ax = b$, with A being a matrix in $\mathbb{R}^{n \times n}$ and x, b vectors in $\mathbb{R}^n$. Here, pending a more complete explanation of parallel com-putation and iterative methods in Chapter 4, the term *parallel computer* is used to denote a machine that has more than one processor or computer, working in parallel, and devoted to the solution of a single problem. The term *computer* is used to mean either an independent machine or a proces-sor within a larger machine. This usually presupposes that: (i) the problem has been decomposed into as many subproblems as there are computers; and (ii) that there exists some method of putting together the solutions to the subproblems in order to get the solution to the original undecomposed problem. A parallel iterative method is one that realizes both these tasks in a straightforward manner. Such methods are typically used when the dimension n of the matrix A is large and the matrix A has some special characteristics that make the use of the conventional Gaussian elimination method inefficient or slow in parallel computers.

The standard method of setting up an iterative method is to split the matrix A into two matrices $A = M - N$, where M is invertible and has a particularly simple structure, for example, is diagonal. The original lin-ear equation can then be rewritten as $Mx = Nx + b$, and given an initial guess $x(0)$ for the solution, it is reasonable to update this guess by setting $Mx(1) = Nx(0) + b$. This is a linear system in the unknown vector $x(1)$, and, since M has been assumed to have a simple structure, it is easier to solve for $x(1)$ than it was to solve the original system. This iteration can be continued and, in order to analyze it, let it be assumed that convergence to the vector x^* occurs. This means that $Mx^* = Nx^* + b$, which means that $(M - N)x^* = Ax^* = b$, i.e., x^* is a solution to the original equation. Sub-tracting this equation from the equation for the general iteration leads to $M(x(k+1) - x^*) = N(x(k) - x^*)$, so that if the error vector $e(k) = x(k) - x^*$ is introduced, one is led to the *error equation* $e(k+1) = M^{-1}Ne(k)$. Clearly, the convergence of the general iteration is implied by the asymptotic sta-bility of the zero solution to the error equation and this, in turn, occurs if and only if the *iteration matrix* $H := M^{-1}N$ has spectral radius less than unity [Ort72].

With this concise background, now suppose that the problem size n is large enough that in order to compute the fixed point or solution vector x^*, it is either necessary or desirable to decompose the vector x into subvectors

x_i, with the iteration corresponding to each subvector being computed separately. Assume that these separate computations are actually carried out in separate computers or processors and, for added generality, that the decompositions are also allowed to vary with each iteration (i.e., $A = M(k) - N(k)$). In order to write down a mathematical model that represents the evolution of the error of such an iterative method, one starts with a Cartesian product decomposition of $\mathbb{R}^n$, namely:

$$\mathbb{R}^n = \mathbb{R}^{n_1} \times \cdots \times \mathbb{R}^{n_m}, \; n_1 + \cdots + n_m = n. \tag{1.14}$$

Given this decomposition, the vector x and the matrix $M^{-1}N$ are partitioned conformally, to get the m error equations in the form:

$$e_i(k+1) = \Sigma_{j=1}^m H_{ij}(k) e_j(k), \tag{1.15}$$

for $i = 1, \ldots, m$, $k = 1, 2, \cdots$, and where $e_i \in \mathbb{R}^{n_i}$, $H(k) = (H_{ij}(k))$, and for all $k \in \mathbb{N}$, the matrices $H_{ij}(k)$ are in $\mathbb{R}^{n_i \times n_j}$. Note that in the simple case discussed above $H(k) = H = M^{-1}N$.

The notion of asynchronous computation can now be introduced in a natural manner. Note that in equation (1.15) it is being assumed that, for the kth update, the ith computer (that updates $e_i(k)$) has the most recent information from all the other computers (i.e., receives $e_j(k)$ from the jth computer, for all j).

Here asynchronism refers to the possibility of using, in the kth iteration of an iterative method based on (1.15), variables that have been computed in earlier iterations: i.e., in the jth iteration with $j \leq k - 1$. To be more specific, assume that, at update k, the ith computer (which updates e_i) receives information from the jth processor with a time-varying delay of $k - d_{ij}(k)$ units. These delays are caused by communication and memory access times and so on, which are present in any real parallel computing system. This form of asynchronous computation usually leads to significant reduction in the total time required to get a solution, although a nonstandard convergence analysis needs to be performed in order to ensure that this asynchronous computation leads to the same answer as standard synchronous computation. This is the problem touched upon in this section and studied in detail in Chapter 4.

The asynchronous implementation of (1.15) can be written as:

$$e_i(k+1) = \Sigma_{j=1}^m H_{ij}(k) e_j(d_{ij}(k)) \tag{1.16}$$

The assumption that the delays are uniformly bounded by a positive integer d is realistic for all existent parallel computing systems. It is stated formally as a restriction on the range of the positive integer-valued functions $d_{ij}(\cdot)$ below.

Bounded delays assumption

$$\exists \, d \in \mathbb{N}, \forall k \in \mathbb{N}, \forall i, j \in \{1, \ldots, m\}, \; d_{ij}(k) \in \{k, k-1, \ldots, k-d\}. \tag{1.17}$$

Note that the $d_{ij}(k)$s are functions of three variables: update or iteration number k; sending computer j; and receiving computer i.

In order to analyze the convergence of the asynchronous iterative method described by (1.16), the diagonal-type Liapunov function below, based on the delayed variables, was introduced in [KBŠ90].

$$V(k) := \max_{i \in \mathbf{m}, p \in \mathbf{d}} \{w_i^{-1} \|e_i(k)\|, w_i^{-1} \|e_{i,p}(k)\|\} \qquad (1.18)$$

where $w_1, \ldots, w_m$ are positive weights, $\mathbf{m} := \{0, 1, \ldots, m\}$, $\mathbf{d} := \{0, 1, \ldots, d\}$, and $e_{i,p}(k) := e_i(k - p)$ are the delayed variables. The convergence result obtained with this Liapunov function is as follows.

Theorem 1.4.1 *If the bounded delays assumption holds and there exists a closed set $D = D_1 \times \cdots \times D_m \subset \mathbb{R}^n$, such that $H(D) \subset D$, and matrix $U = (u_{ij})$ is Schur diagonally stable, where*

$$u_{ij} := \sup_{i,j \in \mathbf{m}, k \in \mathbb{N}} \{\|H_{ij}(k)\|\}, \qquad (1.19)$$

then the zero solution of the asynchronous iterative method (1.16) is locally exponentially stable, in the domain D.

Two points are important here. First, it is clear that, under the conditions of the above theorem, the local exponential stability of the error variables implies local convergence to the desired solution x^*. In other words, the Schur diagonal stability of the matrix U is a sufficient condition for the asynchronous iteration (1.16) to converge to the same solution as the synchronous iteration (1.15). Second, it should be emphasized that the actual asynchronous computations are based on the iterations $M_i x_i(k+1) = \sum_j N_{ij} x_j(d_{ij}(k)) + b_i$ and do not involve explicit formation either of the inverse of the iteration matrix H or the matrices M_i.

Theorem 1.4.1, its proof, generalizations, and applications are found in Chapter 4. It is worth pointing out that the matrix U is nonnegative and that, furthermore, for this class of matrices, the diagonal stability condition is equivalent to having spectral radius less than one. In other words, a nonnegative matrix, such as U is Schur stable if and only if it is Schur diagonally stable, which, in turn, is directly related to the existence of a diagonal-type Liapunov function (see Chapters 2 and 4 for further explanations). The problem of stability of a linear discrete-time system subject to *time-varying, possibly unbounded* delays was first studied by Chazan and Miranker [CM69] in the context of what they termed *chaotic iterative computation*. The chaos referred to is exactly the arbitrary variation in the time delays that are usually assumed to be either absent or else constant and bounded.

To give an outline of the ideas on which the proof is based, consider the following simple example.

Example 1.4.2 *The following two-by-two system describes an asynchronous iterative method.*

$$x_1(k+1) = h_{11}(k)x_1(d_{11}(k)) + h_{12}(k)x_2(d_{12}(k)) \qquad (1.20)$$
$$x_2(k+1) = h_{21}(k)x_1(d_{21}(k)) + h_{22}(k)x_2(d_{22}(k)) \qquad (1.21)$$

Assume that the uniform bound d on the delays $d_{ij}(\cdot)$ is unity, i.e.,

$$\forall k \in \mathbb{N}, \ \forall i,j \in \{1,2\}, d_{ij}(k) \in \{k, k-1\} \qquad (1.22)$$

Define the binary functions

$$\delta_{11}(k) = \begin{cases} 0, & \text{if} \quad d_{11}(k) = k-1 \\ 1, & \text{if} \quad\quad d_{11}(k) = k \end{cases} \qquad (1.23)$$

and

$$\gamma_{11}(k) = 1 - \delta_{11}(k) \qquad (1.24)$$

and similarly for $\delta_{12}(\cdot), \gamma_{12}(\cdot), \delta_{21}(\cdot), \gamma_{21}(\cdot), \delta_{22}(\cdot)$, and $\gamma_{22}(\cdot)$.
An augmented state-vector is defined as

$$\mathbf{x}(k) := (x_1(k), x_{1,1}(k), x_2(k), x_{2,1}(k))^T,$$

and the equations (1.20), (1.21) can be represented in standard state-space form as:

$$\mathbf{x}(k+1) = H_a(k)\mathbf{x}(k), \qquad (1.25)$$

where the matrix $H_a(k)$ is given by:

$$H_a(k) = \begin{bmatrix} h_{11}(k)\delta_{11}(k) & h_{11}(k)\gamma_{11}(k) & h_{12}(k)\delta_{12}(k) & h_{12}(k)\gamma_{12}(k) \\ 1 & 0 & 0 & 0 \\ h_{21}(k)\delta_{21}(k) & h_{21}(k)\gamma_{21}(k) & h_{22}(k)\delta_{22}(k) & h_{22}(k)\gamma_{22}(k) \\ 0 & 0 & 1 & 0 \end{bmatrix}.$$

$$(1.26)$$

Note that, for all k, the entries of $H_a(k)$ are either zeros, ones, or entries $h_{ij}(k)$ of the iteration matrix $H(k)$.

Once the asynchronous iteration is written in the form (1.25), it can be seen that the diagonal-type Liapunov function (1.18) is just the weighted infinity norm of the augmented state vector $\mathbf{x}$. This Liapunov function is, however, only nonincreasing along the trajectories of the discrete-time dynamical system (1.16), so that some additional technicalities are needed to show that the system is in fact exponentially stable and therefore converges to the desired solution (see Chapter 4 for a more detailed discussion as well as the proof of exponential stability). It should be pointed out the term convergence has been loosely used synonymously with exponential stability in this section. Strictly speaking, the latter implies the former and not vice-versa. A good discussion of the relations between different stability and convergence notions is in [Ort73].

The subject of asynchronism and numerical algorithms that make use of it are treated in [BT89] and, in the Russian language, in [AKKK92].

1.5   Global Stability of Neural Networks

In a series of influential papers, Hopfield and Tank [Hop82, Hop84, HT85, TH86] introduced a dynamical system model that now bears their name. This dynamical system has a simple physical interpretation as an electrical circuit containing nonlinear amplifiers, interconnected through a network of linear gains (= real numbers). This circuit was referred to as a neural circuit or network (often, and more accurately, as an *artificial* neural network) in view of the fact that each nonlinear amplifier could be seen as a model of a neuron activation function, and the *interconnection network* of linear gains as synaptic weights. Another way to regard such a dynamical system is to consider it as an analog computer to solve certain types of problems. In this book, the term neural network, now widely accepted to mean artificial neural network in the engineering community, will be used and the Hopfield–Tank neural network is the one being considered throughout. The importance of these neural networks for the solution of many optimization problems in real time was discussed in [TH86, CU93] and also in [AFMM91] for the specific problem of analog-to-digital converter design. Within this context it is required that the neural network have a unique equilibrium point, stable and globally attractive, in order to avoid so-called spurious solutions.

The continuous-time mathematical model of the Hopfield–Tank neural network is given by the equations [Hop84, HT86]:

$$C_i \frac{du_i}{dt} = \sum_{j=1}^{n} t_{ij} v_j - \frac{u_i}{R_i} + I_i \tag{1.27}$$

$$v_i = g_i(u_i), \tag{1.28}$$

for $i = 1, 2, \ldots, n$, and where the scalars $C_i > 0$ represent the neuron amplifier input capacitances and the scalars $R_i > 0$ the resistances; $T = (t_{ij})$ is the $n \times n$ real constant network interconnection matrix; I_i are the (constant) external current inputs to the ith neuron; $u_i = u_i(t) \in \mathbb{R}$ are the neural voltages; and $v_i = v_i(t) \in \mathbb{R}$ are the neuron output voltages.

Under the standard assumptions of Lipschitz continuity, monotonicity and first-quadrant–third-quadrant confinement on the neuron activation functions, the following result on global stability is obtained, resorting, once again, to a diagonal-type Liapunov function for the stability proof.

Theorem 1.5.1 *Given the Hopfield–Tank model subject to the assumptions above, if the interconnection matrix $T = (t_{ij})$ is diagonally stable,*

*then the neural network has a unique isolated equilibrium point which is
globally asymptotically stable.*   □

The existence of an equilibrium depends on some technical notions that
are discussed in detail in Chapter 5, so only a sketch of the global stability proof is given to indicate how the concept of diagonal stability and a
diagonal-type Liapunov function enter the scene.

By a shift of the origin to the (assumed) unique equilibrium point u^*,
equations (1.27) and (1.28) can be written in the following equivalent form:

$$\frac{dx_i}{dt} = \frac{1}{C_i} \left[\sum_{j=1}^{n} t_{ij}\phi_j(x_j) - \frac{1}{R_i}x_i \right], \tag{1.29}$$

for $i = 1, 2, \ldots, n$, where $x_i := u_i - u_i^*$; $\phi_i(x_i) = g_i(x_i + u_i^*) - g_i(u_i^*)$, and
the functions $\phi_i(x_i)$ continue to satisfy the assumptions mentioned above.

Denoting $\Phi : \mathbb{R}^n \to \overline{H} \subset \mathbb{R}^n : \overline{H} = \{(\overline{h}_1, \ldots, \overline{h}_n)^T : \overline{h}_i \in (-2, 2)\}$; and
$[\Phi(x)]_i = \phi_i(x_i)$, equation (1.29) can be written in the form:

$$\frac{dx}{dt} = C^{-1}T\Phi(x) - (RC)^{-1}x, \tag{1.30}$$

where $C := \text{diag}\,(C_1, \ldots, C_n)$ is a positive diagonal constant $n \times n$ matrix;
and $R := \text{diag}\,(R_1, \ldots, R_n)$ is a positive diagonal constant $n \times n$ matrix.
This turns out to be a slightly modified version of the model (1.5) for which
the Persidskii-type Liapunov function (see Theorem 3.2.3) can still be used
and has the following form:

$$V(x) = 2 \sum_{i=1}^{n} p_i \int_{0}^{x_i} \phi_i(\tau)d\tau, \tag{1.31}$$

where the p_i are positive constants that are identified below. Along the
trajectories of (1.30), the time derivative of $V(x)$ is given by

$$\frac{dV(x)}{dt} = \Phi^T(x)\left[P(C^{-1}T) + (C^{-1}T)^T P\right]\Phi(x) - 2\Phi^T(x)P(CR)^{-1}x. \tag{1.32}$$

In Chapter 2, Lemma 2.1.4, it is shown that the set of diagonally stable matrices is closed under the operation of premultiplication by positive diagonal
matrices. Under the condition that T is diagonally stable, it follows that
$(C^{-1}T)$ is diagonally stable as well and thus there exists a positive diagonal
matrix $P = \text{diag}\,(p_1, \ldots, p_n)$ such that the matrix $P(C^{-1}T) + (C^{-1}T)^T P$
and hence dV/dt are negative definite. Since the $\phi_i(\cdot)$s are also monotonic,
and confined to the first and third quadrants, the condition

$$\lim_{|x_i| \to \infty} \int_{0}^{x_i} \phi_i(\tau)d\tau = \infty \tag{1.33}$$

holds, ensuring radial unboundedness of the Liapunov function $V(x)$ and
guaranteeing that $x = 0$ (or equivalently u^*) is globally asymptotically

stable. A detailed discussion and refinements of this result may be found in Chapter 5.

Stability issues in neural networks are treated in detail in [Gol96] and briefly in [Hay99], which also contains much additional material on neural networks as well as an extensive bibliography.

1.6 Variable Structure Systems

Dynamical systems in which controls are discontinuous functions of the states (for example) are known as *variable structure* systems. Such systems were first studied in the phase plane (state space of a vector differential equation in $\mathbb{R}^2$). The objective is to make suitable changes in the control, as a discontinuous function of the current state, so as to modify the state trajectories in a given manner (usually to achieve a stable equilibrium). The problem of choosing such controls is reducible to that of the selection of surfaces in the state space on which the controls have discontinuities.

A basic idea is to ensure that, in the vicinity of a discontinuity surface, the state trajectories are always directed back to this surface. Once on the surface, the state cannot move along any adjacent trajectory over any period, however short. Indeed, in response to any perturbation that takes the state off such a surface, a motion that returns the state to the surface always starts, since the surface has been chosen to have the system trajectories pointing toward it. Thus, in the variable structure system, the state can only move along the discontinuity surface. This motion is referred to as a *sliding mode*. The term variable structure arises because the system can be thought of as consisting of several continuous subsystems, also referred to as structures. At the times when the structures change, the right-hand sides of the differential equations describing the system motion have discontinuities on certain surfaces in the state space of the system.

Consider a linear control system affine in the control given below.

$$\dot{x} = Ax + Bu; \quad s = Gx \qquad (1.34)$$

where, A, B, and G are real matrices. The control u is discontinuous on the surfaces $s_i = 0$, $i = 1, \cdots, m$. The vector s is defined as $(s_1, \cdots, s_m)^T$, so that $s = 0$ defines the intersection manifold of all discontinuity surfaces. The equation of the controlled system in the so-called homogeneous case is [Utk92]

$$\dot{s} = Ku, \quad \text{where} \quad K = GB. \qquad (1.35)$$

The problem is to establish conditions under which globally stable motion with respect to the desired sliding manifold $s = 0$ is produced, i.e., according to definition given in [Utk92, p. 49], the origin of the s-space is globally asymptotically stable in the usual sense of Liapunov.

For systems with no uncertainty in the matrix K, it was shown in [Utk92] that, by appropriate choices of the amplitudes α_i of the sign functions (defined as $\text{sgn}(x) = +1$ if $x > 0$, $\text{sgn}(x) = -1$, if $x < 0$, and $\text{sgn}(0)$ is some real number in the interval $[-1, 1]$), it is possible to ensure the global asymptotic stability of $s = 0$.

If this procedure is followed for a nominal value K_0, with F denoting the diagonal matrix $\text{diag}\,(-\alpha_1, \cdots, -\alpha_m)$, the matrix K can be defined as $K_0 F$. Thus, the problem is reduced to determining conditions for the global asymptotic stability of $s = 0$ for the system

$$\dot{s} = K\,\text{sgn}(s). \tag{1.36}$$

If K is uncertain, then it is of interest to find stability conditions that require K to belong to some open set of matrices, since this means that, even if the actual system matrix K differs a little from the nominal one the stability of the system (1.36) is still ensured. In other words, the system is required to be robustly stable.

The crucial observation is that the dynamical system (1.36) has the form of (1.5) with a discontinuous sign nonlinearity on the right-hand side. This section shows that a stability result can be derived for the system (1.36) using a Persidskii-type function, similar to the one used in the previous section for continuous systems. The difference is that, in this case, since the Liapunov function is applied to systems with discontinuous right-hand side of the type (1.36), it is also chosen to be nonsmooth. The main objective here is to point out that diagonal stability also has an important role in the analysis of discontinuous systems with nonsmooth diagonal-type Liapunov functions, and the reader is assumed to have a working knowledge of variable structure systems at the level of [Utk92] so that not all terms are completely defined here.

Solutions of systems with discontinuous right-hand side, of the type (1.36), are understood as defined by Filippov [Fil64]. Roughly speaking, a solution $x(t)$ is called a *Filippov solution* if (i) it is an absolutely continuous function of t, thus almost everywhere differentiable, i.e., $\dot{x}(t)$ is almost everywhere defined, and (ii) it is identical to the usual definition of solution of differential equations, except on a sliding mode on some discontinuity manifold.

The following definition simplifies the discussion that follows.

Definition 1.6.1 [HC96] *Consider a Filippov solution $x(t)$ of system (1.36), defined for t in a certain interval T. Then, the* extended equivalent control $u_{eq}(t)$ *is an integrable function, defined almost everywhere in T, given by*

$$u_{eq}(t) = -K^{-1}[GAx - \frac{d}{dt}s(x(t), t)]. \tag{1.37}$$

Extended equivalent control is identical to conventional equivalent control when the sliding mode $s(t) \equiv 0$ occurs $(ds/dt \equiv 0)$ in some open interval of time. However, it gives a complete description of an equivalent control signal which is also valid when no sliding or only partial sliding occurs, i.e., only some elements of $s(t)$ are zero. This concept was also used in [Utk78] to describe real sliding modes the presence of practical imperfections of the switching control mechanism. The concept is crucial in the proof of the following stability theorem.

Theorem 1.6.2 *Consider the discontinuous system (1.36). Then $s(t) = 0$ is a globally asymptotically stable solution if K is a diagonally stable matrix. Moreover, in this case the origin is reached in finite time.*

Proof. Consider the following diagonal-type Liapunov function:

$$V(s) = 2 \sum_{i=1}^{m} p_i \int_0^{s_i} \text{sgn}(z)dz = 2 \sum_{i=1}^{m} p_i |s_i|, \tag{1.38}$$

where $P = \text{diag}(p_1, \cdots p_m)$ is the diagonal solution to the Liapunov equation in K. Since $V(s)$ is a Lipschitz-continuous function and $s(t)$ is absolutely continuous, the time function $V(s(t))$ is also absolutely continuous [Roy68]. Thus, the time-derivative of $V(s(t))$, denoted $\dot{V}$ and defined as dV/dt, along a Filippov solution $s(t)$ of (1.36) exists almost everywhere. If the function $V(s)$ were continuously differentiable, then its time derivative would be given by

$$\dot{V} = \frac{dV}{dt} = \left(\frac{\partial V}{\partial s}\right)^T (t) \, \dot{s} = \left(\frac{\partial V}{\partial s}\right)^T (t) \, K u_{eq}(t). \tag{1.39}$$

However, the above expression cannot be directly used with the function V defined by (1.38) since $(\frac{\partial V}{\partial s_i})(t) = 2p_i \, \text{sgn}(s_i(t))$ is not defined whenever $s_i(t)=0$, for some $i = 1, \cdots, m$: this occurs, for example, during sliding modes. In order to calculate $\dot{V}$ along a solution $s(t)$, we show that

$$\left(\frac{\partial V}{\partial s}\right)(t) = 2P u_{eq}(t). \tag{1.40}$$

Indeed, first rewrite $\dot{V}$ as

$$\dot{V}(t) = \sum_{i=1}^{m} \left(\frac{\partial V}{\partial s_i}\right)(t)\dot{s}_i(t). \tag{1.41}$$

If $s_i \neq 0$, $i = 1, \cdots, m$, then

$$\left(\frac{\partial V(s)}{\partial s_i}\right)(t) = 2p_i \, \text{sgn}(s_i(t)) = 2p_i \, u_{ieq}(t). \tag{1.42}$$

Now, if for some open interval of t, there exists i such that $s_i(t) \equiv 0$, then $\dot{s}_i(t) \equiv 0$ in that interval. Thus clearly, the relation

$$\left(\frac{\partial V(s)}{\partial s_i}\right)(t)\dot{s}_i(t) \equiv 2p_i u_{ieq}(t)\dot{s}_i(t) \equiv 0$$

holds in the given interval. Since all other possible situations involve some $s_i = 0$ only in sets of measure zero, it follows that the equality (1.40) holds. Then, from (1.36), (1.41), and the diagonal stability of K, one gets

$$\dot{V} = -u_{eq}^T(K^T P + PK)u_{eq} = -u_{eq}^T Q u_{eq}. \tag{1.43}$$

Moreover, the vector $u_{eq}(t)$ is null only when $s(t) = 0$. In fact, $\|\text{sgn}(s)\|_2 \geq 1$, for all $s \neq 0$. Hence, $\dot{V} \leq -\lambda_{min}^{1/2}(Q)$, for all $s \neq 0$. Hence $V(s(t))$ must converge to zero in finite time and the time needed to reach the sliding mode $s = 0$ is less than $V_0/(\lambda_{min}^{1/2}(Q))$, V_0 being the initial value of V. It has thus been proved that the origin is globally attractive. Stability follows from the positive definiteness of $V(s)$, which is not differentiable but is continuous and lower bounded by a class $\mathbf{K}$ function (see Definition 2.6.5), and that (essentially) $dV(s(t))/dt < 0$, $\forall s \neq 0$.   ∎

The Liapunov theory of nonsmooth systems developed in [SP94], which uses the machinery of generalized gradients, can also be used to justify the use of the nonsmooth diagonal-type Liapunov function above in order to give an alternate proof of the above theorem. Details are given in [HKB99]. The reader will note that Theorem 1.6.2 has been given a complete proof unlike the other results in this chapter. The reason for this is that, unlike the other topics treated in this introductory chapter, variable structure systems and nonsmooth Liapunov functions will not be pursued in the chapters that follow. Thus readers interested in variable structure systems should consult the references given at the end of this section.

Example 1.6.3 *Consider the following matrix.*

$$K = \begin{bmatrix} -0.5 & 3.5 \\ -0.5 & -0.5 \end{bmatrix}$$

A particular case of the diagonal stability condition, namely that the identity matrix be a diagonal solution of the Liapunov equation in K, i.e., that the matrix $K + K^T$ be negative definite, was given in [Utk92]. In this example, however, $K + K^T$ is not negative definite, although $P = \text{diag}(1,7)$ is easily seen to be a diagonal solution of the Liapunov equation in K. Thus for this choice of K, the system (1.36) is seen to be asymptotically stable, from theorem 1.6.2, but not from the result in [Utk92].

Theorem 1.6.2 can be used to characterize matrices K of dimension two for which the system (1.36) is stable.

Theorem 1.6.4 [HKB99] *For $K \in \mathbb{R}^{2 \times 2}$, the zero solution of $\dot{s} = K \, sgn(s)$ is globally asymptotically stable if and only if the diagonally scaled matrix DK is Hurwitz, where $D = \text{diag}\,((1/|k_{12}|),(1/|k_{21}|))$, if the off-diagonal entries are both nonzero, and $D = I$, otherwise.*   □

Recalling the definition that a matrix A is *Hurwitz D-stable* if DA is Hurwitz for all positive diagonal matrices D, the following corollary can be stated.

Corollary 1.6.5 [HKB99] *If $K \in \mathbb{R}^{2 \times 2}$ is D-stable, then the zero solution of (1.36) is globally asymptotically stable.*   □

It is known that a matrix $A \in \mathbb{R}^{2 \times 2}$ is D-stable if and only if $a_{ii} \leq 0$, $i = 1, 2$, at least one a_{ii} is negative, and $\det(A) > 0$ [Cro78] and D-stability of K is implied by diagonal stability of K (Lemma 2.1.4).

Note that Corollary 1.6.5 requires the matrix K to be Hurwitz, while Theorem 1.6.2 does not: it is enough to have a positive diagonal scaling that results in a Hurwitz matrix. For example, the matrix $K_2 = \begin{bmatrix} -0.5 & 1 \\ -5 & 1 \end{bmatrix}$ is not Hurwitz and hence not D-stable, yet when diagonally scaled by the reciprocals of the moduli of its off-diagonal elements, it becomes Hurwitz. Thus, the theorem ensures stability of the zero solution of (1.36), although the corollary does not.

Variable structure systems are given a comprehensive treatment in [Utk78, Utk92].

1.7 Existence of Diagonal-Type Liapunov Functions

Consider a diagonally stable matrix A and a diagonal solution P to the corresponding Liapunov matrix equation (1.3) in A, or the Stein equation (1.4). Given a nonsingular matrix S, now consider $B := SAS^{-1}$ similar to the matrix A. There is no obvious way to find a solution to the equations (1.3) or (1.4) in B from a knowledge of the matrices P and S and no reason to suppose that this new solution will be diagonal. On the other hand, suppose that A is not diagonally stable. It may be asked if there exists a nonsingular matrix T such that the matrix TAT^{-1} is diagonally stable. An example that serves to illustrate the above question is as follows.

Example 1.7.1 *Observe that the matrix*

$$A = \begin{bmatrix} 1 & 2 \\ -2 & -2 \end{bmatrix}$$

is not diagonally stable, since it has a positive diagonal element (see condition (2.42)). However,

$$\text{if } T = \begin{bmatrix} 1 & -1 \\ 1 & 1 \end{bmatrix}, \text{ then } C := TAT^{-1} = \begin{bmatrix} -0.5 & 3.5 \\ -0.5 & -0.5 \end{bmatrix}$$

is diagonally stable, with, for example, a diagonal solution $P = \text{diag}\,(1,7)$ to the Liapunov equation: i.e., $C^T P + PC = Q$, where Q is the negative definite matrix diag $(-1,-7)$.

In other words, diagonal stability is clearly not a coordinate-free property.

On the other hand, if a given system is not in a form that allows the use of a diagonal-type Liapunov function and the associated matrix diagonal stability concept, there may exist a change of state variables that makes it possible to do this. The example in Section 1.3 illustrated one such (nonlinear) change of variables. More generally, for certain classes of systems, the new "good" state variables, corresponding, for example, to a suitable similarity transformation, can be determined indirectly through realization procedures (see Chapter 5). To illustrate this in a simple example, consider a nonlinear dynamical system represented by the equations below.

$$\begin{aligned}
\dot{x}_1 &= x_1 + 2x_2 + 2f(x_2) \\
\dot{x}_2 &= -2x_1 - 2x_2 - 2f(x_2)
\end{aligned} \tag{1.44}$$

The above equations constitute a special case of (1.5) since they can be written as as follows:

$$\begin{pmatrix} \dot{x}_1 \\ \dot{x}_2 \end{pmatrix} = \begin{bmatrix} 1 & 2 \\ -2 & -2 \end{bmatrix} \begin{pmatrix} x_1 \\ x_2 + f(x_2) \end{pmatrix} =: A \begin{pmatrix} \phi_1(x_1) \\ \phi_2(x_2) \end{pmatrix}.$$

Note that if the scalar function $f(\cdot)$ is confined to the first and third quadrants, meaning that $xf(x) > 0$, $f(0) = 0$, then clearly so is $\phi_2(x_2) := x_2 + f(x_2)$. Observe that the matrix A is not diagonally stable, as pointed out in the example above. This rules out the use of Persidskii diagonal-type Liapunov functions of the kinds used in Sections 1.3, 1.5, and 1.6. Furthermore, if the coordinate change $z = Tx$, with T defined as in the example above, is made in the system (1.44), the resulting representation of this system is:

$$\dot{z} = Cz + \begin{bmatrix} 4 \\ 0 \end{bmatrix} f((z_1 - z_2)/2),$$

which clearly is no longer in the form (1.5), because the variables z_1 and z_2 in the function f are not separable.

However, anticipating the developments in Chapter 3, it turns out that a different representation of this system, together with a suitable coordinate transformation, does indeed permit the use of a diagonal Liapunov function. The system (1.44) is first noticed to be in the Lur'e form (discussed in Section 3.2):

$$\begin{pmatrix} \dot{x}_1 \\ \dot{x}_2 \end{pmatrix} = \begin{bmatrix} 1 & 2 \\ -2 & -2 \end{bmatrix} \begin{pmatrix} x_1 \\ x_2 \end{pmatrix} + \begin{bmatrix} 2 \\ -2 \end{bmatrix} f(x_2) =: Ax + bf(x_2). \tag{1.45}$$

Using the coordinate transformation given by $w_1 = x_2$ and $w_2 = -2x_1 - 2x_2$, system (1.45) becomes:

$$\begin{pmatrix} \dot{w}_1 \\ \dot{w}_2 \end{pmatrix} = \begin{bmatrix} 0 & 1 \\ -2 & -1 \end{bmatrix} \begin{pmatrix} w_1 \\ w_2 \end{pmatrix} + \begin{bmatrix} -2 \\ 0 \end{bmatrix} f(w_1) =: A_1 w + b f(w_1), \quad (1.46)$$

which is still in Lur'e form.

In this representation, A_1 is not diagonally stable, but $A_1 - \epsilon I$ is diagonally stable for all $\epsilon > 0$: such a matrix A_1 is said to be *additively diagonally stable* (see Section 2.8). This being the case, it is known that there exists a diagonal Liapunov function (see Section 3.2) which ensures absolute stability of the zero solution of (1.46) and consequently of the system (1.44):

$$V(w) = w_1^2 + \frac{1}{2}w_2^2, \quad \text{for which}$$

$$\dot{V}(w_1, w_2) = -4f(w_1)w_1 - w_2^2,$$

which is clearly negative definite, by the assumption on the function f. Note also that the matrix A is not additively diagonally stable.

The discussion above emphasizes the importance of a suitable mathematical representation insofar as the existence of diagonal-type Liapunov functions is concerned and also emphasizes the fact that diagonal stability and additive diagonal stability are not invariant under coordinate changes. Furthermore, convenient changes of variables clearly should not be left to one's ingenuity, especially for systems with larger dimensions. Some systematic ways of calculating similarity transformations that lead to diagonally stable structures can be found in Chapter 3.

To conclude this chapter, it is opportune to say that the ultimate goal of the use of diagonal-type Liapunov functions is to derive nonconservative stability results, associated to the simplicity of the functions and their manipulation, and indeed this is what happens in the examples presented in this chapter as well as in many others in the rest of this book.

Chapters 2 and 3 expose a set of results that provide a mathematical framework that makes the role of matrix diagonal stability more understandable and susceptible to further analyses. Additional examples, as well as those of this chapter, are treated in more detail in Chapters 4, 5, and 6 in order to reinforce the general conclusion that matrix diagonal stability is an important topic in the understanding, as well as in the analysis and design problems of many types of dynamical systems.

1.8   Notes and References

Prerequisites for this book
Linear algebra and difference and differential equations prerequisites at the level needed for this book can be found in [Str88] and [HS74]. Matrix theory is comprehensively covered in [HJ85, HJ91, LT85] as well as the classic [Gan59].

Prerequisites from linear system and control theory are covered in [Kai80, Del88, Son98]. The books [Kha92, Vid93] and the classic [Hah67] contain all that is needed from nonlinear systems and continuous-time Liapunov stability theory, while [LaS76, LaS86, Ela96] cover discrete-time Liapunov stability theory.

2
Matrix Diagonal and D-Stability

This chapter presents the matrix theory concepts of diagonal stability and also of D-stability and gives the properties of these classes of matrices in various cases that are used extensively in the remaining chapters of this book as well as related results that are important in themselves. It is thus intended as a reference chapter for the whole book. The reader does not have to read this whole chapter before reading the rest of the book, but rather consult it when necessary. There is, of course, no claim to completeness, in terms of theoretical results on matrix diagonal and D-stability, but the Notes and References section at the end of the chapter points out where many additional results can be found.

2.1 Matrix Stability Concepts

Around 1890, the Russian mathematician Liapunov showed that the differential equation

$$\frac{d}{dt}x(t) = Ax(t) \qquad (2.1)$$

is asymptotically stable (i.e., all trajectories converge to zero) if and only if there exists a positive definite matrix P such that the matrix

$$A^T P + PA \text{ is negative definite.}$$

Around the same time Hurwitz showed that stability could also be characterized in terms of the eigenvalues of the matrix A that occurs on the right-hand side of the differential equation above. This fact is commemorated today by defining a real square matrix to be Hurwitz stable if all of its eigenvalues lie in the open left half of the complex plane. Actually Liapunov also showed that one can choose any symmetric positive definite matrix Q and then solve the linear matrix equation

$$A^T P + P A = -Q, \tag{2.2}$$

known as the *Liapunov equation*, for the unique matrix P which is guaranteed to be positive definite if the system (2.1) is asymptotically stable. The importance of this result is well captured in the following quote: "Somehow the location of eigenvalues of a troublesome general matrix (A) is related to the location of the eigenvalues of a much more friendly symmetric matrix (P) via a linear matrix equation." [Joh98]

In the discrete-time case one considers the difference equation

$$x(k + 1) = A x(k). \tag{2.3}$$

The analogous facts for this equation are as follows. The zero solution is stable if and only if there exists a positive definite matrix P such that

$$A^T P A - P \text{ is negative definite.}$$

Once again, stability can be characterized in terms of the eigenvalues of the matrix A, which is said to be Schur stable if all eigenvalues lie in the interior of the unit disk in the complex plane. Similarly, in discrete-time, A is Schur stable if and only if, for any choice of positive definite matrix Q, the *Stein equation*

$$A^T P A - P = -Q \tag{2.4}$$

admits a unique positive definite solution P.

It should be noted that Hurwitz stability is related to the Liapunov equation while Schur stability is related to the Stein equation.

The *quadratic form*, $x^T P x$, where $x \in \mathbb{R}^n$, associated to the positive definite matrix P is known as a *quadratic Liapunov function* for the respective dynamical system. The Stein equation is sometimes also referred to as the discrete-time Liapunov equation. If it is desired only to find a positive definite matrix P that makes the matrices $A^T P + P A$ or $A^T P A - P$ negative definite, then one speaks of finding feasible solutions to the *Liapunov and Stein linear matrix inequalities (LMIs)* $A^T P + P A < 0$ and $A^T P A - P < 0$. For a review of concepts related to the Stein and Liapunov equations as well as a review of Liapunov stability theory, the reader may consult the appendices to Chapters 2 and 3.

The concept of stability and asymptotic stability of a compact set of matrices was introduced in [BC73, Vid78, BT79], and a particular case of this concept later studied in [BY89]. Following these references, we say that a set $\{A_1, \ldots, A_n\}$ of matrices is called *simultaneously stable* if there exists a single Liapunov function $V(\cdot)$ that establishes the stability of the matrices $A_1, \ldots, A_n$; in this case $V(\cdot)$ is called a *simultaneous Liapunov function (SLF)*. Note that in the definition in [BY89], V was restricted to be a quadratic Liapunov function: Following [BT79] this is not required here. Also, in [NT73, Vid78], a simultaneous Liapunov function was referred to as a *common* Liapunov function. See Section 2.6 for further details and definitions.

It should be pointed out here that the stability definitions adopted above are motivated by the study of the behavior of solutions to the dynamical systems of, for example, the type (1.1). Roughly speaking, the equilibrium solution, which is the zero solution, is asymptotically stable if any solution starting from some suitably close initial condition tends to it. In the continuous-time case this means that all eigenvalues of the matrix A must lie in the open left half plane. From the matrix theoretic and purely mathematical point of view, one could equally well choose to define the concept of *positive stability* and require all eigenvalues to belong to the open right half plane. This choice has the advantage of being able to discuss stability exclusively in terms of positive definite matrices and, in fact, the matrix theory literature often deals with positive stability (see [HJ91] for example). An important warning must be issued to the reader here: In some proofs, it is convenient to use positive stability and this will be done with suitable notice whenever necessary. Clearly a matrix A is Hurwitz stable if and only if $-A$ is positive stable, so that it is easy to translate results on one type of stability to results on the other.

For the purposes of this book, the central stability concept is that of *diagonal stability*. A matrix A is said to be Hurwitz diagonally stable if there exists a positive diagonal matrix P that makes $A^T P + PA$ negative definite, and A is Schur diagonally stable if there exists a positive diagonal P that makes $A^T PA - P$ negative definite. The notation that is used throughout the book denotes the class of Hurwitz diagonally stable matrices by $\mathcal{D}_c$ and the class of Schur diagonally stable matrices by $\mathcal{D}_d$. In both cases, diagonal stability is a stronger requirement than just stability, since the solution to the Liapunov equation must be diagonal and positive definite, as opposed to merely positive definite. Note that Hurwitz diagonal stability is also known by several other names in the literature: Volterra stability, Volterra–Liapunov or VL-stability [Cro78, HS98], dissipativeness [SL83], D^+L-stab [CDHJ98]. The diagonal solution to the Liapunov equation is sometimes referred to as a Volterra multiplier [Red85a, Red85b] or scaling factor [BS91], since the earliest application of this class was in the Liapunov stability analysis of the Lotka–Volterra system of equations of population biology [Vol31].

A related concept is that of D-stability: A matrix A is Hurwitz D-stable if for every positive diagonal matrix D, the matrix DA is Hurwitz stable. Similarly, a matrix A is Schur D-stable if for every diagonal matrix with all diagonal elements in the closed interval $[-1, 1]$, the matrix DA is Schur stable. Diagonal stability is a sufficient condition for D-stability (in both the

Hurwitz [Joh74b] and Schur [BK93] cases), but the converse is only true for certain classes of matrices. Ever since the introduction of the concept of D-stability in the context of price adjustment in economic systems in [AM58, Joh77] much has been written on it. Attempts to characterize Hurwitz D-stability have been successful only for certain classes of matrices [Joh74b, Dat78, CDJ82]. The concept of D-stability is also intimately related to robustness of dynamical systems, as will become clear at various places in this book, and can also be examined in the framework of the structured singular value of a matrix [CFY95, MB96].

Theoretical characterizations of the class of diagonally stable matrices have been given in terms of: (a) theorems of the alternative [BBP78] and (b) the Hadamard product [Kra91]. However these characterizations are not computationally effective. Another theoretical approach, based on a generalization of Cross' characterization for three by three matrices [Cro78] is given in [Red85b]: It gives a theoretical solution for four by four matrices and suggests a computational procedure for matrices of larger dimension. On the other hand, various optimization-based numerical algorithms exist to decide whether a given matrix belongs to class $\mathcal{D}_c$ [Kha82, Ger85, Hu87, Hu92, BGFB94]. The reference [BGFB94] discusses efficient polynomial-time interior point algorithms to solve linear matrix inequalities, such as the Liapunov inequality. At present, algebraic characterizations are known only for matrices of dimension less than or equal to three, for both diagonal and D-stability. Section 2.8 contains the algebraic characterizations of diagonal and D-stability for matrices of dimensions two and three, in addition to discussing theoretical and computational aspects.

Although many results are available for Hurwitz diagonal and D-stability, the Schur case has received little attention in the literature. Thus more space will be devoted to the Schur case in this chapter. The qualifiers Schur and Hurwitz will be dropped whenever the context makes it clear which kind of stability is meant. However, to avoid confusion, the subscript d will be used for classes of matrices in the discrete-time or Schur case and the subscript c for the continuous-time or Hurwitz case.

Once again, it should also be pointed out that this chapter is not intended to be a comprehensive survey of the literature on the problems of diagonal and D-stability, but rather a compilation of results that are relevant to the problems treated in this book.

2.1.1   Basic Notation and Terminology

The basic notation and definitions of general classes of matrices that are used throughout this book are established below, although the reader will have noticed that some have already been used informally in the preceding chapter. More specific classes are defined in the text where needed.

Definition 2.1.1 *A matrix* $A = (a_{ij})$ *in* $\mathbb{R}^{n \times n}$ *is called:*

(a) positive *if* $a_{ij} > 0$ *for all* i, j. *This is denoted* $A \succ 0$. *The notation* $A \succ B$ *means that* $A - B \succ 0$.

(b) nonnegative *if* $a_{ij} \geq 0$ *for all* i, j. *This is denoted* $A \succeq 0$. *The notation* $A \succeq B$ *means that* $A - B \succeq 0$. *The term* nonpositive matrix *is defined analogously and denoted* $A \preceq 0$.

(c) positive definite *if* A *is symmetric and all its eigenvalues are positive. This is denoted* $A > 0$. *The term* negative definite *means* $-A > 0$, *and is denoted* $A < 0$.

(d) positive semidefinite *if* A *is symmetric and all its eigenvalues are nonnegative. This is denoted* $A \geq 0$. *The term* negative semidefinite *means* $-A \geq 0$, *and is denoted* $A \leq 0$.

(e) positive diagonal *if* $a_{ii} > 0$ *and* $a_{ij} = 0, i \neq j$. *Denoting* A *as* diag $(a_{11}, \cdots, a_{nn})$, *note that* $A \succeq 0$ *and* $A > 0$ *in this case.*

(f) *a* Z-matrix *(or in the class $\mathcal{Z}$) if* $a_{ij} \geq 0$ *for* $i \neq j$.

(g) *a* nonsingular M-matrix *or in the class* $\mathcal{M}$ *if it is a Z-matrix such that the real part of every eigenvalue is negative. Alternatively, A can be expressed in the form* $A = B - rI$ *for some nonnegative matrix* $B \succeq 0$, *with* $r > \rho(B)$.

(h) *a* singular M-matrix *or in the class* $\mathcal{M}_o$ *if* $a_{ii} \leq 0$, $a_{ij} \geq 0$ *and the real part of every eigenvalue of A is nonpositive.*

(i) *a* nonsingular H-matrix *or in the class* $\mathcal{C}$ *if* $a_{ii} \leq 0$ *and its* comparison matrix $C \in \mathcal{M}$, *where* $C = (c_{ij})$ *is defined as* $c_{ii} = a_{ii}$, $c_{ij} = |a_{ij}|$. *Note that the comparison matrix of a matrix in class $\mathcal{M}$ is equal to itself.*

(j) *a* singular H-matrix *or in the class* $\mathcal{C}_o$ *if its comparison matrix is in* $\mathcal{M}_o$.

(k) *a* quasidominant *matrix, or in the* class *of diagonally row-sum or, respectively, column-sum* quasidominant *matrices if there exists a positive diagonal matrix* $P = $ diag $(p_1, \cdots, p_n)$ *such that* $a_{ii}p_i \geq \sum_{j \neq i} |a_{ij}|p_j, \forall i$, *(respectively)* $a_{jj}p_j \geq \sum_{i \neq j} |a_{ij}|p_i, \forall j$. *If these inequalities are strict, the matrix is referred to as strictly row-sum (respectively column-sum) quasidominant. If P can be chosen as the identity matrix, then the matrix is called row- or column- diagonally dominant.*

Some points are worth noting about Definition 2.1.1 (k). Clearly A is row-sum quasidominant with the scaling factors given by the diagonal entries of the matrix P if and only if AP is row dominant. Actually, row-sum and column-sum quasidominance are equivalent, although strict row and column dominance are not. Of course, there is no requirement that the *same* diagonal matrix work for rows and columns. In other words, if A is quasidominant, there exist positive diagonal matrices P, Q such that AP and QA are row- and column-dominant respectively. Thus, it is usual to refer to a matrix as simply being *quasidominant*. Notice also that the definition only applies to a matrix which has positive diagonal entries. If this is not the case, quasidominance is defined by some authors using the absolute value sign on the diagonal entries a_{ii}, i.e., $d_i|a_{ii}| > \sum_{j \neq i} d_j|a_{ij}|$. This definition is not used in this book.

If a matrix A has nonnegative off-diagonal elements, it is said to have the *sign pattern* of an M-matrix. An M-matrix is also sometimes referred to as a *type K matrix*: The latter term is somewhat more general and is used in the theory of order preserving flows of monotone dynamical systems (see [Smi88] for details). This terminology will not be used in this book, preference being given to the more widespread term M-matrix.

Given a matrix $A = (a_{ij})$, the nonnegative matrix $|A|$ is defined by replacing the elements of A by their moduli, i.e., $|A| = (|a_{ij}|)$. Similarly, for a vector $x \in \mathbb{R}^n$, the vector $|x|$ is defined as $(|x_1|, |x_2|, \cdots, |x_n|)^T \in \mathbb{R}^n_+ \cup \{0\}$.

In what follows, for a matrix (or vector) A, $\rho(A)$ and $\lambda_i(A)$ are the spectral radius of A and the ith eigenvalue of A (in some ordering) respectively.

Consider the vector $x = (x_1, \ldots, x_n)^T \in \mathbb{R}^n$ and the associated vector norms:

$$\|x\|_\infty := \max_i\{|x_i|\}, \quad \|x\|_2 := \sqrt{\sum_i x_i^2}.$$

Given $P = \text{diag}\,(p_1, \ldots, p_n)$, $p_i > 0, \forall i$, diagonal scalings (or weightings) of these norms are as follows:

$$\|x\|_\infty^P := \|P^{-1}x\|_\infty := \max_i\{p_i^{-1}|x_i|\},$$

which induces the matrix norm $\|A\|_\infty^P = \|P^{-1}AP\|_\infty$ and

$$\|x\|_2^P := \|P^{\frac{1}{2}}x\|_2 = \sqrt{\sum_i p_i x_i^2},$$

which induces the matrix norm $\|A\|_2^P = \|P^{\frac{1}{2}}AP^{-\frac{1}{2}}\|_2$, where the matrix $P^{\frac{1}{2}} := \text{diag}\,(\sqrt{p_1}, \ldots, \sqrt{p_n})$ and $P^{-\frac{1}{2}} = (P^{\frac{1}{2}})^{-1}$. All the vector norm functions defined above are often used as diagonal-type Liapunov functions.

It is sometimes convenient to write the positive diagonal entries of a positive diagonal scaling matrix $P = \text{diag}\,(p_1, \cdots, p_n)$ as a positive vector $v = (p_1, \cdots, p_n) \succ 0$, and, in this case, the weighted infinity norm defined above is also written as:

$$\|x\|_\infty^v = \max_i \frac{|x_i|}{v_i}, \tag{2.5}$$

where x_i is the ith component of the vector x. It is clear that

$$\|x\|_\infty^v \leq \beta \quad \text{if and only if} \quad |x| \preceq \beta v. \qquad (2.6)$$

Let $x = (x_1, \cdots, x_m)^T \in \mathbb{R}^N$, where $N = n_1 + \cdots + n_m$, and $x_i \in \mathbb{R}^{n_i}$. The *vectorial norm* of x, denoted $[\![x]\!]$, is defined as:

$$[\![x]\!] := (\|x_1\|, \cdots, \|x_m\|)^T \in \mathbb{R}^m, \qquad (2.7)$$

where the norms of the vectors x_i are arbitrary (i.e., the norms may, in fact, all be different). The symbol $\mathbb{1}$ denotes a square *matrix of ones*, i.e., a matrix of dimension appropriate to the context with all elements equal to 1.

The notation $\langle x, y \rangle := x^T y$ is used for the standard *inner product* on $\mathbb{R}^n$.

Associated to any $n \times n$ matrix $M = (m_{ij})$, with $n \geq 2$, there exists a *directed graph* $G = (N, E)$ with the set of nodes $N = \{1, \ldots, n\}$ and edges $E = \{(i, j) : i \neq j \text{ and } m_{ij} \neq 0\}$.

Definition 2.1.2 *An $n \times n$ matrix A, with $n \geq 2$, is called* irreducible *if for every $i, j \in N$ there exists an oriented path from i to j in graph G and consequently this graph is* strongly connected *[Var62]. For $n = 1$, A is called irreducible if its only element is nonzero.*

Alternatively, A is called *reducible* if there exists a permutation matrix P such that

$$B = P^T A P \quad \text{has the form} \quad \begin{bmatrix} B_{11} & B_{12} \\ 0 & B_{22} \end{bmatrix}.$$

Otherwise A is called *irreducible* or *indecomposable*.

2.1.2   Basic Results on Hurwitz Diagonal and D-Stability

This section contains basic results on various subsets of Hurwitz stable matrices that are used throughout the book.

Definition 2.1.3 *A matrix $A := (a_{ij})$ is said to belong to:*

(a) *the class $\mathcal{H}$ of Hurwitz stable (or $\mathcal{H}$-stable) matrices if all of its eigenvalues have negative real parts.*

(b) *the class $\mathcal{D}_c$ of Hurwitz diagonally stable matrices whenever there exists $P > 0$ diagonal such that $A^T P + P A < 0$. When it is desired to make explicit a diagonal matrix P that makes the matrix $A^T P + P A$ negative definite, the notation $A \in \mathcal{D}_c(P)$ is used.*

(c) *the class $\mathcal{D}_{co}$ of Hurwitz diagonally semistable matrices whenever there exists $D > 0$ diagonal such that $A^T D + D A \leq 0$.*

(d) *the class $\mathbb{D}_c$ of Hurwitz D-stable matrices if AD (or DA) is in $\mathcal{H}$ for any positive diagonal matrix D.*

(e) the class $\mathcal{T}_c$ of totally D-stable *matrices if every principal submatrix of A is Hurwitz D-stable.*

(f) the class $\mathcal{A}_o$ if $(A - \alpha I)$ is in $\mathcal{D}_c$, for every scalar $\alpha > 0$. A matrix in this class is called additively diagonally stable.

(g) the class $\mathcal{P}$ of matrices if all of its principal minors are positive.

(h) the class $\mathcal{P}_o$ if all of its principal minors are nonnegative.

(i) the class $\mathcal{P}_o^+$ if all its principal minors are nonnegative and at least one principal minor of each order is positive. Clearly $\mathcal{P}_o^+ \subset \mathcal{P}_o$.

(j) the class $\mathcal{AD}_c$ of additively D-stable matrices, if $(A - D)$ is Hurwitz stable for any nonnegative diagonal matrix D.

The *stability modulus* of an $n \times n$ matrix A is denoted $s(A)$ and defined as $s(A) = \max \mathrm{Re}(\lambda)$, where $\mathrm{Re}(\lambda)$ denotes the real part of the complex number λ and λ runs over the spectrum of A. Clearly, A is Hurwitz stable if and only if $s(A) < 0$.

An alternative definition of diagonal stability arises in the literature on mathematical population biology. In [SL83, p.132ff.] (also see [Red85a]), motivated by the behavior of the trajectories of the Lotka–Volterra differential equations, the following definitions are introduced: (i) a matrix $A \in \mathbb{R}^{n \times n}$ is called *dissipative* if there exists a positive diagonal matrix $P \in \mathbb{R}^{n \times n}$ such that the quadratic form $x^T P A x$ is negative definite; (ii) a matrix $A \in \mathbb{R}^{n \times n}$ is called *conservative* if there exists a positive diagonal matrix $P \in \mathbb{R}^{n \times n}$ such that the quadratic form $x^T P A x$ is identically zero.

Since the quadratic form $x^T P A x$ is negative definite if and only if the symmetric part of PA $(:= 0.5(A^T P + PA))$ is negative definite, this means that the concepts of dissipativeness and Hurwitz diagonal stability are identical. This way of looking at diagonal stability also motivates the name Volterra multiplier for the positive diagonal scaling matrix P that makes the symmetric part of PA negative definite. Similarly, if a matrix A is conservative, then there exists a positive diagonal scaling of A that makes the scaled matrix PA skew-symmetric.

It has long been known that Hurwitz diagonal stability is sufficient to ensure Hurwitz D-stability. In fact a little more is true:

Lemma 2.1.4 *If $A \in \mathcal{D}_c(P)$ then, for all diagonal matrices $K > 0$, $AK \in \mathcal{D}_c(PK)$, $KA \in \mathcal{D}_c(K^{-1}P)$ and A is additively D-stable, i.e., $A \in \mathcal{AD}_c$ and, in fact, for all nonnegative diagonal matrices $D \geq 0$, $(A-D) \in \mathcal{D}_c(P)$.*

Proof. By the hypothesis $A \in \mathcal{D}_c(P)$, there exists $P > 0$ diagonal such that $PA + A^T P = -Q < 0$. Pre- and post-multiplying by the positive diagonal matrix K yields

$$KPAK + KA^T PK = -KQK < 0,$$

which can be rewritten as $P_1(AK) + (AK)^T P_1 < 0$, where $P_1 := PK = KP > 0$, diagonal. The statement for KA can be proved similarly. To see that A is additively D-stable, consider the following calculation. For any $D \geq 0$ (nonnegative diagonal matrix),

$$(A^T - D)P + P(A - D) = A^T P + PA - 2DP \leq A^T P + PA < 0,$$

showing that $(A - D) \in \mathcal{D}_c(P)$.   ∎

Lemma 2.1.4, in the case of premultiplication, can also be stated as follows: $A \in \mathcal{D}_c$ implies that KA is Hurwitz stable for any positive diagonal K; thus A is D-stable. The implication clearly also holds for postmultiplication.

Lemma 2.1.5 *If $A \in \mathcal{D}_c(P)$ then $A^{-1} \in \mathcal{D}_c(P)$.*

Proof. By the hypothesis $A \in \mathcal{D}_c(P)$, there exists $P > 0$ diagonal such that $PA + A^T P = -Q < 0$ and further, A is nonsingular. Pre-multiplying by the matrix $(A^{-1})^T$ and post-multiplying by the matrix A^{-1} yields

$$
\begin{aligned}
(A^{-1})^T PA(A^{-1}) + (A^{-1})^T A^T P(A^{-1}) &= (A^{-1})^T P + P(A^{-1}) \\
&= -(A^{-1})^T Q(A^{-1}) < 0.
\end{aligned}
$$

and this completes the proof.   ∎

The following example shows that Hurwitz D-stability does not imply Hurwitz diagonal stability, so that the set of D-stable matrices is a strict subset of the set of diagonally stable matrices

Example 2.1.6 *The matrix* $\begin{bmatrix} 0 & -1 \\ 1 & -1 \end{bmatrix}$ *is D-stable. This is because, if it is premultiplied by any positive diagonal matrix $D = \mathrm{diag}\,(d_1, d_2)$, then the trace of the product DA is $-d_2 < 0$ and its determinant is $d_1 d_2 > 0$. This ensures that the degree two monic characteristic polynomial of the matrix DA has positive coefficients and therefore has roots (eigenvalues of A) with negative real parts. It is also easy to see by explicit calculation that the matrix $A^T D + DA$ can never be negative definite for any positive diagonal matrix D since its (1,1) entry is zero. Of course, matrix A is an example on the frontier between diagonal and D-stability, since if the (1,1) entry is made slightly negative then the resulting matrix is clearly diagonally stable. However, it is also possible to give an example of a matrix that is in the interior of the set of D-stable matrices [Har80]. The matrix*

$$B = \begin{bmatrix} -1 & 0 & 50 \\ -1 & -1 & 0 \\ -1 & -1 & -1 \end{bmatrix}$$

can be shown to be D-stable (see Section 2.8), while the following analysis shows that it is not diagonally stable. To simplify matters, consider the matrix $C = -B$, and let the case of positive stability be considered. Let

$D = \mathrm{diag}\ (d_1, d_2, d_3)$, *with* $d_i > 0$. *Then*

$$DC + C^T D = \begin{bmatrix} 2d_1 & d_2 & d_3 - 50d_1 \\ d_2 & 2d_2 & d_3 \\ d_3 - 50d_1 & d_3 & 2d_3 \end{bmatrix}.$$

The matrix $DC + C^T D$ *is positive definite if and only if all its principal minors are positive, although it suffices to work with the leading principal minors. The first minor* $(2d_1)$ *is clearly positive. Thus positivity of the following minors must hold (noting that positivity of (i) and (iii) suffices):*

(i) $4d_1 d_2 - d_2^2$,

(ii) $4d_2 d_3 - d_3^2$, *and*

(iii) $54 d_1 d_2 d_3 - d_1 d_3^2 - d_2^2 d_3 - 2500 d_1^2 d_2$.

Since the $d_i s$ *are positive, the principal minors listed above are positive if:*

(a) $4d_1 > d_2$,

(b) $4d_2 > d_3$, *and*

(c) $54 d_1 d_2 d_3 > d_1 d_3^2 + d_2^2 d_3 + 2500 d_1^2 d_2$.

Substituting (a) and (b) into the left side of (c) yields

$$54 d_1 (4d_1)(4d_2) > d_1 d_3^2 + d_2^2 d_3 + 2500 d_1^2 d_2$$

or

$$864 d_1^2 d_2 > d_1 d_3^2 + d_2^2 d_3 + 2500 d_1^2 d_2.$$

But this inequality does not hold for any positive values of d_1, d_2, *and* d_3, *because the term on the left-hand side is always exceeded by the third term on the right-hand side. Hence* C *is not positive diagonally stable, implying that* B *is not in class* $\mathcal{D}_c$. *Another test that shows this is given in Section 2.8.*

The following lemma relates stability properties of a matrix to those of its transpose and its inverse, thus providing an extension as well as useful alternative view of Lemma 2.1.5.

Lemma 2.1.7 *If a matrix* A *is Hurwitz stable, D-stable, or diagonally stable, then the matrices* A^T *and* A^{-1} *also have the corresponding properties.*

Proof. First note that the spectra of A and A^T coincide. Thus if A is Hurwitz stable, then so is A^T, since Hurwitz stability of a matrix depends on its eigenvalue location. The eigenvalues of A^{-1} are the reciprocals of those of A. This means that if an eigenvalue of A has negative real part: i.e., $\lambda(A) = -\sigma \pm j\omega$, then its reciprocal $1/(\lambda(A)) = (-\sigma/\alpha) \pm j(\omega/\alpha)$, $\alpha = \sigma^2 + \omega^2$ also has negative real part.

Turning now to D-stability, note that, for positive diagonal matrices D, the matrix DA is similar to the matrix $AD = D^{-1}(DA)D$, so that the eigenvalues of DA and AD coincide. Thus if DA is stable for all D, so are AD and its transpose $(AD)^T = DA^T$, showing that if A is D-stable, then so is A^T. By the first paragraph of the proof, AD stable for all positive diagonal D implies $(AD)^{-1}$ stable for all D, which can be written as $D^{-1}A^{-1}$ stable for all D, which, in turn, can be interpreted as saying that A^{-1} is D-stable.

The remaining assertions are proved using the equivalent definition of dissipativity instead of diagonal stability. If A is dissipative, then there exists a positive diagonal matrix P such that $\langle PAx, x \rangle < 0, \forall x$. But $\langle PAx, x \rangle = \langle x, A^T Px \rangle$. Applying the congruence transformation (change of variables) $x = P^{-1}z$, which maintains the negative definiteness of the quadratic form, $\langle P^{-1}z, A^T PP^{-1}z \rangle = \langle z, P^{-1}A^T z \rangle < 0, \forall z$, which shows that A^T is dissipative. Note, by the equivalence of dissipativeness and diagonal stability, that if A is dissipative then it is Hurwitz stable and hence invertible. Writing the definition of dissipativeness as $\langle A^T Px, x \rangle < 0, \forall x$ and applying the congruence transformation $x = A^{-1}z$ leads to $\langle A^T PA^{-1}z, A^{-1}z \rangle = \langle PA^{-1}z, AA^{-1}z \rangle < 0, \forall z$, completing the proof. ∎

In a quadratic form of fixed sign, if some variables are set to zero, the form will still have the same fixed sign with respect to the remaining variables: this is also known as *specializing the quadratic form*. Thus any principal submatrix of a dissipative matrix is itself dissipative and hence, by Lemma 2.1.4, D-stable as well.

The above results lead to the following chain of implications: Diagonal stability implies total stability which implies D-stability which implies stability. The observation that principal submatrices of diagonally stable matrices are also diagonally stable illustrates one of the points being made in this book, which is that diagonal stability has several nice "structural consequences." The principal submatrix property above implies that all the corresponding "principal subsystems" of a given diagonally stable system are diagonally stable, and, as later chapters will show, also enjoy other "structural properties" associated with diagonal stability. An alternative way of stating the observation just made is to say that a property of a matrix A is *hereditary* if every principal submatrix of A shares it. Thus, diagonal stability and diagonal semistability are hereditary properties, while D-stability is not. This discussion is formalized in the following lemma.

Lemma 2.1.8 [Cro78] *If A is in $\mathcal{D}_c$, then so are all principal submatrices of A. If A is D-stable, then all principal submatrices of A are D-semistable, i.e., the matrix DA has nonpositive real parts for all positive diagonal matrices D.*

Proof. Let P be a positive diagonal matrix such that $Q := PA + A^T P < 0$. Since any principal submatrix of Q is negative definite, as can be seen by specialization of the associated quadratic form, the result is immediate. To prove the second assertion, let a matrix with all eigenvalues having nonpositive real part be called semistable. Given P positive diagonal and a submatrix $A_{i_1,\ldots,i_\ell}$ (the subscripts indicate the rows and columns of A that make up the submatrix), let L be the limit of DA as the d_i with $i \notin \{i_1,\ldots,i_\ell\}$ decrease to zero. Every eigenvalue of $D_{i_1,\ldots,i_\ell}A_{i_1,\ldots,i_\ell}$ is an eigenvalue of L. But L is semistable, because A is D-stable and the eigenvalues of DA depend continuously on the d_i.   ∎

Corollary 2.1.9 *If A is in $\mathcal{D}_c$, then A is in $\mathcal{T}_c$.*

Proof. Any principal submatrix of a dissipative matrix is itself dissipative and hence, by Lemma 2.1.4, D-stable as well. It follows that any dissipative (i.e., Hurwitz diagonally stable) matrix is totally D-stable.   ∎

Lemma 2.1.10 [Met45] *A necessary condition for A to be totally D-stable is that every principal minor of even order is positive and every principal minor of odd order is negative.*   □

Theorem 2.1.11 *If $-A$ is quasidominant, then A is in $\mathcal{D}_c$.*

Proof. This theorem is restated and proved as Theorem 2.7.8 below.   □

There are many connections between dominance and diagonal-type Liapunov functions [Wil76].

A useful alternate characterization of the class $\mathcal{P}_o$ is as follows.

Lemma 2.1.12 [SW69, Theorem 1] *A matrix A in $\mathbb{R}^{n \times n}$ is in class $\mathcal{P}_o$ if and only if $\det(P + A) \neq 0$ for every positive diagonal matrix.*   □

Another point worthy of note is that, unlike the case of stability, diagonal stability of a given matrix cannot be decided just by the location of its eigenvalues in the sense that if A and B are two matrices with the same eigenvalues, then it is possible for A to be diagonally stable, even though B is not. For example, $A = -\begin{bmatrix} 1 & 0 \\ 0 & 1 \end{bmatrix}$ is diagonally stable, whereas $B = \begin{bmatrix} 1 & -2 \\ 2 & -3 \end{bmatrix}$ is not, although it has the same eigenvalues as A.

However, the following general fact about the relationship of eigenvalue locations to diagonal stability is true.

Fact 2.1.13 *For all $A \in \mathbb{R}^{n \times n}$, and for any positive diagonal matrix P, there exists $\alpha > 0$ such that $A - \alpha I \in \mathcal{D}_c(P)$.*

Proof. Given any positive diagonal matrix P, observe that

$$\begin{aligned}(A - \alpha I)^T P + P(A - \alpha I) &= (A^T P + PA) - 2\alpha P \\ &= P^{1/2}\left[P^{-1/2}(A^T P + PA)P^{-1/2} - 2\alpha I\right]P^{1/2}\end{aligned}$$

Choosing α larger than the largest of the eigenvalues of the symmetric matrix $S := P^{-1/2}[A^T P + PA]P^{-1/2}$, makes the matrix $S - 2\alpha I$ negative definite. It follows that the congruent matrix $P^{1/2}[S - 2\alpha I]P^{1/2} = A^T P + PA$ is also negative definite. ∎

Fact 2.1.13 can be interpreted in three ways: (i) after an appropriate shift α, a diagonally scaled symmetrized version of A (i.e., S) can be made negative definite; (ii) if the eigenvalues of a matrix are moved sufficiently "deep" into the interior of the left half of the complex plane, then the resulting matrix becomes diagonally stable; (iii) if a matrix has a sufficient degree of diagonal dominance and the diagonal elements are all negative, which is a necessary condition for diagonal stability, then this matrix is diagonally stable. In the example above, note that $B - \alpha I$ is in $\mathcal{D}_c(I)$, for all $\alpha > 1$ and, similarly $B - \text{diag}\,(1 + \epsilon_1, \epsilon_2)$ is also in $\mathcal{D}_c(I)$, for all positive ϵ_1, ϵ_2. In other words, the matrix $B - \text{diag}\,(1, 0)$ is additively diagonally stable. From the proof of Fact 2.1.13, it is now clear that the following fact can also be stated.

Fact 2.1.13a *For all $A \in \mathbb{R}^{n \times n}$, there exists $\alpha > 0$ such that $A - \alpha I$ is in class $\mathcal{A}_o$.* □

More details, references and results involving these definitions can be found in [HJ91, Her92]. Some specific results on these classes of matrices that are used in this book follow.

Lemma 2.1.14 [OT51] *For matrices $A, B, C \in \mathbb{R}^{n \times n}$, if $A = B + C$, where B is symmetric positive definite and C is skew-symmetric, then $\det A > \det B$.* □

Corollary 2.1.15 *If the quadratic form associated to any matrix A, not necessarily symmetric, is positive definite, then it has positive determinant.*

Proof. The matrix A can be written as $B + C$, where $B := \frac{1}{2}(A + A^T)$ is symmetric and $C := \frac{1}{2}(A - A^T)$ is skew-symmetric. Clearly $x^T A x = x^T B x > 0$, for all x, so that B is positive definite. Thus, by Lemma 2.1.14, $\det A > \det B > 0$. ∎

Lemma 2.1.16 [Cro78] *The following facts are true.*

(i) $A \in \mathcal{D}_c$ implies that $-A \in \mathcal{P}$.

(ii) $A \in \mathcal{D}_{co}$ implies that $-A \in \mathcal{P}_o$.

(iii) If $A \in \mathcal{A}_o$, then $-A \in \mathcal{P}_o$.

(iv) If A is D-stable, then $-A \in \mathcal{P}_o^+$.

Proof. Applying Corollary 2.1.15 to the matrix $-PA$, it follows that $\det(-PA) > 0$. But $\det(-PA) = \det P \det(-A)$, and $\det P > 0$, implying that $\det(-A) > 0$. A similar result is obtained for each principal submatrix by specializing the quadratic form, as in Lemma 2.1.8. This gives $-A \in \mathcal{P}$, as required. Applying the result to $A - \epsilon I$ and letting $\epsilon \to 0_+$ shows that $A \in \mathcal{D}_{co}$ implies that $-A \in \mathcal{P}_o$. This argument also proves item (iii). Finally, to prove item (iv), note that by Lemma 2.1.8, every principal sub-matrix $A_{i_1,\ldots,i_\ell}$ is semistable, implying that $(-1)^\ell \det A_{i_1,\ldots,i_\ell} > 0$, since it is the product of eigenvalues of $A_{i_1,\ldots,i_\ell}$, and the complex eigenvalues occur in conjugate pairs. Finally, stability of A implies that every coefficient of its characteristic polynomial (these are sums of principal minors with sign) is positive, and hence $-A \in \mathcal{P}_o^+$.

Lemma 2.1.17 [Cro78] *For normal matrices and within the set $\mathcal{Z}$, D-stability, additive D-stability, and diagonal stability are all equivalent to Hurwitz stability.*

Proof. If a normal matrix A is Hurwitz stable, then it is also diagonally stable, because $A^T + A < 0$, i.e., $A \in \mathcal{D}_c(I)$. By Lemma 2.1.4, this implies additive D-stability as well as D-stability. By a standard result on M-matrices [FP62, Theorem 4.3], Hurwitz stability and the condition $A \in \mathcal{P}$ are equivalent in $\mathcal{Z}$, and if $A \in \mathcal{Z}$ is Hurwitz stable, then AD has positive dominant diagonal for some positive diagonal matrix $D > 0$. Thus, by Theorem 2.1.11, A is in $\mathcal{D}_c$, i.e., a stable $A \in \mathcal{Z}$ is also diagonally stable. The proof is completed by applying Lemma 2.1.4. ∎

However, in general, these stability types are quite different. From the proof of Lemma 2.1.4, it is clear that diagonal stability of A is equivalent to:

For every $D > 0$, there exists a diagonal $P > 0$ such that:

$$(A^T D)P + P(DA) < 0, \tag{2.8}$$

while by the Liapunov theorem, D-stability is equivalent to:

For every $D > 0$, there exists a symmetric $P > 0$ such that (2.8) holds.

An algebraic transformation that is used frequently with Liapunov and Stein equations is postmultiplication by a certain matrix W and pre-multiplication by its transpose, i.e., A is transformed to $W^T A W$. This transformation, applied to a matrix, is known as *congruence*. The associated invariance property that holds is called *Sylvester's law of inertia* and states that the number of positive (p), zero (z), and negative (n) eigenvalues of a symmetric matrix is conserved under a congruence transformation (for a beautiful proof using homotopy ideas, see [Str88, p.341]). The triple (p, z, n) is called the *inertia* of the symmetric matrix. Also note that a congruence transformation has the interpretation of a change of variables in the quadratic form associated to a symmetric matrix: i.e., if A is a symmetric matrix and $x^T A x = \langle x, Ax \rangle$ the corresponding quadratic form, then making the change of variables $x = Wz$, where W is a non-singular matrix leads to $\langle Wz, AWz \rangle = \langle z, W^T AWz \rangle$. In particular, if P

is a symmetric and positive definite matrix, and K is a positive diagonal matrix, then $KPK^T = KPK$ is also positive definite; equivalently, the quadratic forms $\langle x, Px \rangle$ and $\langle x, KPKx \rangle$ are both positive definite. A congruence transformation can also be applied to a nonsymmetric matrix: In this case the *inertia* of a matrix A, denoted $\operatorname{In} A$, is defined as the triple $(n_+(A), n_0(A), n_-(A))$, where $n_+(A)$ is the number of eigenvalues of A in the right open halfplane, $n_0(A)$ is the number of purely imaginary eigenvalues of A, and $n_-(A)$ is the number of eigenvalues in the open left halfplane. Using this terminology, a matrix $A \in \mathbb{R}^{n \times n}$ is Hurwitz stable if $n_-(A) = n$. A key result connecting inertia, stability, and the Liapunov equation was proved independently in [OS62] and [Tau61] and is as follows.

Theorem 2.1.18 *For a given real matrix A, there exists a symmetric matrix P such that $A^T P + PA$ is negative definite if and only if $n_0(A) = 0$ and, in this case, $\operatorname{In} A = \operatorname{In} P$.* $\qquad \square$

A real matrix A is defined to be *inertia preserving* if for every real invertible diagonal matrix D, $\operatorname{In} AD = \operatorname{In} D$. A real matrix A is defined to be *strongly inertia preserving* if for every real diagonal matrix D (not necessarily invertible), $\operatorname{In} AD = \operatorname{In} D$ [BS91]. The following fact is clearly true in view of the above definitions.

Fact 2.1.19 *Inertia preserving matrices are Hurwitz D-stable.* $\qquad \square$

From Theorem 2.1.18 it is not difficult to show:

Theorem 2.1.20 [BS91] *A Hurwitz diagonally stable matrix is strongly inertia preserving.* $\qquad \square$

Thus the class of inertia preserving matrices is a subset of the class of D-stable matrices and contains the diagonally stable matrices.

Some results on additive diagonal and D-stability

The class $\mathcal{A}_o$ (Definition 2.1.3 (f)) of additively diagonally stable matrices admits an alternate characterization.

Fact 2.1.21 *Let the class of matrices $\mathcal{A}_1$ be defined as*

$$\{A \in \mathbb{R}^{n \times n} : (A - D) \in \mathcal{D}_c, \ \forall \text{ diagonal } D > 0\}. \tag{2.9}$$

Then $\mathcal{A}_1 = \mathcal{A}_o$.

Proof. Clearly $\mathcal{A}_1 \subset \mathcal{A}_o$, since it is enough to choose $D = \alpha I$. So it remains to show that if $(A - \alpha I) \in \mathcal{D}_c$, $\forall \alpha > 0$, then $(A - D) \in \mathcal{D}_c$, $\forall D > 0$. For any positive diagonal matrix $D = \operatorname{diag}(d_1, \cdots, d_n)$, the matrix $A - D$ can be written as $A - \alpha I - D_1$, where $\alpha = \min_i\{d_i\} - \epsilon$, and $\epsilon > 0$ is chosen such that $\alpha > 0$, and $D_1 := D - \alpha I > 0$. Then, since $A \in \mathcal{A}_o$, there exists $P_\alpha > 0$ such that $(A - \alpha I)^T P_\alpha + P_\alpha(A - \alpha I) =: Q_\alpha$ is negative definite. Now, $(A - \alpha I - D_1)^T P_\alpha + P_\alpha(A - \alpha I - D_1)$ is equal to $Q_\alpha - 2D_1 P_\alpha$. This is the sum of two negative definite matrices, thus it is negative definite. $\qquad \blacksquare$

Lemma 2.1.22 *If a matrix* $-T$ *is row-sum or column-sum quasidominant then* $T \in \mathcal{A}_o$.

Proof. It suffices to observe that the addition of an arbitrary positive diagonal matrix D to a quasidominant matrix T makes the matrix $-T + D$ strictly quasidominant and so, by Theorem 2.7.8, in class $\mathcal{D}_c$ as well. Thus T belongs to class $\mathcal{A}_o$. ∎

The following lemma relates additive D-stability and D-stability.

Lemma 2.1.23 [Cai76] *A real three by three matrix is D-stable if and only if* $A - D$ *is D-stable for all diagonal matrices* D *with nonnegative diagonal entries.* □

The proof of this lemma depends on the characterization of D-stability that is discussed in Section 2.8 and may be found in [Cai76, p.76].

2.2 Special Classes of Hurwitz Diagonally Stable Matrices

For many purposes, it is of interest to identify special classes of matrices that can be proved to be diagonally stable, often by supplying a diagonal matrix solution of the Liapunov equation. Some examples of such classes are: Hurwitz triangular matrices [BBP78], quasidominant matrices [HNC65], M-matrices [BP94], and sign-stable matrices with nonzero diagonal entries [JKD77, MQ69, QR65]. The paper [Red85a] gives a useful overview of several special classes of diagonally stable matrices.

In this context, the following result for M-matrices is very useful.

Theorem 2.2.1 *Let* A *be an M-matrix. Then the stability modulus* $s(A)$ *is an eigenvalue of* A, *there exists a corresponding eigenvector* $v \succ 0$ *and the following are equivalent:*

(i) A *is Hurwitz stable, i.e.,* $s(A) < 0$.

(ii) *There exists* $u \succ 0$ *such that* $Au \prec 0$.

(iii) $A \in \mathcal{D}_c$.

(iv) $-A^{-1} \succeq 0$.

(v) *The following determinantal conditions hold:*

$$(-1)^k \begin{bmatrix} a_{11} & |a_{12}| & \cdots & & |a_{1k}| \\ |a_{21}| & a_{22} & |a_{23}| & & |a_{2k}| \\ \vdots & & & & \vdots \\ |a_{k1}| & \cdots & & |a_{k,k-1}| & a_{kk} \end{bmatrix} > 0, \quad 1 \le k \le n.$$

□

The Perron–Frobenius theorem is the key to the proof of this theorem—see [FP62, LaS76, BP94, Šil78, OŠ85, Smi88, HS98] for more details.

In what follows, the diagonally stable subset of the class of so-called indecomposable quasi-Jacobi matrices [May66] is characterized algebraically by conditions that are similar to the determinantal conditions (v), also known as the Hawkins–Simon conditions [HS49], that characterize M-matrices in the lemma above. Another subset of $\mathcal{D}_c$, namely the class $\mathcal{D}_{c2}$ defined below, is also characterized.

Definition 2.2.2 *A matrix A is said to belong to class $\mathcal{D}_{c_2}$ if and only if $A^T P + PA = -Q < 0$ is satisfied with $P > 0$ and $Q > 0$ both diagonal.*

For both classes of matrices mentioned above, a pair (P, Q) satisfying the Liapunov equation is explicitly given. The interest in these specific classes of matrices is due to the fact that they arise in many of the applications treated in this book.

Some additional terminology is needed. Given a matrix A, the *digraph* $D(A)$ associated to it contains n vertices $(v_i, i = 1, 2, \ldots, n)$ and a directed edge (v_i, v_j) from vertex v_i to vertex v_j if and only if $a_{ij} \neq 0$, $i \neq j$. Assigning the weight a_{ij} to each edge of $D(A)$ the corresponding *weighted digraph* $D_w(A)$ is obtained. The matrix $U = (u_{ij})$ (of zeros and ones), obtained from the matrix $A = (a_{ij})$ in the following manner: $u_{ij} = 1$, whenever $a_{ij} \neq 0$, and $u_{ij} = 0$, otherwise, is called the *adjacency matrix* associated to the digraph $D_w(A)$. The properties of a digraph associated with a given matrix are referred to as properties of the matrix, and vice versa.

A matrix A is *combinatorially symmetric* if and only if $a_{ij} \neq 0$ implies $a_{ji} \neq 0$ for $i \neq j$; in this case $D(A)$ is a symmetric graph and, consequently, the adjacency matrix is symmetric as well. A *chain of length (or order) r* of A is the product $a_{i_1 i_2} \cdot a_{i_2 i_3} \cdot \ldots \cdot a_{i_{r-1} i_r} \cdot a_{i_r i_{r+1}}$, where $i_1, i_2, \ldots, i_r$ are distinct and all the elements of this product are nonzero. A chain is a *cycle of length (or order) r* if and only if $i_{r+1} = i_1$. If A has no cycles of order $r \geq k(k \geq 1)$, then A is said to be *acyclic-k*. Considering the case of a combinatorially symmetric matrix A, each edge of $D(A)$ is in a 2-cycle and $D(A)$ can be transformed into an undirected graph $G(A)$. The edges of $G(A)$ are denoted by $|v_i, v_j|$. Indecomposable Jacobi (or tridiagonal) matrices and more generally quasi-Jacobi matrices [May66] are acyclic-3, and the corresponding graphs $G(A)$ are trees.

The following lemma characterizes the class of acyclic-3 matrices in structural terms.

Lemma 2.2.3 [KH84b] *An indecomposable matrix A is acyclic-3 if and only if:*

(i) A is combinatorially symmetric, and

(ii) it has exactly $n - 1$ nonzero elements above (and below) the main diagonal.

Proof. (*if*) Condition (i) implies that $D(A)$ is symmetric, and thus one can define the undirected graph $G(A)$. Since A is indecomposable, $G(A)$ is connected. Moreover, from condition (ii) one concludes that $G(A)$ has $n - 1$ edges. Thus, A is an acyclic-3 matrix.

(*only if*) If A is acyclic-3 and indecomposable, then $D(A)$ is symmetric and $G(A)$ has $n - 1$ edges, or equivalently, A has $n - 1$ nonzero elements above (and below) the main diagonal.   ■

A matrix A satisfying $|a_{ij}| = |a_{ji}|$ for all $i \neq j$ is said to be *symmetric in modulus*. Using this term, one can say more about acyclic-3 matrices:

Lemma 2.2.4 *If A is indecomposable and acyclic-3, then there exists a similarity transformation given by a diagonal matrix $D = \text{diag}\,(d_1, \cdots, d_n)$, such that $\overline{A} = DAD^{-1}$ is symmetric in modulus, $\overline{a}_{ij} \cdot \overline{a}_{ji} = a_{ij} \cdot a_{ji}$, and $\overline{a}_{ii} = a_{ii}$; furthermore, the diagonal entries of D can be chosen to be positive, i.e., D is a positive diagonal matrix.*

Proof. Let $\overline{N} = \{(p, q) : p \neq q \text{ and } a_{pq}a_{qp} \neq 0\}$. By Lemma 2.2.3, $\overline{N}$ has $n - 1$ elements. The matrix $\overline{A} = DAD^{-1}$ is symmetric in modulus if

$$\left| \frac{a_{ij}}{a_{ji}} \right| = \frac{d_j^2}{d_i^2}, \qquad (i, j) \in \overline{N}. \tag{2.10}$$

Considering $z_k := d_k^2$ $(k = 1, \ldots, n)$ as unknowns of the algebraic system (2.10), it can be shown that consistent positive $(z_k > 0)$ solutions exist by constructing an undirected graph as follows: (1) associate n vertices to the n values z_k $(k = 1, 2, \cdots, n)$, and (2) associate $n - 1$ edges, $|z_i, z_j|$, connecting vertices z_i and z_j to the ratios $|a_{ij}/a_{ji}|$. Assign an arbitrary positive value to an arbitrary vertex z_r. Since the graph is connected and acyclic, any other vertex can be reached through a unique path. Thus, using the relations (2.10), the (positive) values associated to all vertices can be found. The corresponding (2^n) solutions of (2.10) are given by $d_i = \pm\sqrt{z_i}\,(i = 1, \ldots, n)$. Any of these solutions, in particular the positive one, gives a matrix $D = \text{diag}\,(d_1, \cdots, d_n)$ such that $\overline{A} = DAD^{-1}$ is symmetric in modulus.   ■

Definition 2.2.5 *Let A and B be two acyclic-3 matrices. Then A and B are said to be principally equal (denoted $A \overset{p}{=} B$) if and only if (i) $a_{ii} = b_{ii}$ and (ii) $a_{jk}a_{kj} = b_{jk}b_{kj}$.*

Lemma 2.2.6 [May66] *If $A \overset{p}{=} B$ then A and B have the same spectrum. This is also true for all their corresponding principal submatrices. Furthermore, if $A \overset{p}{=} B$, then their corresponding principal minors are equal.*
                                      □

Given a general indecomposable acyclic-3 matrix A, define a matrix $A^+ = (a_{ij}^+)$ with $a_{ij}^+ = a_{ij}$ if $a_{ij} \cdot a_{ji} > 0$, $a_{ij}^+ = a_{ji}^+ = 0$ if $a_{ij} \cdot a_{ji} \le 0$, and $a_{ii}^+ = a_{ii}$. The following theorem then holds.

Theorem 2.2.7 *An indecomposable acyclic-3 matrix A is in $\mathcal{D}_c$ if and only if*

$$(-1)^i \det_i(A^+) > 0, \qquad i = 1, 2, \dots, n, \tag{2.11}$$

where $\det_i$ denotes the determinant of the ith principal minor. The corresponding diagonal matrix D satisfying $DA + A^T D = -Q < 0$ is given by a positive solution of (2.10).

Proof. (*if*) By Lemma 2.2.4 there exists a positive diagonal matrix D such that $DAD^{-1} = A_s$, where A_s is symmetric in modulus. Obviously A_s^+ is the symmetric part of A_s. Though A_s^+ can be decomposable (with strictly negative diagonal), A_s^+ is combinatorially symmetric and acyclic-3. Furthermore, $A_s^+ \overset{p}{=} A^+$, and by Lemma 2.2.6, $(-1)^i \det_i(A_s^+) = (-1)^i \det_i(A^+) > 0$, i.e., A_s^+ is negative definite and thus

$$I A_s + A_s^T I = 2 A_s^+ < 0. \tag{2.12}$$

This relation can be written as

$$D^2 A + A^T D^2 = 2 D A_s^+ D < 0. \tag{2.13}$$

Hence, $A \in \mathcal{D}_c$ and a corresponding positive diagonal matrix which satisfies the Liapunov equation is given by D^2.

(*only if*) Supposing $A \in \mathcal{D}_c$, then there exists $P = \mathrm{diag}(p_i) > 0$ such that $PA + A^T P = -R < 0$. This also means that all principal submatrices of A belong to $\mathcal{D}_c$. Hence, by Lemma 2.1.4, all principal submatrices of A are D-stable, i.e., A is totally D-stable. Since A^+ and all its principal submatrices are principal submatrices of A, Lemma 2.1.10 implies the determinantal conditions (2.11). ∎

From a practical point of view, conditions (2.11) can be simplified by applying them to the strongly connected (or indecomposable) principal submatrices $A_k^+ (k = 1, \dots, p)$ of A^+.

The theorem below is a generalization (in the indecomposable case) of the characterization of totally D-stable tridiagonal matrices given in [CDJ82, Her88].

Theorem 2.2.8 *Given an indecomposable acyclic-3 matrix A, the following conditions are equivalent:*

(i) $A \in \mathcal{D}_c$,

(ii) A is totally D-stable,

(iii) A satisfies condition (2.11). □

It is worth noting that even if the conditions (2.11) hold for a given matrix, this does not ensure that all its principal minors satisfy the necessary condition of Lemma 2.1.10. However, for indecomposable acyclic-3 matrices (as a consequence of Theorem 2.2.8 and Lemma 2.1.10), the latter condition is equivalent to the simplified conditions (2.11). In this case, instead of requiring sign conditions on all principal minors of A (as in [BH83, CDJ82]), only the leading principal minors of A^+ have to be considered.

The characterization of the class $\mathcal{D}_{c_2}$ is much simpler and the theorems given below are stated without proofs (which are straightforward).

Introducing the notation $\tilde{A} = A - \text{diag}(a_{11}, \cdots, a_{nn})$, a matrix A is said to be *D-skew-symmetric* if and only if there exists a diagonal matrix $D > 0$ such that $D\tilde{A}$ is skew-symmetric. The main results for the class $\mathcal{D}_{c_2}$ are the following.

Theorem 2.2.9 *A matrix A is in class $\mathcal{D}_{c2}$ if and only if: (i) A has strictly negative diagonal and (ii) A is D-skew-symmetric. The diagonal matrix $D > 0$ that skew-symmetrizes $\tilde{A}$ satisfies the Liapunov equation, $DA + A^T D = -Q$, where Q is a positive diagonal matrix as well.* □

In graph theoretic terms, an alternative characterization is as follows.

Theorem 2.2.10 *A matrix A is of class $\mathcal{D}_{c2}$ if and only if: (i) $a_{ii} < 0$; (ii) $\tilde{A}$ is sign skew-symmetric (i.e., sign $\tilde{a}_{ij} = -$sign $\tilde{a}_{ij}$, $i \neq j$), and (iii) for all cycles of length $k \geq 3$ of A the equality $a_{i_1 i_2} a_{i_2 i_3} \cdots a_{i_{k-1} i_k} a_{i_k i_1} = (-1)^k a_{i_1 i_k} a_{i_k i_{k-1}} \cdots a_{i_3 i_2} a_{i_2 i_1}$ holds (for distinct $i_1, i_2, \cdots, i_k$).* □

Corollary 2.2.11 *Let matrix A be indecomposable and acyclic-3. Then $A \in \mathcal{D}_{c2}$ if and only if A has strictly negative diagonal and $\tilde{A}$ is sign skew-symmetric.* □

Theorem 2.2.10 gives a basis for a test to check whether a matrix belongs to the class $\mathcal{D}_{c2}$. First check conditions (i) and (ii) of the theorem. Then, either check (iii) or proceed as follows: Extract an acylic-3 indecomposable submatrix $\hat{A}$ from A by eliminating nonzero pairs (a_{ij}, a_{ji}) from A. In terms of the associated digraph $G(A)$, a *spanning tree* is found by this process. The system $|\hat{a}_{ij}/\hat{a}_{ji}| = q_j^2/q_i^2$, $(i, j) \in \overline{N}$ is then solved for $q_k, k = 1, \cdots, n-1$, as for the system (2.10). Finally check if $D\tilde{A}$, where $D = \text{diag}(q_1^2, \cdots, q_{n-1}^2)$, is skew-symmetric. The following example illustrates this procedure.

Example 2.2.12

$$A = \begin{bmatrix} -1 & +2 & +1 & -2 \\ -6 & -3 & 0 & 0 \\ -10 & 0 & -2 & +5 \\ +4 & 0 & -1 & -1 \end{bmatrix}$$

Conditions (i) and (ii) of Theorem 2.2.10 are verified. Then take

$$\hat{A} = \begin{bmatrix} -1 & +2 & +1 & -2 \\ -6 & -3 & 0 & 0 \\ -10 & 0 & -2 & 0 \\ +4 & 0 & 0 & -1 \end{bmatrix}$$

where $\hat{A}$ is an indecomposable, acyclic-3 matrix obtained from A by zeroing the entries a_{43}, a_{34}. Solve

$$\left|\frac{\hat{a}_{12}}{\hat{a}_{21}}\right| = \frac{1}{3} = \frac{q_2^2}{q_1^2}, \left|\frac{\hat{a}_{13}}{\hat{a}_{31}}\right| = \frac{1}{10} = \frac{q_3^2}{q_1^2}, \left|\frac{\hat{a}_{14}}{\hat{a}_{41}}\right| = \frac{1}{2} = \frac{q_4^2}{q_1^2}.$$

Setting $q_1 = 1$, one obtains $q_2^2 = \frac{1}{3}, q_3^2 = \frac{1}{10}, q_4^2 = \frac{1}{2}$, $D = \mathrm{diag}\left(1, \frac{1}{3}, \frac{1}{10}, \frac{1}{2}\right)$, and

$$DA = \begin{bmatrix} -1 & +2 & +1 & -2 \\ -2 & -1 & 0 & 0 \\ -1 & 0 & -\frac{1}{5} & +\frac{1}{2} \\ +2 & 0 & -\frac{1}{2} & -\frac{1}{2} \end{bmatrix}.$$

Thus A is in class $\mathcal{D}_{c2}$, since $D\tilde{A} = \widetilde{DA}$ is skew-symmetric.

In this example of low dimension, condition (iii) of Theorem 2.2.10 is not difficult to verify: $a_{13}a_{34}a_{41} = -a_{14}a_{43}a_{31} = 20$, and the diagonal pair (P, Q) is given by $P = D$ and $Q = DA + A^T D = -\mathrm{diag}\left(2, 2, \frac{2}{5}, 1\right)$.

A class of matrices that is strongly related to class $\mathcal{D}_{c2}$ is the class of qualitatively stable or sign-stable matrices. The problem of qualitative stability of real $n \times n$ matrices was considered in [QR65, JKD77, JKD87]. These papers provided conditions under which a given matrix can be guaranteed to be stable when only the signs of the matrix entries are known. The problem has its origin in mathematical economics where the stability of a linear dynamical system of the form $dx/dt = Ax$ is studied given only the signs of the entries of the matrix A (see the review paper [MQ69] and corrections in [JKD77]).

Definition 2.2.13 *A real matrix $A = (a_{ij})$ is said to be sign-stable whenever every matrix B which is sign similar to A is Hurwitz stable.*

Using the results above, a sufficient condition for sign-stability can be stated as follows.

Lemma 2.2.14 *A given real $n \times n$ indecomposable matrix $A = (a_{ij})$ is sign-stable if:*

(i) it is acyclic-3,

(ii) $\tilde{A}$ is sign skew-symmetric,

(iii) the diagonal elements a_{ii} are all negative.

Proof. The proof is immediate using Theorem 2.2.7 and Corollary 2.2.11, since preserving the signs of the elements (a_{ij}), conditions (i), (ii), and (iii) above are preserved as well, and the determinantal conditions (2.11) are satisfied for any matrix B which is sign similar to A.

Thus there exists a diagonal solution P for the Liapunov equation $A^T P + PA = Q < 0$ which is given by (2.10) and consequently A is Hurwitz. Since this is true for any matrix which is sign similar to A, the matrix A is sign-stable.   ∎

It should be noticed that in the Lemma 2.2.14; conditions (i) and (ii) are known to be necessary conditions for sign-stability, whereas condition (iii) is not, i.e., not all the diagonal elements have to be negative. Actually, in order to be sign-stable, matrix A cannot have any positive diagonal entries and must have at least one negative entry that has to be conveniently located in order to ensure that Q (diagonal and negative semidefinite) will imply asymptotic stability of A.

Example 2.2.15 *The matrix*

$$A = \begin{bmatrix} -1 & 1 & 0 & 0 & 0 \\ -1 & 0 & 1 & 0 & 0 \\ 0 & -1 & 0 & 1 & 0 \\ 0 & 0 & -1 & 0 & 1 \\ 0 & 0 & 0 & -1 & 0 \end{bmatrix}$$

is sign-stable, although only one diagonal entry is negative. On the other hand, if the only diagonal negative entry is moved from the (1,1) position to the (3,3) position, then the resulting matrix is no longer sign-stable. In fact, this matrix is not even Hurwitz stable [Jef74].

Notice that sign-stable matrices are in the class $\mathcal{D}_{c2}$, i.e., matrix Q is also diagonal. Sign-stable matrices are also known as *qualitatively stable* matrices [Bar71]. For further details and more results on sign-stable matrices, see [QR65, Jef74, JKD77, JKD87].

The discrete-time counterpart of the Hurwitz sign stability result is discussed in Section 2.7.3.

2.3   Similarity for Hurwitz Diagonally Stable Matrices

Example 1.7.1 showed that an appropriate change of coordinates can result in the property of diagonal stability of a system (or matrix). Within this context the following question is natural: Given a Hurwitz (or Schur) stable matrix A, is it possible to find a coordinate (similarity) transformation matrix T such that TAT^{-1} is diagonally stable (or semistable)? The answer to this question is affirmative for so-called *nonderogatory* matrices that have exactly one eigenvector per eigenvalue.

Proposition 2.3.1 *For any real nonderogatory Hurwitz matrix A, there exists a matrix T such that* $TAT^{-1} \in \mathcal{D}_{co}$.

Proof. Let matrix A be nonderogatory. Then it is always possible, using a a suitable similarity transformation, to represent it either in Hessenberg $\left(H = T_{HA}AT_{HA}^{-1}\right)$ or companion $\left(C = T_{CA}AT_{CA}^{-1}\right)$ forms [Dat74, BS70]. Subsequently, matrices H and C may be transformed into the Schwarz form S [Sch56, AJM76] using, for example, the algorithms in [Dat74], where:

$$
S = \begin{bmatrix}
-w_1 & 1 & & & & \\
-w_2 & 0 & 1 & & & \\
& -w_3 & 0 & 1 & & \\
& & \ddots & & \ddots & \\
& & & & & 1 \\
& & & -w_n & 0
\end{bmatrix}.
$$

Thus, $S = T_{SH} H T_{SH}^{-1} = T_{SC} C T_{SC}^{-1}$ and

$$
S = T_{SH} T_{HA} A T_{HA}^{-1} T_{SH}^{-1} = T_{SC} T_{CA} A T_{CA}^{-1} T_{SC}^{-1} = T_{SA} A T_{SA}^{-1}.
$$

Given the tridiagonal structure of matrix S, it is easy to see that the Hurwitz stability condition is given by $w_i > 0$; $i = 1, 2, \cdots, n$ [Par62, NT73, Dat74], and this condition is also equivalent to sign-stability of S (compare with Example 2.2.15). The diagonal Liapunov function that ensures stability (and sign-stability as well) is given by $x^T D x$, $x \in \mathbb{R}^n$, where $D = \text{diag}\,(1, w_2^{-1}, w_3^{-1}, \cdots, w_n^{-1})$ and

$$
DS + S^T D = \text{diag}\,(-2w_1, 0, 0, \cdots, 0).
$$

This shows that $S \in \mathcal{D}_{c2} \cap \mathcal{D}_{co}$. In other words, S is such that there exists a diagonal matrix D such that $DS + S^T D$ becomes negative semidefinite. Thus, with the above diagonal Liapunov function, LaSalle's theorem [LL61] has to be invoked to ensure asymptotic stability. ∎

Another type of diagonal Liapunov function for matrix S as well as other types of transformations T such that TAT^{-1} becomes diagonally stable are discussed in [RKH86]. Other diagonal solutions for the Liapunov equation for a matrix in Schwarz form are in [Dat78].

To determine the range of a parameter α for which the matrix $A(\alpha)$ is stable, it is useful to state the following:

Proposition 2.3.2 $A(\alpha) \in \mathcal{H}$ *if and only if* $S(\alpha)$ *presents a diagonal solution* $D(\alpha)$ *to the linear matrix inequality* $D(\alpha)S(\alpha) + S^T(\alpha)D(\alpha) \leq 0$, *where* $D(\alpha) = \text{diag}\,(1, w_2^{-1}(\alpha), w_3^{-1}(\alpha), \cdots, w_n^{-1}(\alpha))$ *and* $S(\alpha)$ *is the corresponding Schwarz form of* $A(\alpha)$.

Proof. (*if*) Whenever $A(\alpha) \in \mathcal{H}$, the matrix $S(\alpha)$ (in Schwarz form) is such that $w_i(\alpha) > 0$, $i = 1, 2, \cdots, n$ [Sch56] that, on the other hand, ensures sign stability and consequently the existence of a diagonal solution of the form $D(\alpha) = \text{diag}\,(1, w_2^{-1}(\alpha), w_3^{-1}(\alpha), \cdots, w_n^{-1}(\alpha))$.

(*only if*) (by contradiction): Suppose $D(\alpha)$, as above, is not a solution to $D(\alpha)A(\alpha) + A^T(\alpha)D(\alpha) \le 0$ this implies that, for at least one index j, the inequality $w_j(\alpha) > 0$ is not satisfied, and consequently $A(\alpha) \notin \mathcal{H}$ which is a contradiction.   ∎

Thus, the evaluation of Hurwitz stability regions for a given parameter-dependent matrix $A(\alpha)$ is equivalent to finding the region of diagonal semi-stability of the corresponding Schwarz form $S(\alpha)$. This is exactly what happens in the illustrative example in Section 1.2, as well as in many other examples in Chapter 6, in which conditions for diagonal stability coincide with those for Hurwitz stability, and for absolute stability as well.

2.4   Persistence of Diagonal Stability under Perturbations

The set of diagonally stable matrices is an open set, since it is defined by an inequality (linear matrix inequality (LMI)). It is also easy to see that it is a convex set. Thus, given a diagonally stable matrix A,

(i) it is reasonable to look for bounds on perturbations, ΔA, such that $A + \Delta A$ continues to be diagonally stable.

(ii) checking if A is diagonally stable is equivalent to checking for the feasible solution of a convex optimization problem of a special type (LMI).

In order to calculate bounds, it is necessary to specify the type of permissible perturbations. Results are available for two types of perturbations: (i) perturbations bounded in Frobenius norm; and (ii) bounded entrywise perturbations. More specifically, consider the norm bounded and interval bounded perturbation classes defined below.

$$\Delta A_F(\zeta) = \{\Delta A = (\delta a_{ij}) \in \mathbb{R}^{n \times n} : \|\Delta A\|_F \le \zeta, \ \zeta > 0\} \qquad (2.14)$$

$$\Delta A_I(\zeta) = \{\Delta A = (\delta a_{ij}) \in \mathbb{R}^{n \times n} : \Delta A = \zeta \Sigma, \zeta > 0;$$
$$\Sigma = (\sigma_{ij}), \sigma_{ij} \in [\underline{\sigma}_{ij}, \overline{\sigma}_{ij}], \underline{\sigma}_{ij} \le 0, \overline{\sigma}_{ij} \ge 0\} \qquad (2.15)$$

For these classes, the following results were derived in [FT95] for the Hurwitz case.

Theorem 2.4.1 *Let the real $n \times n$ matrix $A \in \mathcal{D}_c(P)$. Then $A + \Delta A \in \mathcal{D}_c(P)$ for all $\Delta A \in \Delta A_F(\zeta)$, if*

$$\zeta < \zeta_F = \frac{1}{\|\mathcal{F}\|_2},$$

where $\mathcal{F} \in \mathbb{R}^{n^3 \times n}$ is the matrix

$$\mathcal{F} = \left[(L^T)^{-1} [Pe_1 e_1^T]^S L^{-1} \cdots (L^T)^{-1} [Pe_i e_j^T]^S L^{-1} \cdots (L^T)^{-1} [Pe_n e_n^T]^S L^{-1} \right].$$

In the formula above, $[M]^S$ denotes the symmetric part of a real matrix M: i.e., $[M]^S = \frac{1}{2}(M + M^T)$, e_i is the ith standard basis vector, and $L \in \mathbb{R}^{n \times n}$ is any factor of $[PA]^S$, i.e.,

$$L^T L = [PA]^S. \tag{2.16}$$

Although the evaluation of the bound ζ_F above only involves the calculation of the 2-norm of a rectangular matrix (maximum singular value), the bound is not tight in general. A tighter bound is obtained for the interval perturbation class (2.15) as follows.

Theorem 2.4.2 [FT95] *Let the real $n \times n$ matrix $A \in \mathcal{D}_c(P)$. Then $A + \Delta A \in \mathcal{D}_c(P)$, for all $\Delta A \in \Delta A_I(\zeta)$ if and only if*

$$\zeta < \zeta_I = \min_{k=1,\ldots,2^{n^2}} \frac{1}{\lambda_{\max}\left\{ (L^T)^{-1} [P\Sigma_k]^S L^{-1} \right\}},$$

where L is defined in (2.16) and Σ_k denotes the kth of the 2^{n^2} $n \times n$ matrices obtained by putting in (2.15) either $\sigma_{ij} = \underline{\sigma}_{ij}$ or $\sigma_{ij} = \overline{\sigma}_{ij}$, for each $i = 1, \ldots, n; j = 1, \ldots, n$.

Note that there is a combinatorial explosion of the required number of maximum eigenvalue computations, so that the applicability of this tight bound is limited to low dimensions (small n). Another relevant observation is that both results assume that the matrices A and $A + \Delta A$ share the same diagonal solution P, i.e., are in $\mathcal{D}_c(P)$. The following example shows that other possibilities exist.

Example 2.4.3 *Consider the matrices:*

$$A = \begin{bmatrix} -0.5 & 3.5 \\ -0.5 & -0.5 \end{bmatrix}; \quad \Delta A = \begin{bmatrix} -0.4556 & -0.9695 \\ -0.6024 & 0.4936 \end{bmatrix}$$

as well as the matrices $P = \mathrm{diag}\,(1, 7)$ and $P_1 = \mathrm{diag}\,(1, 2.3483)$. Then $A \in \mathcal{D}_c(P)$, but $A \notin \mathcal{D}_c(P_1)$, and $(A + \Delta A) \in \mathcal{D}_c(P_1)$ but $(A + \Delta A) \notin \mathcal{D}_c(P)$. Thus the bounds in Theorems 2.4.1 and 2.4.2 are quite conservative, since a perturbation may change the particular diagonal solution associated to a matrix without affecting the property of diagonal stability of the latter.

2.5   Basic Results on Schur Diagonal and D-Stability

The concept of discrete-time or Schur stability arises in the study of discrete-time dynamical systems of the form $x(k + 1) = Ax(k)$ [LaS76, LaS86].

This book is concerned mainly with matrix diagonal stability and some variants, and this section discusses the basic definitions and properties in the Schur case. The utilization of these stability concepts and results in different classes of nonlinear dynamical systems is the subject of the other chapters.

The qualifier discrete-time or Schur will often be dropped in the following, since this section is devoted entirely to the discrete-time case. However, as a reminder, a subscript "d" (for discrete-time) will be used for various sets.

Definition 2.5.1 *The class of* Schur stable *matrices is defined as the class of matrices that have all eigenvalues within the interior of the unit circle in the complex plane.*

Definition 2.5.2 *A matrix $A \in \mathbb{R}^{n \times n}$ is said to be* Schur diagonally stable *(alternatively, in class $\mathcal{D}_d$ or, simply, in $\mathcal{D}_d$) if and only if there exists a positive diagonal matrix P such that $A^T PA - P$ is negative definite. When it is desired to make explicit a matrix P that satisfies the above condition, the notation $A \in \mathcal{D}_d(P)$ is used.*

Definition 2.5.3 *A matrix $A \in \mathbb{R}^{n \times n}$ is said to be* discrete-time *or* Schur D-stable *(alternatively, in class $\mathbb{D}_d$ or, simply, in $\mathbb{D}_d$), if AD is Schur stable for all real diagonal matrices D whose elements are all less than or equal to unity in absolute value, i.e., $|D| \leq I$.*

As in the Hurwitz case, Schur diagonal stability of a given matrix cannot be decided just by the location of its eigenvalues in the sense that if A and B are two matrices with the same eigenvalues, then it is possible for A to be diagonally stable, even though B is not. However, from the definition of Schur diagonal stability, the following fact, a discrete-time analog of Fact 2.1.13, is true.

Fact 2.5.4 *For all matrices $A \in \mathbb{R}^{n \times n}$, there exists an $|\alpha| < 1$ such that $\alpha A \in \mathcal{D}_d$.*

Proof. The quadratic form corresponding to the left-hand side of the Stein equation (2.4) in the matrix αA is $\alpha^2 (x^T A^T PAx) - x^T Px$ and the Stein equation will be satisfied if this quadratic form can be made negative definite for some positive definite matrix P. By the Rayleigh theorem [Str88, p.349], for any positive definite matrix P (and therefore, in particular, for any positive diagonal matrix P), the choice $\alpha^2 < (\lambda_{\min}(P))/(\lambda_{\max}(A^T PA))$ suffices to do this.   ∎

Fact 2.5.4 can be interpreted as saying that if the eigenvalues of a matrix are moved sufficiently "deep" into the interior of the unit circle, then it becomes Schur diagonally stable.

The following calculation also leads to another way of looking at Schur diagonal stability. Since $A \in \mathcal{D}_d(P)$ means that there exists some diagonal $P > 0$ such that the matrix $A^T P A - P$ is negative definite, i.e.,

$$A^T P A - P = -Q \quad \text{for some } Q > 0.$$

Using the congruence transformation defined by the positive definite matrix $P^{-1/2}$ on both sides of this equation, yields

$$\left(P^{-1/2} A^T P^{1/2}\right) \left(P^{1/2} A P^{-1/2}\right) - I = -P^{-1/2} Q P^{-1/2}$$

i.e., $B^T B - I$ is required to be negative definite, where $B = P^{1/2} A P^{-1/2}$ and this will occur if and only if

$$\lambda_{\max}(B^T B) = \sigma_{\max}(B) = \|P^{1/2} A P^{-1/2}\|_2 < 1. \tag{2.17}$$

Thus requiring that $A \in \mathcal{D}_d(P)$ is equivalent to requiring that a weighted 2-norm of A be less than unity (the weighting is two-sided and the weights are given by the square roots of the positive diagonal elements of the matrix P).

It is useful to associate two polytopes of matrices to a given matrix and to do this the following *class* $\mathbb{K}^{[-1,1]}$ of diagonal matrices is defined.

Definition 2.5.5 *The subset $\mathbb{K}^{[-1,1]}$ of the set of diagonal matrices is defined as follows.*

$$\mathbb{K}^{[-1,1]} := \{K = \text{diag}\,(k_1, \ldots, k_n) : \quad -1 \le k_i \le 1, \forall i\}, \tag{2.18}$$

also written as $\{K : |K| \le I\}$. The set of 2^n vertices of $\mathbb{K}^{[-1,1]}$ is also called the class of signature matrices, denoted $\mathbb{K}_{\text{sgn}}$, and defined by $\mathbb{K}_{\text{sgn}} := \{K : |K| = I\} = \{K : K = \text{diag}\,(\pm 1, \ldots, \pm 1)\}$.

Definition 2.5.6 *Given a matrix $A \in \mathbb{R}^{n \times n}$, the polytopes $\mathcal{R}(A)$ and $\mathcal{L}(A)$ are defined as:*

$$\mathcal{R}(A) := \{AK : K \in \mathbb{K}^{[-1,1]}\}; \quad \mathcal{L}(A) := \{KA : K \in \mathbb{K}^{[-1,1]}\}.$$

It can easily be shown that $\mathcal{R}(A)$ and $\mathcal{L}(A)$ are polytopes that can be written explicitly as the convex hull of a set of vertex matrices. For example,

$$\mathcal{R}(A) = \mathbf{Co}\{AK_i, i = 1, \ldots, 2^n; \; K_i \in \mathbb{K}_{\text{sgn}}\}.$$

Using this definition, a matrix A is Schur D-stable if and only if every matrix B in the polytope $\mathcal{R}(A)$ is stable. More succinctly, A is Schur D-stable if and only if $\mathcal{R}(A)$ is stable. From the identity $\det(\lambda I - AK) = \det(\lambda I - KA)$, it follows that $\mathcal{R}(A)$ is stable if and only if $\mathcal{L}(A)$ is stable, so that D-stability may also be defined in terms of left multiplication by diagonal matrices in $\mathbb{K}^{[-1,1]}$. An important difference between the Hurwitz and Schur

D-stability is that in the latter, the diagonal matrices that multiply the given matrix may be singular.

The following useful fact is a consequence of a geometrical property of the ellipsoidal level sets of the quadratic form $x^T P x$.

Fact 2.5.7 *Let P be a positive definite matrix in $\mathbb{R}^{n \times n}$. For all $x \in \mathbb{R}^n$, for all $K \in \mathbb{K}^{[-1,1]}$*

$$x^T K P K x \leq x^T P x \quad \text{if and only if } P \text{ is diagonal.} \tag{2.19}$$

Proof. (*if*) Immediate, by calculation.

(*only if*) By contradiction. Only an outline of the proof is given. The basic idea is to show that if P is not diagonal, then given, for example, the level curve $L_1 := \{x : x^T P x = 1\}$, it is always possible to find a point x^o on L_1 such that if the coordinates of x^o are reduced in absolute value to get a new point $K x^o$, then this new point lies on another level curve $L_c := \{x : x^T P x = c\}$, where $c > 1$. To keep notation simple and illustrate the general idea, consider the nondiagonal positive definite matrix $P = \begin{bmatrix} 2 & 1 \\ 1 & 2 \end{bmatrix}$. Given $x = (x_1 \ x_2)^T \in \mathbb{R}^2$, the equation of the level curve L_1 is

$$2x_1^2 + 2x_1 x_2 + 2x_2^2 = 1 \tag{2.20}$$

and represents an ellipse in the x_1-x_2 plane. Now consider the line $x_2 = b$ parallel to the x_1 axis. To find the value of b for which this line touches the ellipse L_1, set $x_2 = b$ in equation (2.20) and require that the discriminant of the resulting quadratic equation be zero. This gives the point $x^o = (0.5\sqrt{2/3} \ \sqrt{2/3})^T$. Let $K = \text{diag}\,(\alpha, 1)$ and substitute $K x^o$ on the left-hand side of (2.20) in order to calculate the value of the right-hand side. This will be the value of c, as a function of α, which defines the new level curve L_c to which the point $K x^o$ belongs. This gives $3c = \alpha^2 - 2\alpha + 4$, showing that the minimum value of c is 1, attained for $\alpha = 1$, and that $c > 1$ for all $\alpha < 1$. In other words, as claimed, for this nondiagonal positive definite matrix P, a point x^o has been found such that $x^{oT} K P K x^o > x^{oT} P x^o$ for a matrix $K \in \mathbb{K}^{[-1,1]}$. It is geometrically clear that this construction will work for any nondiagonal positive definite matrix of any dimension. ∎

The reader's attention is called to the related Lemmas 3.4.5 and 3.4.8 in the next chapter.

Schur diagonal stability of a given matrix implies its Schur D-stability, as in the Hurwitz case (see Lemma 2.1.4).

Proposition 2.5.8 *If P is a diagonal solution to the Stein equation associated to a matrix A, then it is also a diagonal solution to the Stein equation associated to the matrix AK, for every $K \in \mathbb{K}^{[-1,1]}$. In symbols, $A \in \mathcal{D}_d(P)$ implies $AK \in \mathcal{D}_d(P), \forall K \in \mathbb{K}^{[-1,1]}$.*

Proof. Since $|K| \leq I$, K may be singular, and two cases have to be considered:

• $x \in$ nullspace(K). In this case, $x^T K^T A^T P A K x - x^T P x$ is negative for all $x \neq 0$, since the first term is zero, and the second negative definite.

• $x \notin$ nullspace(K). Then, on pre- and postmultiplying the left-hand side of the Stein equation by $K = K^T$, one gets the quadratic form $x^T (K^T A^T P A K - K P K)x$, which can be written as $z^T (A^T P A - P)z$, where $z = Kx \neq 0$ for $x \neq 0$. The latter quadratic form is negative definite, since $A \in \mathcal{D}_d(P)$. By Fact 2.5.7, for P diagonal and $|K| \leq I$, $x^T K P K x \leq x^T P x$; thus $x^T K^T A^T P A K x - x^T P x$ is negative for all $x \neq 0$. In other words, $AK \in \mathcal{D}_d(P)$, $\forall K \in \mathbb{K}^{[-1,1]}$. ∎

In polytope language, Proposition 2.5.8 may also be stated as follows. If any one vertex, namely A, of the polytope $\mathcal{R}(A)$ is diagonally stable, with solution P, then the entire polytope $\mathcal{R}(A)$ is diagonally stable with the same diagonal solution P. In this case, the quadratic Liapunov function is said to be a *simultaneous Liapunov function (SLF)* for the set of matrices $\mathcal{R}(A)$, which in turn is said to be *simultaneously stable*. More on these concepts appears in Section 2.6 below.

The left multiplication analog of the above proposition is as follows.

Proposition 2.5.9 $A \in \mathcal{D}_d(P)$ *implies* $KA \in \mathcal{D}_d(P)$, $\forall K \in \mathbb{K}^{[-1,1]}$.

Proof. $A \in \mathcal{D}_d(P)$ implies that

$$\exists P > 0, P \text{ such that } A^T P A - P < 0. \tag{2.21}$$

Adding and subtracting the matrix $A^T K^T P K A$ from the left of (2.21) gives:

$$A^T P A - A^T K^T P K A + A^T K^T P K A - P < 0,$$

which can be written as

$$[A^T (P - K^T P K)A] + [A^T K^T P K A - P] < 0. \tag{2.22}$$

Since P is diagonal and $|K| \leq I$, it follows from Fact 2.5.7 that the matrix $P - K^T P K$ is positive semidefinite, implying that the first matrix in square brackets in (2.22) is positive semidefinite. Since the sum of the two matrices in (2.22) is negative definite, it follows that the second matrix in square brackets in (2.22) must be negative definite, i.e.,

$$\text{for all } K \text{ diagonal, } |K| \leq I, (KA)^T P(KA) - P < 0,$$

i.e., $KA \in \mathcal{D}_d(P)$, $\forall K$ diagonal, $|K| \leq I$. ∎

The obvious question that arises at this point is whether there exist some converses of Propositions 2.5.8 and 2.5.9. There are two candidates for a converse: (i) simultaneous stability of the polytope $\mathcal{R}(A)$ implies diagonal stability of A; (ii) stability of the polytope $\mathcal{R}(A)$ implies diagonal stability of the matrix A.

In regard to (i), the following converse of proposition 2.5.9 is true.

Proposition 2.5.10 *If the matrix polytope $\mathcal{L}(A)$ admits a quadratic, not necessarily diagonal SLF, $V(x) = x^T P x$, then it also admits a diagonal quadratic SLF, i.e., the matrix A is Schur diagonally stable.*

Proof. Since it can be shown that the quadratic form $x^T(B^T P B - P)x$ assumes its maximum value at some vertex B_1 of the polytope $\mathcal{L}(A)$ [BK93], rename, without loss of generality, this vertex as A; consequently

$$\text{for all } K \in \mathbb{K}^{[-1,1]}, \quad x^T(A^T K P K A - P)x \leq x^T(A^T P A - P)x < 0$$

whence

$$x^T(A^T K P K A - A^T P A)x \leq 0. \tag{2.23}$$

If x is in the nullspace of A, then (2.23) is satisfied for any choice of matrix P. In particular, any choice of diagonal matrix P is allowable. When x is not in the null space of A, setting $Ax = z$, the above equation can be written as

$$\text{for all } K \in \mathbb{K}^{[-1,1]}, \quad z^T K P K z - z^T P z \leq 0,$$

which can also be satisfied by a diagonal matrix. Indeed, if A is nonsingular, then Fact 2.5.7 shows that P must be diagonal. ∎

Note that (ii) is a weak form of (i) and is the exact discrete-time analog of the well-known continuous-time problem of whether D-stability implies diagonal stability (see Example 2.1.6). Not surprisingly, in the Schur case too, D-stability does not imply diagonal stability and this is shown by the following counterexample.

$$A = \begin{bmatrix} 0.97701 & -0.32047 & -0.08986 \\ 0.07282 & 0.66868 & 0.389559 \\ 0.15153 & -0.27886 & 0.697892 \end{bmatrix}. \tag{2.24}$$

The matrix A can be shown to be Schur D-stable (in fact, it belongs to the interior of the set of Schur D-stable matrices, see Section 2.8), but it is not Schur diagonally stable, for reasons that are discussed in the next section.

In other words, the class $\mathcal{D}_d$ of diagonally stable matrices is strictly contained in the class of all D-stable matrices. Section 2.6 contains more material on the relationship between these two classes.

2.6 A Matrix Polytopic View of Schur Diagonal and D-Stability

An interesting perspective on the relationship between Schur diagonal and D-stability may be obtained from an abstract approach to the stability of sets of matrices. The definitions of stability and asymptotic stability of a set of matrices M [BC73, BT79, BT80] are stated next.

Definition 2.6.1 *A set of matrices* M *is* stable, *if for each neighborhood U of the origin 0 in* $\mathbb{R}^n$, *there exists another neighborhood V of the origin such that,* $\forall\Pi \in$ M′, $\Pi V \subset U$, *where* M′ *is the set of all finite product of matrices in* M.

Definition 2.6.2 *A set of real* $n \times n$ *matrices* M *is* asymptotically stable *if there exists* $\alpha > 1$ *such that* αM *is stable.*

The following facts, proved for example in [BC73, BT80], can also be adopted as alternative definitions of stability and asymptotic stability.

Fact 2.6.3 *A set of real* $n \times n$ *matrices* M *is* stable, *if and only if there exists a bounded* M-*invariant neighborhood,* W, *of the origin; i.e.,*

$$\forall A \in M, \ AW \subset W.$$

Furthermore W *can be chosen to be closed, convex, and balanced (*W *is balanced if* $z \in W$ *implies* $\alpha z \in W$, *for all* $|\alpha| \leq 1$).

Fact 2.6.4 *A set of* $n \times n$ *real matrices* M *is* asymptotically stable *if and only if there exists* $\alpha > 1$ *and a neighborhood* W *of the origin such that*

$$\forall A \in M, \ \alpha AW \subset W.$$

In order to give the formal definition of a simultaneous Liapunov function, some definitions from Liapunov stability theory are required [Vid93] (also see the Appendix to Chapter 3).

Definition 2.6.5 *A function* $\phi : \mathbb{R}_+ \to \mathbb{R}_+$ *is of* class **K** *if it is continuous, strictly increasing and* $\phi(0) = 0$. *A function* $V : \mathbb{R}^n \times \mathbb{R}_+ \to \mathbb{R}$ *is said to be* positive definite *if (i) it is continuous; (ii)* $V(0,t) = 0, \forall t \geq 0$; *and (iii) there exist a constant* $r > 0$ *and a function* ϕ *of class* **K** *such that*

$$\phi(\|x\|) \leq V(x,t) \quad \forall x \in \mathbb{R}^n. \tag{2.25}$$

The function V *is called* radially unbounded *if (2.25) is satisfied for some continuous function* ϕ, *not necessarily of class* **K**, *with the additional property that* $\phi(r) \to \infty$ *as* $r \to \infty$.

Definition 2.6.6 *A continuous, real-valued, positive definite and radially unbounded function* $v_M : \mathbb{R}^n \to \mathbb{R}_+$ *is called a* simultaneous Liapunov function (SLF) *for a set of matrices* M, *if*

$$\forall A \in M, \ \Delta v_M := v_M(x) - v_M(Ax) \text{ is positive definite.}$$

Definition 2.6.7 *A simultaneous Liapunov function that is also a norm on* $\mathbb{R}^n$ *is called a* norm SLF. *An SLF that is a convex function on* $\mathbb{R}$ *is called a* convex SLF. *Of course, all norm SLFs are convex SLFs.*

The main equivalence result is stated as follows.

Theorem 2.6.8 [BM94] *If* $\mathsf{M} \subset \mathbb{R}^{n \times n}$ *is compact then the following are equivalent:*

(i) *The set of matrices* M *is asymptotically stable, in the sense of Definition 2.6.2;*

(ii) *Given any induced matrix norm* $\| \cdot \|$, *there exist a finite* $k \in \mathbb{N}$ *and a real number* $b \in (0, 1)$ *such that*

$$\forall \Pi_k \in \mathsf{M}^k, \ \|P_{k^*}\| \leq b < 1; \tag{2.26}$$

(iii) *There exist an* M-*invariant neighborhood of* 0, W, *and a real number* $c \in (0, 1)$ *such that for all finite* $j \in \mathbb{N}$

$$\forall \Pi_j \in \mathsf{M}^j, \ \rho(\Pi_j) \leq c < 1; \tag{2.27}$$

(iv) *There exists a norm SLF for* M.

(v) *The difference inclusion* $x(k+1) \in \mathsf{M} \, x(k)$ *is globally exponentially stable (for definitions see Section 3.4.1).*   □

It is important to emphasize the difference between the equivalences (ii) and (iii) of the above theorem. In (ii), it is enough that k-products of matrices from M have norm strictly less than unity for all k *larger* than some integer value. In (iii) the spectral radii of all finite products must be strictly less than unity. In fact, if *any finite product* of matrices from M has spectral radius greater than or equal to unity, then the set of matrices M is not asymptotically stable.

Details of the proofs of these and other equivalences may be found in [BM94]. The objective here is to apply these results to the polytope $\mathcal{R}(A)$, to arrive at the following hierarchy.

A is diagonally stable (i.e., $\exists$ *diagonal, quadratic SLF for* $\mathcal{R}(A)$ *and any* $\Pi_k \in \mathcal{R}(A)^k$ *is stable for* k *finite)* is, by Proposition 2.5.10, equivalent to

A is quadratically stable (i.e., $\exists$ *quadratic SLF for* $\mathcal{R}(A)$ *and any* $\Pi_k \in \mathcal{R}(A)^k$ *is stable for* k *finite)* implies that

A is simultaneously stable (i.e., $\exists$ *norm SLF for* $\mathcal{R}(A)$ *and any* $\Pi_k \in \mathcal{R}(A)^k$ *is stable for* k *finite)* implies that

A is D-stable (i.e., $B \in \mathcal{R}(A)$ *is stable).*

In Section 2.5, the following facts were shown.

Fact 2.6.9 *If a matrix* A *is Schur diagonally stable, then the polytope of matrices* $\mathcal{R}(A)$ *is simultaneously asymptotically stable (and hence* A *is D-stable);*

Fact 2.6.10 *The fact that a matrix* A *is Schur D-stable does not imply that it is Schur diagonally stable. A counterexample, (2.24), was given.*

A new perspective on these facts can be obtained from the main results in this section. Fact 2.6.9 asserts that the polytope $\mathcal{R}(A)$ has a special property: Namely, if $A \in \mathcal{D}_d(P)$, then $B \in \mathcal{D}_d(P)$ for all $B \in \mathcal{R}(A)$. In other words, if any vertex of this polytope is in $\mathcal{D}_d(P)$, then every matrix in the polytope is in $\mathcal{D}_d(P)$. It is of interest to ask whether there exist other examples of such polytopes; the authors do not know of any other nontrivial examples.

To understand the "gap" between diagonal and D-stability, reconsider the matrix A in (2.24): in particular, the vertices A and $B :=$ $A \, \text{diag} \, (-1, 1, 1)$ of the associated polytope $\mathcal{R}(A)$. The spectral radius of the product of the vertex matrices A and B, $\rho(AB) = 1.0001 > 1$. From the equivalence of items (iii) and (iv) of Theorem 2.6.8, this means that the set of matrices $\mathcal{R}(A)$ does not admit any convex SLF. In particular, no quadratic diagonal SLF exists. Thus A is not Schur diagonally stable. This argument also shows that finding an unstable k-product of matrices in the polytope $\mathcal{R}(A)$ is one (rather roundabout) way of concluding that the matrix A does not admit a diagonal solution to the Stein equation.

In other words, the polytope $\mathcal{R}(A)$ is such that all 1-products (i.e., elements of $\mathcal{R}(A)$) are Schur stable, but not all n-products, $n \geq 2$ are. Thus the failure of D-stability to imply diagonal stability can be understood as the failure of the asymptotic stability of 1-products alone to imply simultaneous stability.

With regard to the conjectured characterization of D-stability (2.7.21) in Section 2.7.1, a perspective is obtained by first formulating the following corollary to Theorem 2.6.8.

Corollary 2.6.11 *The matrix polytope* $\mathcal{R}(A)$, $A \in \mathbb{R}^{n \times n}$ *is simultaneously stable if and only if the set of its vertices is simultaneously stable.*

Proof. (*only if*) Obvious.

(*if*) Observe that any matrix B in the polytope can be written as a convex combination of the 2^n vertices. Writing this as $B = \sum_{i=1}^{2^n} \lambda_i A_i$, where the A_i are the vertex matrices, and using the norm SLF (denoted $\|\cdot\|_S$ whose existence is guaranteed by Theorem 2.6.8, it follows that $\|B\|_S = \|\sum_{i=1}^{2^n} \lambda_i A_i\|_S \leq \sum_{i=1}^{2^n} \lambda_i \|A_i\|_S < \sum_{i=1}^{2^n} \lambda_i = 1.$ ∎

Thus Conjecture 2.7.21 refers to the possibility of deleting both occurrences of the word "simultaneously" from the above corollary. For the case of a quadratic SLF, an independent proof of Corollary 2.6.11 was given in [BK93].

In closing, it should be pointed out that it is also possible to view Hurwitz diagonal stability from the matrix polytopic point of view of this section as follows. Since, for any positive scalar α, the eigenvalues of A satisfy the relation $\alpha \lambda_i(A) = \lambda_i(\alpha A)$, for $i = 1, \cdots, n$, it follows that Hurwitz D-stability (i.e., stability of the matrix AD) can be tested for with D positive diagonal with $\text{trace}(D) = 1$. Thus one can consider the set of vertices

$\{V_i = \text{diag}\,(\epsilon/(n-1),\cdots,1-\epsilon,\epsilon/(n-1))\}$, where $\epsilon > 0$ is arbitrarily small, $1-\epsilon$ occurs in the ith position, $i = 1,\cdots,n$ and $n \geq 2$. Then, as $\epsilon \to 0$, D-stability of A is approximated by the Hurwitz stability of the polytope $\mathbf{Q} = \mathbf{co}\,(AV_1,\cdots,AV_n)$. Such an approach, based on [BK93], is pursued in [GOH98], which shows how the positive real concept can be used to study the stability of sets of products of matrices of the polytope $\mathbf{Q}$, in terms of LMIs, leading to a numerical test for Hurwitz D-stability. In [OG99], the same idea leads to a numerical test for Schur D-stability (also see Section 2.8).

2.7 Special Classes of Schur Diagonally Stable Matrices

It is useful to identify classes of matrices that are diagonally stable, since Schur diagonal stability for a matrix of dimension greater than two has not been characterized algebraically. By Proposition 2.5.8, such classes also have the property of D-stability, which has not been characterized either.

Proposition 2.7.1 $\{Schur\ stable\ symmetric\ matrices\} \subset \mathcal{D}_d$.

Proof. Since A is Schur stable $(\rho(A) < 1)$ and symmetric, $A^T A = A^2$ and $\rho(A^2) = (\rho(A))^2 < 1$ so that $A^T A - I$ is negative definite, which shows that $A \in \mathcal{D}_d(I)$. ∎

Definition 2.7.2 *A matrix A is said to be* d-symmetrizable *if there exists a nonsingular diagonal matrix T such that $T^{-1}AT$ is symmetric.*

A generalization of Proposition 2.7.1, based on the above Definition, is as follows.

Proposition 2.7.3 $\{Schur\ stable\ d\text{-}symmetrizable\ matrices\} \subset \mathcal{D}_d$.

Proof. By the definition of d-symmetrizability, there exists a nonsingular diagonal matrix T such that $T^{-1}AT$ is symmetric. By the proof of the previous proposition, this means that

$$(T^{-1}AT)^T(T^{-1}AT) - I < 0$$

$$\Leftrightarrow T^T[A^T(T^{-1})^T(T^{-1})A - (T^{-1})^T T^{-1}]T < 0$$

$$\Leftrightarrow A^T P A - P < 0, \text{ where } P = (T^{-1})^T(T^{-1}) = T^{-2}, \text{ by congruence}$$

i.e., $A \in \mathcal{D}_d(P)$, since $P = T^{-2}$ is diagonal. ∎

The following lemma is needed to prove some useful facts about quasidominance.

Lemma 2.7.4 [Nik68, Corollary 1, p.364] *If all principal minors of A are positive, then there exists a vector $x \succ 0$ such that $Ax \succ 0$.* □

Theorem 2.7.5 [Moy77] *All principal minors of A are positive if and only if, for all $K \in \mathbb{K}_{\mathrm{sgn}}$, there exists a vector $x \succ 0$ such that $KAKx \succ 0$.*

Proof. (*only if*) Since the principal minors of KAK are identical to those of A, the claim follows by Lemma 2.7.4.

(*if*) The proof is by induction on the size of principal minors of A. Suppose that for each K there exists $x \succ 0$ such that $KAKx \succ 0$, and suppose also that all $m \times m$ principal minors of A are known to be positive. (The case $m = 1$ is trivially handled). Let A be partitioned as

$$A = \begin{bmatrix} A_{11} & A_{12} & a_{13} \\ A_{21} & A_{22} & a_{23} \\ a_{31} & a_{32} & a_{33} \end{bmatrix},$$

where $a_{23}, a_{32}^T \in \mathbb{R}^{m \times 1}$ and let $K = \mathrm{diag}\,(k_1, k_2, 1)$ be chosen such that $(a_{32}A_{22}^{-1}A_{21} - a_{31})k_1 \succeq 0$ and $(-a_{32}A_{22}^{-1})k_2 \succeq 0$. By the induction hypothesis, $\det A_{22} > 0$ and A_{22} is invertible. Now, by assumption, there exists

$$x = \begin{pmatrix} x_1 \\ x_2 \\ x_3 \end{pmatrix} \succ 0 \text{ such that}$$

$$KAKx = \begin{bmatrix} k_1 A_{11} k_1 & k_1 A_{12} k_2 & k_1 a_{13} \\ k_2 A_{21} k_1 & k_2 A_{22} k_2 & k_2 a_{23} \\ a_{31} k_1 & a_{32} k_2 & a_{33} \end{bmatrix} \begin{pmatrix} x_1 \\ x_2 \\ x_3 \end{pmatrix} =: \begin{pmatrix} y_1 \\ y_2 \\ y_3 \end{pmatrix} \succ 0.$$

Elimination of x_2 leads to

$$(a_{33} - a_{32}A_{22}^{-1}a_{23})x_3 = y_3 + (a_{32}A_{22}^{-1}A_{21} - a_{31})k_1 x_1 - a_{32}A_{22}^{-1}k_2 y_2,$$

so that $a_{33} - a_{32}A_{22}^{-1}a_{23} > 0$. This means that

$$\det \begin{bmatrix} A_{22} & a_{23} \\ a_{32} & a_{33} \end{bmatrix} = (a_{33} - a_{32}A_{22}^{-1}a_{23}) \det A_{22} > 0.$$

In other words, this particular $(m+1) \times (m+1)$ principal minor is positive. The proof is completed by permuting the rows and columns of A, so that the above argument applies to all minors of size $(m + 1)$.   ∎

Theorem 2.7.6 [Moy77] *A square matrix A is quasidominant if and only if there exists a vector $x \succ 0$ such that $KAKx \succ 0$ for every $K \in \mathbb{K}_{\mathrm{sgn}}$.*

Proof. The condition $KAKx \succ 0$ may be written as

$$a_{ii}x_i > -\sum_{j \neq i} k_{ii} k_{jj} a_{ij} x_j,$$

where $k_{ii} k_{jj} = \pm 1$. Since the inequality must hold for every choice of the "signs" $k_{ii} k_{jj}$, it is clearly equivalent to the condition

$$a_{ii}x_i > \sum_{j \neq i} |a_{ij}| x_j,$$

which is the quasidominance condition.   ∎

Corollary 2.7.7 [Tau49] *All principal minors of a quasidominant matrix are positive. In particular, every symmetric quasidominant matrix is positive definite.* □

Theorem 2.7.8 [Moy77] *If A is quasidominant, there exists a diagonal $P > 0$ such that $PA + A^T P > 0$.*

Proof. Since quasidominance of A is equivalent to quasidominance of A^T, by Theorem 2.7.6 it follows that there exist $x \succ 0$ and $y \succ 0$ such that $KAKx \succ 0$ and $KA^T Ky \succ 0$ for any signature matrix K. Let P be a diagonal matrix with positive diagonal entries $p_i := y_i/x_i$. Then:

$$K(PA + A^T P)Kx = PKAKx + KA^T KPx = PKAKx + KA^T Ky \succ 0,$$

which means that by Theorem 2.7.6 $PA + A^T P$ is quasidominant. Since it is symmetric, it is positive definite, by Corollary 2.7.7. ∎

Proposition 2.7.9 $\{A : I - |A| \text{ is quasidominant}\} \subset \mathcal{D}_d.$

Proof. From Theorem 2.7.26, it follows that there exists a diagonal $P > 0$ such that $P - |A|^T P |A|$ is an M-matrix (and therefore quasidominant). A comparison of the entries of $P - |A|^T P |A|$ and $P - A^T PA$ then shows that this latter matrix is also quasidominant, for the same P and the proof is completed by Corollary 2.7.7. ∎

Lemma 2.7.10 [Reg92] *Let A be a matrix in companion form:*

$$A = \begin{bmatrix} a_1 & a_2 & \cdots & a_{n-1} & a_n \\ 1 & & & & \\ & 1 & & & \\ & & \ddots & & \\ & & & 1 & 0 \end{bmatrix}, \qquad (2.28)$$

and let $a^T := [a_1 \ a_2 \ \cdots \ a_{n-1} \ a_n]$. Then $A \in \mathcal{D}_d(P)$ if and only if

$$\sum_{k=1}^{n} |a_k| = \|a\|_1 \leq 1. \qquad (2.29)$$

Moreover, the diagonal elements p_i of P must satisfy:

$$p_1 \geq p_2 \geq \cdots \geq p_n > 0. \qquad (2.30)$$

Proof. (*only if*) Assume that $A \in \mathcal{D}_d(P)$, with $P = \text{diag}(p_1, \cdots, p_n)$ positive diagonal. Since the Stein equation (2.4) is linear in the solution P, the latter may be scaled by any positive constant. So it may be assumed without loss of generality that P is scaled so that $p_1 = 1$. Writing the negative definite matrix $-Q$ that occurs on the right-hand side of the Stein equation (2.4) as $-CC^T$, it is easy to verify that (2.4) can be written as:

$$aa^T + \text{diag}(p_2, \cdots, p_n, 0) - \text{diag}(1, p_2, \cdots, p_n) = -CC^T,$$

which can be rearranged as

$$[a \ C]\begin{bmatrix} a^T \\ C^T \end{bmatrix} = aa^T + CC^T = \text{diag}\,(1 - p_2, p_2 - p_3, \cdots, p_{n-1} - p_n, p_n).$$

$$(2.31)$$

By positivity of the left-hand side, the inequality (2.30) results. Now equation (2.31) holds if and only if there exists an orthogonal matrix U such that

$$\begin{bmatrix} a^T \\ C^T \end{bmatrix} = \text{diag}\,(\sqrt{1 - p_2}, \cdots, \sqrt{p_{n-1} - p_n}, \sqrt{p_n}) = U\text{diag}\,(\sigma_1, \cdots, \sigma_n),$$

$$(2.32)$$

for some orthogonal matrix U; the elements σ_i clearly being singular values. Equation (2.32) constrains the coefficients a_i to satisfy

$$a_1 = u_{11}\sigma_1 \quad \text{where} \quad \sum_{k=1}^{n} u_{1k}^2 = 1 \quad (U \text{ is orthogonal})$$
$$a_2 = u_{12}\sigma_2$$
$$\vdots \qquad\qquad\qquad\qquad\qquad\qquad\qquad\qquad (2.33)$$
$$a_n = u_{1n}\sigma_n \quad \text{and} \quad \sum_{k=1}^{n} \sigma_k^2 = 1 \quad (\text{by inspection of (2.32)}).$$

Hence

$$\left(\sum_{k=1}^{n} |a_k|\right)^2 = \left(\sum_{k=1}^{n} |u_{1k}| \, |\sigma_k|\right)^2 \leq \left(\sum_{k=1}^{n} u_{1k}^2\right)\left(\sum_{k=1}^{n} \sigma_k^2\right) = 1,$$

which gives inequality (2.30).

(*if*) Assume that (2.29) holds. It is always possible to find a set of positive constants σ_k and a second set of constants u_{1k} which satisfy (2.33). The reader is asked to take this statement on faith or else refer to [Reg92]. The constants u_{1k} can be taken as the first row of a matrix which is completed to an orthogonal matrix U. Solving for a set of positive constants $1 = p_1 \geq p_2 \geq \cdots \geq p_n$, one can then construct the matrix in (2.32). The steps leading up to (2.32) are now reversed to construct an equation in the form (2.4) with A in companion form (2.28) and P diagonal. ■

An alternate proof can be found in [Wim98]. The condition (2.29) was proved earlier in the context of stability of polynomials in [MK87]. It is well known that if the coefficients of a polynomial are put into a matrix in companion form, then the roots of the polynomial coincide with the eigenvalues of the matrix in companion form, so that the result in [MK87] is equivalent to (2.29). An immediate corollary of Lemma 2.7.10, obtained in [Ber82] for polynomials and restated here for matrices in companion form, using the same notation as above, is as follows.

Corollary 2.7.11 [Ber82] *The matrix A in (2.28) is Schur stable if* $\|a\|_2 < 1/\sqrt{n}$.

Proof. By Lemma 2.7.10, $A \in \mathcal{D}_d(P)$, if and only if (2.29) is satisfied. By the equivalence of norms, if $\|a\|_2 < 1/\sqrt{n}$, then $\|a\|_1 \leq 1$. ■

Corollary 2.7.12 {*Schur stable triangular matrices*} $\subset \mathcal{D}_d$.

Proof. The argument for a three by three upper triangular matrix is readily generalized to the $n \times n$ upper or lower triangular case. Hence, let $A = (a_{ij}) \in \mathbb{R}^{3 \times 3}$ be upper triangular, then so is $B = I - |A| = (b_{ij})$. It is easy to choose $d_i, i = 1, 2, 3$ to show that B is row quasidominant, as follows. Let $d_3 = 1$; then any $d_2 > |b_{23}|/|b_{22}|$ is a viable choice. Define $m_1 = \max\{d_2|b_{12}|, |b_{13}|\}$. Then any $d_1 > 2m_1/|b_{11}|$ is a viable choice of d_1. ∎

Note that, without using Corollary 2.7.12, it can be seen that the class of triangular matrices is also D-stable, since the absolute values of eigenvalues (on the diagonal) are not increased by postmultiplication by $K \in \mathbb{K}^{[-1,1]}$. This corollary is the Schur counterpart of the Hurwitz case in [BBP78].

The important class of M-matrices is a subclass of the class of quasidominant matrices. Note that, if A has the sign pattern required of an M-matrix, then it is quasidominant if and only if it is an M-matrix. Hence:

Proposition 2.7.13 [Moy77] {A: $I - |A|$ is an M-matrix} $\subset \mathcal{D}_d$. Equivalently, {A: $|A|$ is Schur stable } $\subset \mathcal{D}_d$. □

The above proposition has been exploited in the context of M-matrix-structure-preserving connective stability of large scale systems [Šil78] and robust stability [KB93b]. In this connection, it is interesting to point out that there are diagonally Schur stable matrices A such that $I - A$ is not an M-matrix. For example,

$$A = \begin{bmatrix} -0.5 & 0.75 \\ -0.75 & -0.5 \end{bmatrix} \in \mathcal{D}_d(I),$$

but $|A|$ is not Schur stable, so that $|A| \notin \mathcal{D}_d$ and $I - |A|$ is not an M-matrix. Note also that $I - A$ is not an M-matrix, since it is not Schur stable.

Definition 2.7.14 [Bau63] *If there exist $K_1, K_2 \in \mathbb{K}_{sgn}$ such that $|K_i| = I, i = 1, 2$ and $K_1 A K_2 = |A|$, then A is called* checkerboard. *The class of* stable checkerboard *matrices is defined as $\{A : A$ checkerboard and $|A|$ stable $\}$.*

From the above definition it follows that the class of stable checkerboard matrices is a subset of {A: $|A|$ Schur stable} so that the following proposition is a special case of Proposition 2.7.13. This class of matrices, originally defined in the context of optimal scaling of matrices [Bau63], interestingly enough, also crops up in frequency-domain robustness analysis of continuous-time systems [QD86]. All two by two triangular matrices are checkerboard. Qualitatively stable matrices, defined in Section 2.7.3 below are also a subclass of the class of Schur stable checkerboard matrices.

Proposition 2.7.15 {*Schur stable checkerboard matrices*} $\subset \mathcal{D}_d$.

Proof. By Definition 2.7.14, there exist $K_1, K_2 \in \mathbb{K}_{\text{sgn}}$ such that $K_1 A K_2$ is a nonnegative stable matrix and hence in $\mathcal{D}_d$ by Proposition 2.7.13. Thus there exists a positive diagonal matrix P such that

$$(K_2 A^T K_1) P (K_1 A K_2) - P < 0,$$

which implies that $K_2 A^T P A K_2 - P < 0$ (since $K_1 P K_1 = P$, for $|K_1| = I$) which implies, in turn, that $A^T P A - P < 0$ (pre- and postmultiply by K_2). ∎

2.7.1 Schur Diagonal and D-Stability for Two-by-Two Matrices

For two-by-two matrices it is possible to characterize diagonal stability and, furthermore, show that, in this case, D-stability is actually equivalent to diagonal stability. This section proves these facts with the additional objective of formulating a conjecture for matrices of dimension greater than two.

Schur stability of real two by two matrices is characterized in the following lemma.

Lemma 2.7.16 $A \in \mathbb{R}^{2 \times 2}$ *is Schur stable if and only if* $|\text{trace}(A)| < 1 + \det(A)$ *and* $|\det(A)| < 1$.

Proof. Follows from the Schur–Cohn conditions in [Mar66], or may be worked out from the fact that $\det(A)$ is the product of the eigenvalues of A and $\text{trace}(A)$ is their sum. □

Given the results of the previous section, the only additional fact needed to prove the equivalence of diagonal and D-stability for two by two matrices is the following.

Fact 2.7.17 $A \in \mathbb{R}^{2 \times 2}$ *is d-symmetrizable if and only if* $a_{12} a_{21} > 0$.

Proof. Immediate, by calculation. □

Proposition 2.7.18 $A \in \mathbb{R}^{2 \times 2}$ *is D-stable if and only if* $A \in \mathbb{R}^{2 \times 2}$ *is diagonally stable.*

Proof. (*if*) Proposition 2.5.8.
(*only if*) Let $A = (a_{ij})$ be D-stable. There are three possibilities for the product of the off-diagonal elements: (i) $a_{12} a_{21} = 0$; (ii) $a_{12} a_{21} > 0$; or, (iii) $a_{12} a_{21} < 0$. If (i) holds, then A is triangular and by Corollary 2.7.12, $A \in \mathcal{D}_d$. If (ii) holds, then A is d-symmetrizable, hence $A \in \mathcal{D}_d$. If (iii) holds, then either for $B_1 = A \, \text{diag}(1, -1)$ or for $B_2 = -B_1$, it must hold that $b_{12} b_{21} > 0$. Since the D-stability of A implies that both B_1 and B_2 are stable, it follows that one of them is d-symmetrizable and hence in $\mathcal{D}_d$, and now, by the proof of Proposition 2.5.8, $A \in \mathcal{D}_d$. ∎

Proposition 2.7.19 *$A \in \mathbb{R}^{2 \times 2}$ is in $\mathbb{D}_d$ if and only if each vertex of the polytope $\mathcal{R}(A)$ is stable.*

Proof. (*only if*) Immediate from the definition of D-stability.
(*if*) Consider the vertices A and $B = A \operatorname{diag}(1, -1)$ of $\mathcal{R}(A)$ which, by hypothesis, are both stable. Then one of the following three cases must hold.

$a_{12}a_{21} > 0$. By Proposition 2.7.3, $A \in \mathcal{D}_d$ and by Proposition 2.7.18, this is equivalent to $A \in \mathbb{D}_d$.

$b_{12}b_{21} > 0$. As in the previous case, it can be seen that $B = AK \in \mathcal{D}_d$, where $K = \operatorname{diag}(1, -1)$. By Proposition 2.5.8, $BK \in \mathcal{D}_d$. But $BK = AK^2 = A$.

$a_{12}a_{21} = 0$. This means that A is triangular and by Corollary 2.7.12, the proof is complete.   ■

The following characterization of Schur D-stability for two by two matrices is, by Proposition 2.7.18, also valid for diagonal stability. In the latter case, the condition was first proved, in an equivalent form, in [MMR78].

Proposition 2.7.20 *$A \in \mathbb{R}^{2 \times 2}$ is in $\mathbb{D}_d$, or, equivalently, in $\mathcal{D}_d$ if and only if $|\det(A)| < 1$ and the inequalities*

$$\begin{aligned} |a_{11} + a_{22}| &< 1 + \det(A) \\ |a_{11} - a_{22}| &< 1 - \det(A) \end{aligned} \qquad (2.34)$$

are satisfied.

Proof. By the previous proposition, $A \in \mathbb{D}_d$ if and only if PA is Schur stable for the four matrices $P = \operatorname{diag}(\pm 1, \pm 1)$. By Lemma 2.7.16, given that $|\det(A)| < 1$, condition (2.34) holds if and only if $\rho(PA) < 1$ for $P = I$ (for the $+$ sign) and $P = \operatorname{diag}(1, -1)$ (for the $-$ sign). Since $\rho(-PA) = \rho(PA)$, stability is also ensured for the other two choices of P, namely $-I$ and $\operatorname{diag}(-1, 1)$.   ■

Thus Propositions 2.7.18, 2.7.19, and 2.7.20 show that, for two by two matrices, the concepts of diagonal, vertex, and D-stability are all equivalent. Since the matrix (2.24) shows that diagonal and D-stability are not equivalent in general, this leads to the following conjecture, made in [BK93].

Conjecture 2.7.21 *$A \in \mathbb{R}^{n \times n}$ is Schur D-stable (i.e., the polytope $\mathcal{R}(A)$ is Schur stable) if and only if each vertex of the polytope $\mathcal{R}(A)$ is Schur stable.*

An alternative formulation of this conjecture is as follows: $A \in \mathbb{D}_d$ if and only if all matrices of the form $A \operatorname{diag}(\pm 1, \cdots, \pm 1)$ are Schur stable.

For additional perspectives on this conjecture, see the Notes and References.

2.7.2 Nonnegative Matrices

There is a vast literature and a great wealth of results on nonnegative matrices [BP94, Sen72, Var62, MM64, Min88]. In fact, it is possible to prove several additional facts about nonnegative diagonally stable matrices; for instance, an important fact is that a nonnegative matrix is Schur stable if and only if it is diagonally stable. This section states, without proof, the bare minimum of relevant facts about nonnegative matrices, referring the reader to the cited literature for details.

The best-known and most important result for nonnegative matrices is as follows.

Theorem 2.7.22 *(Perron–Frobenius) Let A be an $n \times n$ real nonnegative matrix. The spectral radius, $\rho(A)$, is an eigenvalue of A and possesses a corresponding nonnegative eigenvector. Furthermore, if A is irreducible, the corresponding eigenvector, v, unique up to scaling, is positive $(v \succ 0)$ and if V is a diagonal matrix with diagonal elements equal to the components of v, then*

$$\|VAV^{-1}\|_\infty = \rho(A),$$

which can also be written as
$\forall \epsilon > 0, \ \exists v \succ 0 \ \text{such that} \ \rho(A) \leq \|A\|_\infty^v \leq \rho(A) + \epsilon.$ ☐

For a proof of this theorem, see [Var62, BP94, BT89]. Simple bounds are available for this maximal eigenvalue, also known as the Perron–Frobenius eigenvalue.

Theorem 2.7.23   *Let A be an irreducible nonnegative $n \times n$ matrix, and let $\mathcal{K}_{\text{pos}}$ be the set of positive diagonal matrices.*
Then, for any $K \in \mathcal{K}_{\text{pos}}$, either

$$\min_{1 \leq i \leq n} \sum_{j=1}^{n} (K^{-1}AK)_{ij} < \rho(A) < \max_{1 \leq i \leq n} \sum_{j=1}^{n} (K^{-1}AK)_{ij}$$

or

$$\sum_{j=1}^{n} (K^{-1}AK)_{ij} = \rho(A) \ \text{for all} \ 1 \leq i \leq n,$$

where $K = \text{diag}(k_1, \ldots, k_n)$ and $(K^{-1}AK)_{ij} = (a_{ij}k_j)/k_i$.
If A is reducible, the strict inequalities above are replaced by nonstrict inequalities. Furthermore,

$$\rho(A) = \sup_{K \in \mathcal{K}_{\text{pos}}} \left\{ \min_{1 \leq i \leq n} \left\{ \sum_{j=1}^{n} (K^{-1}AK)_{ij} \right\} \right\}$$

$$= \inf_{K \in \mathcal{K}_{\text{pos}}} \left\{ \max_{1 \leq i \leq n} \left\{ \sum_{j=1}^{n} (K^{-1}AK)_{ij} \right\} \right\}. \quad ☐$$

Theorem 2.7.23 is of great value since it shows that simple arithmetic methods provide upper and lower bounds for the maximal eigenvalue of a nonnegative matrix. Note also that if this theorem is applied with $K = I \in \mathcal{K}_{\text{pos}}$, then it states that the maximal eigenvalue of a nonnegative matrix lies between its minimum and maximum row-sums. This is clearly also valid if row-sums are replaced by column-sums, since a matrix and its transpose have the same spectral radius. The following variant of Theorem 2.7.23 is also useful.

Theorem 2.7.24 [SW62, Theorem 2, p. 161] *Given a positive matrix A, let $\lambda_{PF}(A)$ denote its Perron–Frobenius eigenvalue. The following holds*

$$\lambda_{PF}(A) = \min_{D \succeq 0} \|D^{-1}AD\|$$

where D is a nonnegative diagonal matrix, and $\|\cdot\|$ any matrix norm induced by a (vector) Hölder norm. □

The following lemma gives various equivalent characterizations of Schur stable nonnegative matrices.

Lemma 2.7.25   *Let $A = (a_{ij})$ be a nonnegative matrix. The following are equivalent:*

(i) There exists a positive diagonal D such that $\|D^{-1}AD\|_\infty < 1$.

(ii) $I - A$ is quasidominant.

(iii) $A \in \mathcal{D}_d$.

(iv) The spectral radius, $\rho(A)$, is strictly less than unity.

(v) $I - A$ is a nonsingular M-matrix.

(vi) There exist $\lambda < 1$ and $d \succ 0$ such that $Ad \preceq \lambda d$.

Scheme of proof:

$$
\begin{array}{ccccccccc}
(iv) & \Leftarrow & (iii) & \Leftrightarrow & (i) & \Leftrightarrow & (ii) & \Rightarrow & (iii) \\
\Downarrow & & \Updownarrow & & & & \Updownarrow & & \Downarrow \\
(ii) & \Rightarrow & (v) & & & & (v) & \Leftarrow & (iv) & \Leftrightarrow (vi)
\end{array}
$$

Proof.   (i) ⇔ (ii): By definition $I - A$ is quasidominant if and only if there exist $d_i > 0$, $i = 1, \ldots, n$ such that

$$d_i(1 - a_{ii}) > \sum_{j \neq i} d_j a_{ij} \quad \text{for } i = 1, \ldots, n$$

$$\Leftrightarrow \quad 1 > a_{ii} + \sum_{j \neq i} \frac{d_j}{d_i} a_{ij} \quad \text{for } i = 1, \ldots, n$$

$$\Leftrightarrow \quad 1 > \sum_{j=1}^{n} \frac{d_j}{d_i} a_{ij} \quad \text{for } i = 1, \ldots, n \tag{2.35}$$

$$\Leftrightarrow \quad \max_{1 \leq i \leq n} \sum_{j=1}^{n} \frac{d_j}{d_i} a_{ij} = \|D^{-1} A D\|_\infty < 1,$$

where $D := \operatorname{diag}(d_1, \ldots, d_n)$.

(ii) $\Rightarrow$ (iii) is Proposition 2.7.9 specialized to nonnegative matrices.

(iii) $\Rightarrow$ (iv) is a classical result of Liapunov theory. A short algebraic proof is as follows. From remark 2.17 we know that $A \in \mathcal{D}_d \Leftrightarrow \|P^{1/2} A P^{-1/2}\|_2 < 1$. Since $\rho(A) = \rho(P^{1/2} A P^{-1/2}) \leq \|P^{1/2} A P^{-1/2}\|_2$, this completes the proof.

(iv) $\Rightarrow$ (v) follows from the definition of a nonsingular M-matrix.

(v) $\Leftrightarrow$ (ii) It is observed in [Moy77, p. 56] that if a matrix has the sign pattern required of an M-matrix, then it is quasidominant if and only if it is an M-matrix. Since A is nonnegative, clearly $I - A$ has all off-diagonal entries nonpositive and the equivalence is now obvious.

(iv) $\Rightarrow$ (vi) follows from the Perron–Frobenius theorem.

(vi) $\Rightarrow$ (iv) The hypothesis can be written as $\|A\|_\infty^d \leq \lambda < 1$. But $\rho(A) \leq \|A\|_\infty^d$, completing the proof. ∎

Thus Lemma 2.7.25 says that, within the class of nonnegative matrices, the stability of a matrix is equivalent to its diagonal stability and, in turn, equivalent to the existence of a weighted infinity norm in which the norm of the matrix is less than unity, as opposed to the requirement (see remark 2.17) that a weighted two-norm be less than unity. For a continuous-time version of Lemma 2.7.25, see [OŠ85], and for different proofs of some equivalences in the discrete-time case, see [BT89].

A generalization of the equivalence between items (iii) and (v) above is as follows.

Theorem 2.7.26 [Ara75] *Let R be a positive diagonal matrix and let A be a nonnegative matrix. Then $R - A$ is an M-matrix if and only if there is a positive definite diagonal matrix P such that $Q := RPR - A^T P A$ is positive definite.*

Proof. The matrix Q is positive definite if and only if the matrix

$$R^{-1} Q R^{-1} = P - (AR^{-1})^T P (AR^{-1})$$

is positive definite. On the other hand, the matrix $R - A$ is an M-matrix if and only if the matrix $I - AR^{-1}$ is an M-matrix [AK72, Theorem 5]. Thus it is enough to show that the theorem holds in the case $R = I$.

(*only if*) By a property of an M-matrix [FP62, BP94], there are positive vectors $u = (u_1 \cdots u_n)^T$ and $v = (v_1 \cdots v_n)^T$ such that

$$y = (I - A)u \succ 0, \qquad z = (I - A)^T v \succ 0.$$

Define the diagonal entries of the matrix P by $p_i = v_i/u_i$, where the subscript i identifies the ith entry. Then,

$$\begin{aligned} Qu &= Pu - A^T PAu = v + A^T P(y - u) \\ &= z + A^T P \succ 0. \end{aligned}$$

Since the diagonal elements of Q are nonpositive, the matrix Q is an M-matrix [BP94]. Since Q is symmetric, Q is positive definite (by the definition of an M-matrix and Sylvester's test).

(*if*) Let λ_{PF} be the Perron–Frobenius eigenvalue of A and $u \succeq 0$ the corresponding eigenvector. Since Q is positive definite:

$$\begin{aligned} 0 < u^T Qu &= u^T(P - A^T PA)u \\ &= (1 - \lambda_{PF}^2)u^T Pu. \end{aligned}$$

By assumption $u^T Pu > 0$. Thus $|\lambda_{PF}| < 1$ and so $(I - A)$ is an M-matrix by Lemma 2.7.25. ∎

Some useful corollaries of Lemma 2.7.25 for general matrices are as follows.

Corollary 2.7.27 [Moy77] *For any $A \in \mathbb{R}^{n \times n}$, if the spectral radius of the modulus matrix of A is less than 1 (i.e., $\rho(|A|) < 1$), then A is in $\mathcal{D}_d$.*

Proof. By Lemma 2.7.25, $\rho(|A|) < 1$ if and only if $I - |A|$ quasidominant and the proof is complete, by Proposition 2.7.9. ∎

The converse of Corollary 2.7.27 is false. Consider, for example,

$$A = \begin{bmatrix} -0.5 & 0.75 \\ -0.75 & -0.5 \end{bmatrix} \in \mathcal{D}_d(I),$$

since $I - A^T A$ is positive definite, but, by Theorem 2.7.23, $\rho(|A|) = 1.25 > 1$.

Corollary 2.7.28 *For any $A \in \mathbb{R}^{n \times n}$ if the 2-norm of A is less than one, then $A \in \mathcal{D}_d$.*

Proof. The 2-norm or maximum singular value is the square root of the largest eigenvalue of the positive semidefinite matrix $A^T A$. Thus, from the hypothesis, $A^T A - I$ is negative definite, i.e., $A \in \mathcal{D}_d(I)$. ∎

Corollary 2.7.29 [BT89, Corollary 6.2, p. 150] *For any $A \in \mathbb{R}^{n \times n}$, there exists a positive diagonal matrix K such that $\|K^{-1}AK\|_\infty < 1$ if and only if $\rho(|A|) < 1$.*

Proof. (*if*) Immediate from 2.7.25 and the absoluteness of the infinity norm.

(*only if*) $\|K^{-1}|A|K\|_\infty = \|K^{-1}AK\|_\infty$, since the infinity norm is an absolute norm. Thus, $\rho(K^{-1}|A|K) < 1$. But $\rho(K^{-1}|A|K) = \rho(|A|)$.   ∎

Corollary 2.7.30 [BT89, Corollary 6.3, p. 151] *For any* $A \in \mathbb{R}^{n \times n}$, $\rho(A) \leq \rho(|A|)$.   □

A useful result on diagonal stability for partitioned matrices is as follows:

Lemma 2.7.31 *Given a block partitioned matrix* $A = (A_{ij}) \in \mathbb{R}^{N \times N}$ *where* $N = \sum_{i=1}^n m_i$, $A_{ij} \in \mathbb{R}^{m_i \times m_j}$, *let* $H = (\|A_{ij}\|) \in \mathbb{R}^{n \times n}$. *Then* $\rho(H) < 1$ *implies* $A \in \mathcal{D}_d$.

Proof. Since H is a nonnegative Schur stable matrix, by Lemma 2.7.25, H is diagonally stable, i.e., in $\mathcal{D}_d(P)$ for some $P = \mathrm{diag}\,(p_1, \ldots, p_n)$. Also define $\mathbb{P}$ as $\mathrm{diag}\,(p_1 I_{m_1}, \ldots, p_n I_{m_n})$ and consider the quadratic form $V(x) = x^T \mathbb{P} x$. It is enough to show that the quadratic form $\Delta V(x) := x^T (A^T \mathbb{P} A - \mathbb{P})x$ is negative definite. Noting that P and $\mathbb{P}$ are diagonal matrices, by the properties of norms,

$$
\begin{aligned}
\Delta V(x) &= x^T (A^T \mathbb{P} A - \mathbb{P})x \\
&\leq [\![x]\!]^T H^T PH [\![x]\!] - [\![x]\!]^T P [\![x]\!] \\
&= [\![x]\!]^T (H^T PH - P)[\![x]\!] \\
&< 0, \quad \forall x \neq 0 \in \mathbb{R}^N,
\end{aligned}
$$

which completes the proof.   ∎

It is clear that in the case of a single partition ($n = 1, m_1 = N$) and using the 2-norm, the above lemma reduces to Corollary 2.7.28. This result can be interpreted as follows. Schur stability (equivalently, by Lemma 2.7.25, Schur diagonal stability) of the matrix of norms of the blocks, H, implies Schur diagonal stability of the "parent" unpartitioned matrix A. Related results on the relations between norms of unpartitioned and partitioned matrices can be found in [Ost61, KLS89].

In the large scale systems approach to dynamical systems, instead of analyzing a large scale system in its full dimension, states of the system are grouped together or aggregated, and the resulting aggregate system is of smaller dimension (see Section 6.1). Under certain conditions, qualitative properties, such as stability, of the aggregate system are shared by the large system.

In terms of the lemma above, the aggregate matrix H corresponding to the block partitioned system matrix $A = (A_{ij})$, which is clearly always nonnegative, since its elements are norms of submatrices, is often required to be an M-matrix ([Šil78]). By Lemma 2.7.25, this is equivalent to requiring that it be Schur diagonally stable as well. In addition, Lemma 2.7.31 shows that the original nonaggregated system matrix must also have been Schur diagonally stable. This does not seem to have been noted in the literature on the large scale systems approach, and it accounts for the conservativism of this approach in the circumstances described above. See Chapter 6 for more details on these topics.

So far this chapter has discussed subclasses of the set of Schur stable matrices all of whose members remain in it under all transformations of a given type (for example, pre- or postmultiplication by a diagonal matrix with diagonal elements in the interval $[-1, 1]$). In the same spirit, another type of stability that arises in the context of stability of certain classes of discrete-time systems and matrices (see Section 3.4.1) is defined in terms of the Hadamard product of matrices.

Definition 2.7.32 *The Hadamard product of* $A = (a_{ij}) \in \mathbb{R}^{n \times n}$ *and* $B = (b_{ij}) \in \mathbb{R}^{n \times n}$ *is written* $A \circ B$ *and defined by the elementwise product* $A \circ B = (a_{ij} b_{ij}) \in \mathbb{R}^{n \times n}$.

Note that the Hadamard product is commutative, i.e., $A \circ B = B \circ A$. For other properties of the Hadamard product, see [JE87, HJ91].

For a matrix $B = (b_{ij})$, write $|B| \preceq \mathbb{1}$, whenever, for all pairs i, j, one has $|b_{ij}| \leq 1$.

Definition 2.7.33 *The class* $\mathcal{HS}$ *of Hadamard stable matrices is:*

$$\mathcal{HS} := \{A \in \mathbb{R}^{n \times n} : A \circ H \quad \text{is Schur stable for all,} \, |H| \preceq \mathbb{1}\}.$$

The following result holds.

Theorem 2.7.34 *The following statements are equivalent:*

(i) A *is Hadamard stable.*

(ii) $|A|$ *is Schur diagonally stable.*

(iii) $|A|$ *is Schur stable.*

(iv) $I - |A|$ *is an M-matrix.*

Proof. In view of Lemma 2.7.25, it is enough to prove that item (i) implies any one of the other items and that any one of the other items implies (i).

$[(i) \Rightarrow (iii)]$. If $A = (a_{ij}) \in \mathcal{HS}$, define $H = (h_{ij})$, as follows $h_{ij} :=$ sgn(a_{ij}), where sgn$(x) = +1$, if $x > 0$; sgn$(x) = -1$, if $x < 0$; and sgn$(0) = 0$; consequently $|H| \preceq \mathbb{1}$ and $A \circ H = |A|$ is Schur stable.

$[(ii) \Rightarrow (i)]$. $|A|^T P |A| - P < 0$, for $P > 0$ diagonal. Adding and subtracting the term $(|A|^T \circ H) P (H \circ |A|)$ on the left-hand side gives, for all $H \preceq \mathbb{1}$:

$$[(|A|^T \circ H^T) P (H \circ |A|) - P] + [|A|^T P |A| - (|A|^T \circ H^T) P (H \circ |A|)] < 0. \tag{2.36}$$

Since P is positive diagonal, the following matrix inequality holds:

$$(|A^T| \circ H^T) P (H \circ |A|) \leq (|A^T| \circ |H^T|) P (|H| \circ |A|)$$

Thus the second term in square brackets in (2.36) can be manipulated as follows:

$$|A^T|\, P\, |A| - (|A^T| \circ H^T)P(H \circ |A|) \geq$$
$$|A^T|\, P\, |A| - (|A^T| \circ |H|^T)P(|H| \circ |A|) \geq 0 \tag{2.37}$$

Since the sum of the two quadratic forms (each within square brackets) in (2.36) is negative definite and it has just been shown that the second is positive semidefinite, therefore the first must be negative definite. In other words, for all $|H| \preceq \mathbb{1}$, $(|A|^T \circ H^T)P(H \circ |A|) - P < 0$, i.e., for all $|H| \preceq \mathbb{1}$, $(|A| \circ H) \in \mathcal{D}_d(P)$. $\blacksquare$

The class of Hadamard stable matrices introduced above appears in Section 3.4.1 in the contexts of the stability of nonlinear interval systems and also of qualitative Schur stability (considered in the next section).

Polytopes of nonnegative matrices

Theorem 2.6.8 says that a polytope of matrices is stable if and only if there exists a norm in which all vertex matrices have norm strictly bounded below unity.

For polytopes of nonnegative matrices, it is possible to go further and generalize Theorem 2.7.26 to show that if simultaneous 1- and infinity-norms exist, then a simultaneous 2-norm exists as well and can be calculated from the former. Here, since the norms are thought of as Liapunov functions for the polytopes, the term simultaneous is being applied to them as well, in a slight extension of Definition 2.6.6 in Section 2.6

Proposition 2.7.35 *Given a polytope of nonnegative matrices, $\mathcal{A} = \mathrm{Co}\{A_1, \ldots, A_n\}$, (i) implies (ii), where statements (i) and (ii) are as follows:*

(i) There exist positive diagonal matrices D_r and D_c such that for $i = 1, \ldots, n$,

$$\|D_r^{-1} A_i D_r\|_\infty < 1, \tag{2.38}$$

$$\|D_c^{-1} A_i^T D_c\|_\infty = \|D_c A_i D_c^{-1}\|_1 < 1, \quad i = 1, \ldots, n. \tag{2.39}$$

(ii) There exists a positive diagonal matrix P such that, for $i = 1, \ldots, n$,

$$\|P^{\frac{1}{2}} A_i P^{-\frac{1}{2}}\|_2 < 1. \tag{2.40}$$

Furthermore, $P = D_c D_r^{-1}$.

Proof. If equations (2.38) and (2.39) hold, then, since the matrices A_i are nonnegative, this is equivalent to asserting that the vectors d_r and d_c, which are the diagonals of the matrices D_r and D_c, respectively, are such that, for $i = 1, \ldots, n$,

$$\begin{aligned} u_i &= (I - A_i)d_r \succeq 0, \\ v_i &= (I - A_i)^T d_c \succeq 0. \end{aligned}$$

Let $P = D_c D_r^{-1}$ as claimed. Then, for $i = 1, \ldots, n$,

$$
\begin{aligned}
(P - A_i^T P A_i) d_r &= P d_r - A_i^T P A_i d_r \\
&= d_c - A_i^T P(d_r - u_i) \\
&= (d_c - A_i^T d_c) + A_i^T P u_i \\
&= v_i + A_i^T P u_i.
\end{aligned}
$$

But $v_i \succ 0$ and since $A_i^T \succeq 0, P > 0$, therefore $A_i^T P u_i \succeq 0$. Thus $(P - A_i^T P A_i) d_r \succ 0$, i.e. $Q_i := P - A_i^T P A_i$ is an M-matrix, for all i. Since Q_i is symmetric, it must be positive definite, for all i (cf. Theorem 2.7.26). ∎

This proposition says that the existence of simultaneous row *and* simultaneous column diagonal scalings such that every vertex has infinity- and 1-norm strictly less than unity implies the existence of a simultaneous diagonal scaling in which every vertex has 2-norm strictly less than unity. Alternatively, for polytopes of nonnegative matrices, the existence of *two scaled infinity- and 1-norm diagonal-type simultaneous Liapunov functions implies the existence of a single quadratic diagonal SLF*. It should be noted that the proof of this proposition is partly a restatement of the classical relation $\|A\|_2^2 \leq \|A\|_1 \|A\|_\infty$ [HJ85, Exercise 5.6.21].

Proposition 2.7.35 guarantees that the only possibility that is ruled out is the existence of scalings D_r and D_c together with the inexistence of a scaling P. Alternatively, if P does not exist, it may be affirmed that at most one of D_r and D_c exists.

If the polytope in Proposition 2.7.35 consists of a single matrix (i.e., $n = 1$), the result still holds. In fact, by Lemma 2.7.25, the existence of D_r is equivalent to the existence of a positive diagonal matrix P that satisfies the Stein equation.

2.7.3 Qualitatively Schur Stable Matrices

Another class of matrices that is diagonally stable is that of qualitatively Schur stable matrices first introduced and characterized in [KB89].

Consider the dynamical system described by a difference equation

$$
x(k+1) = Ax(k), \qquad k = 0, 1, \cdots . \tag{2.41}
$$

In the continuous-time case, sign-stability of a matrix A is determined using qualitative data obtained by comparing the values of the entries of A with the zero element. Therefore, the entries are considered to be either exactly zero (*hard zero*), greater than zero (positive entries), or less than zero (negative entries). It is natural to proceed in a similar manner in the discrete-time case, comparing the absolute values of the entries with unity. Given this change in the comparison, it is reasonable to speak of *qualitative stability* in the discrete-time case.

In order to define discrete-time qualitative stability, it is necessary to introduce four sets of real numbers. The entries of real matrices are then classified into one of these four sets. The sets are defined as follows. $C_1 :=$ $\{z \in \mathbb{R} : 0 < |z| < 1\}$, $C_2 := \{z \in \mathbb{R} : |z| > 1\}$, $C_3 := \{z \in \mathbb{R} : |z| = 1\}$, $C_4 := \{z \in \mathbb{R} : z = 0\}$.

The problem of qualitative stability is to derive conditions under which $A = (a_{ij})$ has all its eigenvalues within the unit circle, provided that the only available information about each entry a_{ij} is the set C_k to which it belongs. Note that only three sets (positive, negative and zero entries) are needed in the definition of sign-stability (the continuous-time analog) [QR65, JKD77, JKD87, MQ69, GM81]. This is discussed further below.

The real $n \times n$ matrices $A = (a_{ij})$ and $B = (b_{ij})$ are said to be m-equivalent (modulus-equivalent) if and only if for every pair (i, j), a_{ij} and b_{ij} belong to the same class C_k, $k = 1, \ldots, 4$. This is denoted by $A\|B$.

Definition 2.7.36 *A real $n \times n$ matrix $A = (a_{ij})$ is said to be* qualitatively stable *if and only if all matrices belonging to the set $m(A) := \{B : B\|A\}$ are Schur stable, and equivalently the system (2.41) is said to be* qualitatively stable.

Note that, strictly speaking, the matrix A is qualitatively Schur stable, but, for reasons of linguistic convenience, the simpler locution qualitatively stable is used below.

It is easy to see the need for the four classes C_k, since, for example, if one defines $\widetilde{C} := C_1 \cup C_3$ then the matrices

$$A := \begin{bmatrix} 0 & 1 \\ 1 & 0 \end{bmatrix} \quad \text{and} \quad B := \begin{bmatrix} 0 & 1 \\ a & 0 \end{bmatrix}, \quad \text{where } |a| < 1,$$

become m-equivalent, although A is not Schur stable and B is stable. In fact, it will transpire that B is qualitatively stable, while A is not. The class C_4 of hard zeros is introduced to account for known absences of interaction. For instance, if it is known that the evolution of state x_i is not affected by that of state x_j, then the element a_{ij} is a hard zero. For discussion of the significance of the class C_3 of hard or "true" ones, see [Wil86].

The definition of qualitative stability above clearly implies the following.

Fact 2.7.37 *A given matrix A is qualitatively stable if and only if $|A|$ is qualitatively stable.* □

This fact is helpful since some well-known results on nonnegative matrices can be used in order to obtain the following characterization of qualitative stability.

Theorem 2.7.38 [KB89] *An irreducible $n \times n (n \geq 2)$ real matrix $A = (a_{ij})$ is qualitatively stable if and only if:*

(i) there is only one cycle in A, and it is of length n.

(ii) $\forall i, j, \ |a_{ij}| \leq 1$.

(iii) *the absolute value of the product of all the matrix entries in the cycle*
is strictly less than unity.   □

Equivalent restatements of conditions (i) and (iii) are as follows:

(i') *A has exactly one nonzero entry per row and per column.*

(iii') *there exists at least one entry which is strictly less than one in*
modulus (clearly (iii) implies (iii') and, given (ii), (iii') implies (iii)).

The discrete-time analogue of sign-similarity is called m-equivalence and
was arrived at by mimicking the continuous-time case. In light of the characterization of m-stability, one sees that, despite the close analogy between
definitions of sign similarity and m-equivalence, the discrete case is much
simpler. The notion of m-equivalence induces a strong restriction on the
class of qualitatively or m-stable matrices—such matrices are permitted to
have only one cycle, of gain strictly less than unity. This implies that an
m-stable matrix of order n must have $n^2 - n$ hard zeros.

It is interesting to note that in [Cai90], the characterization above is
restated as follows:

Theorem 2.7.38a *Let* $n \geq 2$ *and A be irreducible. Then every matrix*
which is modulus equivalent to A is asymptotically stable if and only if
$A = PD$, *where P is an irreducible permutation matrix and D is a diagonal*
matrix satisfying $0 < |d_{ii}| \leq 1$ *with* $|\det(D)| < 1$.   □

This characterization is also easily seen to be true for complex matrices,
since the proof involves only the absolute value (modulus) of each entry of
the matrix.

Qualitative stability implies diagonal stability as shown by the following
lemma, thus providing a link with the subject of this chapter.

Lemma 2.7.39 $A \in \mathbb{R}^{n \times n}$ *is qualitatively stable implies that A is*
diagonally stable, i.e., $A \in \mathcal{D}_d$.

Proof.   Since a qualitatively stable matrix A has one element per row and
column, there always exist diagonal signature matrices K_1 and K_2 such
that $K_1 A K_2 = |A|$; for example $K_2 = I$ and K_1 is a matrix with diago-
nal elements $+1$ corresponding to the row for which the existing nonzero
element (a_{ij}) is positive, and -1 for the case in which the only nonzero element is negative. Therefore a qualitatively stable matrix A is checkerboard
and, by Fact 2.7.37, this implies that the matrix $|A|$ must be Schur stable.
Thus, by Lemma 2.7.15, A is diagonally stable.   ■

Actually, it is easy to verify that $V(x) = x^T I x$ is a simultaneous diagonal
Liapunov function for this class of matrices. This implies that qualitatively
stable matrices are D-stable, which is to be expected since the stability
of these matrices should not be affected by pre- or postmultiplication by a
diagonal matrix $|D| \leq I$. Qualitatively stable matrices are also easily shown
to be: (i) Hadamard stable; and (ii) in class $\mathcal{D}_{d2}$, defined as the class of

matrices that possess a diagonal solution to the Stein equation with right-hand side diagonal as well, by analogy with the continuous-time definition of class $\mathcal{D}_{c2}$.

2.8 Testing for Diagonal and D-Stability

It is important to make a distinction between the problem of finding a matrix that satisfies the Liapunov or Stein inequality and that of finding one that is a solution to the Liapunov or Stein equation. For example, given a matrix A, two questions may be asked: (i) Does there exist a positive diagonal matrix P such that the matrix $A^T P + PA$ is negative definite? (ii) Given a negative definite matrix Q, does there exist a diagonal solution to the equation $A^T P + PA = Q$? In the first case, a feasible solution to the Liapunov linear matrix inequality is sought, while in the second case, a solution to the linear Liapunov equation is sought. Depending on the context, the answers to both questions are of interest, as seen elsewhere in this book. Note also that (i) may have an affirmative answer, while, for a given matrix Q, (ii) may not. For example, if A is a negative diagonal matrix, then it is easy to see it has a diagonal solution to the Liapunov equation (for example, the identity matrix). However, if a nondiagonal matrix is chosen as the right-hand side of the Liapunov equation, then the latter cannot be satisfied for any diagonal matrix, since the left-hand side is diagonal and the right-hand side is not.

Thus one can distinguish between theoretical approaches to testing for diagonal stability, which usually answer the feasibility question of whether a given matrix belongs to the class of diagonally stable matrices, and numerical approaches, which can usually solve both the feasibility problem as well as the problem of finding a solution for a specific right-hand side Q.

This section briefly presents available results in the categories listed above and also gives necessary conditions of the type that can be easily checked by inspection.

Theoretical characterizations
Theoretical characterizations (i.e., answers to the feasibility question) of the class of Hurwitz diagonally stable matrices have been given in terms of (a) algebraic relations between the elements of the matrix $A \in \mathbb{R}^{n \times n}$, for $n = 2, 3$ [Cro78]; (b) cones of matrices [BBP78]; and (c) the Hadamard product [Kra91]. For $n = 4$, a theoretical solution to the problem of finding a diagonal solution to the Liapunov equation is given in [Red85b]. It leads to a computational procedure that is compared with the LMI method below.

For certain classes of matrices it is possible to be more specific about diagonal solutions to the Stein and Liapunov equations. For example, the paper [Moy77] gives solutions to these equations for quasidominant matrices, as discussed in Section 2.7; for the Liapunov equation and acyclic-3 matrices,

diagonal solutions are constructed in [BH83, KH84b, Red85a]; Hurwitz
D-stability of tridiagonal matrices is characterized in [CDJ82, Car84].

Algebraic characterizations of Hurwitz diagonal and D-stability
For matrices of dimensions 2 and 3, some algebraic characterizations of
diagonal and D-stability are available [Cai76, Cro78, Log87] and are listed
below.

Hurwitz diagonal stability

Fact 2.8.1 [Cro78] *A matrix* $A = (a_{ij}) \in \mathbb{R}^{2\times 2}$ *is in class* $\mathcal{D}_c$ *if and only
if* $-A$ *is in* $\mathcal{P}$, *i.e.,* $a_{ii} < 0$, $i = 1, 2$ *and* $\det(A) > 0$.

Fact 2.8.2 [Cro78] *A matrix* $A \in \mathbb{R}^{3\times 3}$ *is in class* $\mathcal{D}_c$ *if and only if* $B = -A$ *has the following properties.*

  (i) B *is in class* $\mathcal{P}$; *and*

  (ii) *Each of the four numbers* $1, \omega_1, \omega_2, \omega_3$ *is smaller than the sum of the
       other three (or, equivalently,* $\max(1, \omega_1, \omega_2, \omega_3) < \frac{1}{2}(1 + \omega_1 + \omega_2 + \omega_3)$,
       where $\omega_i := \sqrt{b_{ii}(B^{-1})_{ii}}$, $i = 1, 2, 3$.

The formulation of item (ii) above is from [Kra91], although derived earlier
in another equivalent formulation in [Cro78].

Hurwitz D-stability
    Given a matrix $A = (a_{ij})$, let m_{ii} denote the principal minor associated
to a_{ii}. For example, for $A \in \mathbb{R}^{3\times 3}$, $m_1 = a_{22}a_{33} - a_{23}a_{32}$.

Fact 2.8.3 [Joh74a] *A matrix* $A \in \mathbb{R}^{2\times 2}$ *is D-stable if and only if* $-A$ *in*
$\mathcal{P}_o^+$, *i.e.,* $a_{ii} \le 0$, $i = 1, 2$, *at least one* a_{ii} *is negative and* $\det(A) > 0$.

Fact 2.8.4 [Cro78] *A matrix* $A \in \mathbb{R}^{3\times 3}$ *is D-stable if and only if* $-A \in \mathcal{P}_o^+$
and

  (i) $\sqrt{\dfrac{a_{11}m_{11}}{\det(A)}} + \sqrt{\dfrac{a_{22}m_{22}}{\det(A)}} + \sqrt{\dfrac{a_{33}m_{33}}{\det(A)}} - 1 > 0$; *or*

  (ii) $\sqrt{\dfrac{a_{11}m_{11}}{\det(A)}} + \sqrt{\dfrac{a_{22}m_{22}}{\det(A)}} + \sqrt{\dfrac{a_{33}m_{33}}{\det(A)}} - 1 = 0$ *and there exists an index*
       $j \in \{1, 2, 3\}$ *such that* $a_{jj}m_{jj} = 0$, *with* $a_{jj} \ne 0$ *or* $m_{jj} \ne 0$.

Examples

  1. The matrix $A = \begin{bmatrix} 0 & 1 \\ -1 & -1 \end{bmatrix}$ is D-stable but not diagonally stable,
     since $-A$ is not in $\mathcal{P}$. It is, however, clearly diagonally semistable
     (i.e., in $\mathcal{D}_{co}$).

2. Example 2.1.6 is reexamined here and, once again, it is convenient to work with $C = -B$. For the matrix

$$C = \begin{bmatrix} 1 & 0 & -50 \\ 1 & 1 & 0 \\ 1 & 1 & 1 \end{bmatrix},$$

$$C^{-1} = \begin{bmatrix} 1 & -50 & 50 \\ -1 & 51 & -50 \\ 0 & -1 & 1 \end{bmatrix}$$

so that $\omega_1 = \sqrt{1} = 1, \omega_2 = \sqrt{51} = 7.1414, \omega_3 = \sqrt{1} = 1$, so that $\max(1, \omega_1, \omega_2, \omega_3) = 7.1414$ while $\frac{1}{2}(1 + \omega_1 + \omega_2 + \omega_3) = 5.0707$ so that the second condition for diagonal stability is violated, indicating that $B \notin \mathcal{D}_c$.

On the other hand, the matrix C is clearly in $\mathcal{P}_o^+$, with minors $m_{11} = 1, m_{22} = 51, m_{33} = 1$ and $\det(C) = 1$. Thus $\sqrt{1} + \sqrt{51} + \sqrt{1} - 1 > 0$ so that condition (i) for D-stability is met, and the matrix B is D-stable. In fact, since $\omega_1 + \omega_2 + \omega_3 > 1$, it actually follows that B is in the topological interior of the set of D-stable matrices (see [Har80, Kra91] for details).

3. It is stated in [Log87] that the set of totally D-stable matrices is a strict superset of the set of Hurwitz diagonally stable matrices, and the following matrix,

$$A = \begin{bmatrix} -1 & 4 & -5 \\ -0.2 & -1 & 0.5 \\ 0.25 & -2 & -1 \end{bmatrix},$$

is claimed to be totally D-stable but not Hurwitz diagonally stable. This is false, because it can be checked by the reader that the matrix $-A$: (a) is in class $\mathcal{P}$ and (b) $\omega_1 = 0.6003, \omega_2 = 0.6367, \omega_3 = 0.5695$, so that $\max(1, \omega_1, \omega_2, \omega_3) = 1 < \frac{1}{2}(1 + \omega_1 + \omega_2 + \omega_3) = 1.4033$, so that the conditions for diagonal stability are satisfied. The matrix is therefore totally D-stable as well by Corollary 2.1.9. The matrix is, of course, also D-stable, as can be checked independently by the calculation $\sqrt{2/5.55} + \sqrt{2.25/5.55} + \sqrt{1.8/5.55} - 1 = 0.8605 > 0$.

Schur diagonal and D-stability

The characterizations of Schur D-stability and diagonal stability in the two by two case [MMR78, BK93] are as follows:

Fact 2.8.5 [BK93] *A matrix $A \in \mathbb{R}^{2 \times 2}$ is Schur diagonally stable if and only if it is Schur D-stable.*

Fact 2.8.6 [MMR78, BK93] *A matrix $A \in \mathbb{R}^{2 \times 2}$ is Schur D-stable or equivalently diagonally stable if and only if the matrices A and $B =$*

A diag $(1, -1)$ *are both Schur stable which means that:* $|\det(A)| < 1$, $|trace(A)| < 1 + \det(A)$ *and* $|trace(B)| < 1 + \det(B)$.

Some necessary conditions that can be checked by inspection
Hurwitz case

$$A = (a_{ij}) \in \mathcal{D}_c \text{ only if } a_{ii} < 0, \text{ for all } i. \qquad (2.42)$$

Schur case

Two easily checked necessary conditions for Schur diagonal stability are given in the following proposition.

Proposition 2.8.7 [Heb94] *If* $A = (a_{ij}) \in \mathbb{R}^{n \times n}, n > 1$ *is Schur diagonally stable, then (i) for all* i, $|a_{ii}| < 1$; *(ii)* $z \le z_{\max} := n - |trace(A)|$, *where* z *is the number of zero-valued diagonal elements of* A.

Proof.   Assume that $A \in \mathcal{D}_d(P)$, with $P = \text{diag}(p_1, \ldots, p_n)$. This means that $p_i > 0$ and there exists a positive definite matrix Q such that $P - A^T P A = Q$. By calculation, the ith diagonal element of Q is $q_{ii} = p_i - \sum_{k=1}^{n} p_k a_{ki}^2$ and must be positive, in order for Q to be positive definite. Rearranging the above equation results in

$$(1 - a_{ii}^2)p_i = \sum_{k=1, \, k \neq i}^{n} a_{ki}^2 p_k. \qquad (2.43)$$

Now, in order to get a contradiction, assume that, for some i, $|a_{ii}| \ge 1$. This assumption implies that, a_{ii} cannot be the only nonzero element in the ith row, for this would contradict stability of A. Since p_i and the right-hand side of (2.43) are positive, it must be true that $|a_{ii}| < 1$, in order for the left-hand side to be positive as well. This contradiction completes the proof of item (i).

For the proof of item (ii), first observe that if $z = n$, then trace$(A) = 0$ and the conclusion $z = z_{\max}$ is obvious. Assuming $z < n$, define $j_i, i = 1, \ldots, n - z$ to be the index of the ith nonzero diagonal entry of A. From the definition of the trace of a matrix and the triangle inequality, it follows that

$$|\text{trace}(A)| = \left| \sum_{i=1}^{n-z} a_{j_i j_i} \right| \le \sum_{i=1}^{n-z} |a_{j_i j_i}|.$$

By item (i), $|a_{j_i j_i}| < 1$, which bounds the right-hand term by $n - z$. This implies that $|\text{trace}(A)| < n - z$; thus $z < n - |\text{trace}(A)| = z_{\max}$.   ∎

Matrices in companion form
As pointed out in Section 2.7, a matrix in the companion form is Schur diagonally stable if and only if the sum of the absolute values of the entries in the nontrivial row (or column) (i.e., the row or column with the entries equal to the coefficients of the characteristic polynomial) is strictly

less than one and this implies that the diagonal entries of P can be put in nonincreasing order. An algorithm to calculate these entries algebraically is given in [BCZ95], which also gives a symbolic algorithm to calculate solutions of the Liapunov equation in a companion form matrix. Note that it is shown in [Wim98] that a diagonal solution cannot exist for dimension greater than or equal to two, i.e., companion form matrices are never Hurwitz diagonally stable!

Matrices known to be in $\mathcal{D}_c$ or $\mathcal{D}_d$ by inspection
For the purposes of this book, it is of interest to identify special classes of matrices that can be proved to be diagonally stable, often with a specification of the diagonal matrix solution of the Liapunov equation. Some examples of such classes are: Hurwitz triangular matrices [BBP78], quasidominant matrices [HNC65], M-matrices [BP94], and sign-stable matrices with nonzero diagonal entries [JKD77, MQ69, QR65, GM81].
Tridiagonal matrix
Let $T = (t_{ij})$ be a tridiagonal matrix: i.e., $t_{ij} = 0$, if $|i - j| \geq 2$. Then $T \in \mathcal{D}_{c2}$ if and only if $t_{ii} < 0$, for $i = 1, \cdots, n$ and $t_{i,i+1}t_{i+1,i} < 0$, for $i = 1, 2, \cdots, n - 1$, and the solution to the Liapunov equation is given by the relationship (2.10). In fact, the explicit solutions are as follows [BS70]:

$$P = \text{diag} \left(1, -\frac{t_{12}}{t_{21}}, \frac{t_{12}t_{23}}{t_{21}t_{32}}, \cdots, (-1)^{n-1}\frac{t_{12}t_{23}\cdots t_{n-1,n}}{t_{21}t_{32}\cdots t_{n,n-1}} \right)$$

and

$$Q = -2 \,\text{diag} \left(t_{11}, -\frac{t_{12}t_{22}}{t_{21}}, \cdots, (-1)^{n-1}\frac{t_{12}t_{23}\cdots t_{n-1,n}t_{nn}}{t_{21}t_{32}\cdots t_{n,n-1}} \right).$$

If the conditions on the entries t_{ij} are satisfied, then P and Q are positive definite and satisfy the Liapunov equation $PT + T^T P = -Q$, showing that $T \in \mathcal{D}_c(P)$.

Cones of positive definite matrices and diagonal stability
Although stable matrices do not form a convex set and it seems, at first sight, that there is no connection between matrix stability and conic structure, Liapunov's basic stability theorem can be stated in terms of cones. Let a positive definite matrix P such that $PA + A^T P$ is negative definite be called a *stability factor for A*. Note that a diagonal stability factor (also called a *scaling factor*) is also referred to as a *Volterra multiplier*. Let *cone* H_A be defined as the set of all stability factors of a given matrix A. In this terminology, Liapunov's theorem asserts that the matrix A is Hurwitz stable if and only if the cone H_A is nonempty. Let H_A^d be the cone of positive semidefinite diagonal matrices D such that the matrix $DA + A^T D$ is negative definite and let C_A^d be the cone of positive semidefinite matrices H such that the diagonal elements of the matrix HA are nonpositive. With this terminology, the following theorem characterizes diagonal stability.

Theorem 2.8.8 [BBP78] *For $A \in \mathbb{R}^{n \times n}$, the following are equivalent:*

(i) A is in $\mathcal{D}_c$.

(ii) The cone H_A^d is nonempty.

(iii) $C_A^d = \{0\}$.   □

A simpler way of stating the equivalence between items (i) and (iii) is as follows:

Theorem 2.8.9 [BBP78] *A matrix A is in $\mathcal{D}_c$ if and only if for every nonzero positive semidefinite matrix B, the matrix BA has a negative diagonal element.*   □

Proofs of the above theorems are omitted here since they require more background of the reader than is being assumed. More details on this approach and its relation to theorems of the alternative can be found in [BBP78, CHS97, Her98]. An easy corollary of the above theorem is that the set $\mathcal{D}_c$ is open [Har80].

Characterization of the class $\mathcal{D}_c$ using the Hadamard product
Theorem 2.8.9 is the basis for the following characterization of diagonal stability.

Theorem 2.8.10 [Kra91] *The following statements are equivalent for a matrix A:*

(i) A is in $\mathcal{D}_c$.

(ii) $A \circ S$ is in class $\mathcal{P}$, for all $S = S^T \succeq 0, s_{ii} \neq 0$.

(iii) $A \circ S$ is in class $\mathcal{P}$, for all $S = S^T \succeq 0, s_{ii} = 1$.   □

Optimization-based procedures for finding diagonal solutions to the Stein and Liapunov equations or LMIs
Various optimization-based numerical algorithms exist to decide whether a given matrix belongs to class $\mathcal{D}_c$ [Kha82, Ger85, Hu87, Hu92, BGFB94]. The reference [BGFB94] discusses efficient polynomial-time interior point algorithms to solve linear matrix inequalities, such as the Liapunov inequality.

The efficient interior point methods in [BGFB94] have been coded into a MATLAB Toolbox called LMI Control Toolbox, which can be used to get a convenient and fast method to check for Hurwitz and Schur diagonal stability, in addition to having the capacity to deal with most LMI-related problems. The LMI Lab environment of this toolbox comes with an interface called the LMI Editor that makes it easy to use the capabilities of the toolbox.

Using LMI Lab to find a diagonal solution

The following MATLAB M-file uses the commands of the LMI Lab environment (contained in the LMI Control Toolbox) to test if a matrix A is in $\mathcal{D}_c(P)$ for some diagonal matrix $P > 0$ and to give this P whenever it exists.

```
function P = hdstab(A)
%This function P = hdstab(A) uses the LMI Lab
%to calculate the diagonal solution
%to the Liapunov equation in A. It returns P = []
%if there is no solution.
[n,n] = size(A);
setlmis([]);
np = ones(n,2);
np(:,2) = zeros(n,1);
P = lmivar(1,np);
lmiterm([-1 1 1 P],A',-1,'s'); %LMI #1 -A'*P - P*A > 0
lmiterm([-2 1 1 P],1,1);       %LMI #2  P > 0
liap = getlmis;
[tmin, P1] = feasp(liap);
if tmin < 0
   Pd = (1/P1(1,1))*P1;
   P = diag(Pd);
else
   P = [];
end
```

The MATLAB code above can be compared to the theoretical method worked out on the following example in [Red85b, p.620].

Example 2.8.11 *Consider the matrix*

$$A = \begin{bmatrix} 2 & 3 & 2 & 1 \\ 1 & 2 & 1 & 3 \\ 1 & 2 & 2 & 3 \\ -1 & 1 & 1 & a_{44} \end{bmatrix}.$$

The problem is to find the ranges of a_{44} for which $-A \in \mathcal{D}_c$. In [Red85b], the values found are as follows: $a_{44} \leq 12$ implies $-A \notin \mathcal{D}_c$; $a_{44} \geq 12.6$ implies $-A \in \mathcal{D}_c$; and $a_{44} = 12.6$ implies $-A \in \mathcal{D}_c(P)$, with $P = \text{diag}(1, 2, 4, 2.28)$.

The use of the MATLAB program hdstab.m *removes the gap between the values of a_{44} for nonexistence (≤ 12) and existence (≥ 12.6) found by the theoretical analysis cited above. In fact, $a_{44} < 12.433$ implies $-A \notin \mathcal{D}_c$, $a_{44} \geq 12.433$ implies $-A \in \mathcal{D}_c$, and for comparison, $a_{44} = 12.6$ implies $-A \in \mathcal{D}_c(P)$, with P calculated as* $\text{diag}(1, 1.9705, 4.5842, 2.2683)$.

In order to test for Schur diagonal stability, the following MATLAB program can be used.

```
function P = sdstab(A)
%The function P = sdstab(A)
%uses the LMI Lab to test the matrix A
%for Schur diagonal stability
```

```
%and returns the diagonal matrix P if yes, [] if no.
  [n,nn]  = size(A);
  setlmis([])
  np = ones(n,2);
  np(:,2) = zeros(n,1);
  P = lmivar(1,np);
  lmiterm([1 1 1 P], A', A)      %LMI # 1, A'*P*A
  lmiterm([1 1 1 P], -1 , 1)     %LMI # 1, -P
  lmiterm([-2 1 1 P], 1, 1)      %LMI # 2, P>0
  lmis = getlmis;
  [tmin,xfeas] = feasp(lmis);
  if tmin < 0
      P = dec2mat(lmis,xfeas,P);
  else
      P = []
  end
```

As mentioned above, other numerical methods for finding diagonal solutions are given in [Kha82, Ger85, Red85b, Hu87, Hu92], although commercial software for using these algorithms is not yet available.

Finally, sufficient conditions for Hurwitz D-stability and Schur D-stability stated in terms of LMIs are given below.

Theorem 2.8.12 [OGH99] *The matrix A is Schur D-stable if there exist positive definite matrices P_i, $i = 1, \cdots, n$ and matrices G and H satisfying the LMI*

$$\begin{bmatrix} GK_i + K_i^T G^T - P_i & K_i^T H^T - G \\ HK_i - G^T & A^T P_i A - H - H^T \end{bmatrix} < 0 \quad i = 1, \cdots, n, \quad (2.44)$$

where $K_i \in \mathbb{K}_{\text{sgn}}$.

This result, although only a sufficient condition, can be used in conjunction with the LMI Toolbox to get a numerical test for Schur D-stability. For the matrix (2.24) in Section 2.5 it turns out that the LMI (2.44) is feasible, showing that the matrix is Schur D-stable.

The analogous result that gives an LMI-based sufficient condition for Hurwitz D-stability is the following.

Theorem 2.8.13 [GOH98] *The matrix A is Hurwitz D-stable if there exist positive definite matrices P_i, $i = 1, \cdots, n$ and matrices G and H satisfying the LMI*

$$\begin{bmatrix} GB_i + B_i^T G^T & P_i A - G + B_i^T H^T \\ A^T P_i - G^T + H B_i & -H - H^T \end{bmatrix} < 0 \quad i = 1, \cdots, n, \quad (2.45)$$

where $B_i := \text{diag}\,(\varepsilon/(n-1), \cdots, 1-\varepsilon, \cdots, \varepsilon/(n-1))$, $i = 1, \cdots, n$ and $\varepsilon > 0$ is arbitrarily small.

2.9   Notes and References

Diagonal and D-stability

More material on diagonal and D-stability can be found in the following references. Hurwitz D-stability [Joh74b, Joh75, Cai76, Cro78, Har80, Cai84, YF90, HJ91, Her92, CM94, CFY95, Her98]; Hurwitz diagonal stability [BBP78, BW78, Cro78, BH83, HS85a, HS85b, HS88, CHS92, Red85a, Red85b, Log87, Log89, Kra91].

The elementary observation that if $A \in \mathcal{D}_c(P)$, then $A^{-1} \in \mathcal{D}_c(P)$ (Lemma 2.1.5) leads to the concept of a *convex invertible cone* of matrices. An investigation of the links between convexity and matrix stability, in terms of convex invertible cones, is carried out in the series of papers [CL93, CL97a, CL97b, Lew99] and briefly surveyed in [Her98].

An application of diagonal stability to get conditions for the stability of linear hyperbolic first order partial differential equations can be found in [Zió90].

The Schur case has been much less studied, usually it is mentioned as a side remark in papers devoted to the Hurwitz case (see, e.g., [BBP78]). In fact, Schur D-stability was first introduced in [BK93] and the 2×2 case was characterized there. Some papers that have followed up on this paper are [CDHJ98], where both Schur and Hurwitz diagonal and D-stability are considered, in the finite and infinite-dimensional cases, and [FGL$^+$98] where the 2×2 characterization of Schur D-stability in [BK93] is rederived, together with some other results.

Acyclic-3 matrices

The class of acyclic-3 indecomposable matrices was initially studied in [May66], where they were called "quasi-Jacobi" (the locution "matrix whose digraph is a tree" has also been used [Wie82]). Some results in [May66] were used in the simple characterization of Theorem 2.2.4.

Earlier more complicated proofs of the "only if" part of Lemma 2.2.3 were given in [QR65] and [May66, May74]. Symmetrization ideas and graph techniques similar to those in the proof of Lemma 2.2.4 were used in [CDJ82, May74, SB82, Red85a].

The equality in condition (iii) of Theorem 2.2.10 appeared in [May74] as a condition for "pseudoskew-symmetry," which is defined by the existence of a real diagonal matrix T such that $T^{-1}\tilde{A}T$ is skew-symmetric. It is easy to verify that the two definitions (D-skew symmetry and pseudoskew-symmetry) are equivalent.

Stability of sets of matrices and simultaneous Liapunov functions

The type of simultaneous Liapunov function (SLF) defined in Section 2.6 is sometimes called a *strict simultaneous Liapunov function*, since Δv_M is positive definite and proves global asymptotic stability of the set M. It is possible to define nonstrict SLFs that relax the conditions of this definition in various ways: for instance, by removing the restriction that v_M be radially unbounded or time-invariant. The concept of a simultaneous or common Liapunov function was studied in depth in the continuous case for a finite set M in [NT73, BY89] and later in the discrete-time context in [BK91] and [BM94].

Results on stability of infinite products of matrices (using the concept of joint spectral radius) and relations to the stability of difference inclusions can be found in [MP89, BW92, DL92, Els95, Gur95, LW95, EF97, Shi99].

Also, it should be pointed out that several stability results are proved in this chapter for both the continuous- and discrete-time cases independently. This was done mainly for reasons of clarity and in order to maintain some independence between the sections. In fact, given a continuous-time stability result, the corresponding discrete-time stability result can be deduced from the following fact that appears implicitly in [Tau64].

Let P be an arbitrary prespecified positive definite matrix. Then, a matrix A is such that $-(A^T P + PA)$ is positive definite if and only if $(P - \hat{A}^T P \hat{A})$ is positive definite, with the same P, where $\hat{A}$ is the image of A under the Cayley transform, i.e., $\hat{A} := (A + I)(A - I)^{-1}$.

In spite of the above fact, there are differences between the stability of difference and differential inclusions. This is pointed out here as a warning, even though the latter are not studied in this book. For instance, in order to compare with Theorem 2.6.8:

Consider the following statements for $\mathcal{A} = \{A_1, A_2, \cdots\}$:

(i) *Every matrix in the convex hull of $\mathcal{A}$ is Hurwitz stable.*

(ii) *The differential inclusion $\dot{x} \in \mathcal{A}x$ is exponentially stable.*

(iii) *For all $A_j \in \mathcal{A}$, there exist a positive definite matrix P and a positive scalar ϵ, such that $-(A_j^T P + PA_j + \epsilon I)$ is positive semidefinite.*

(iv) *For all $A_j \in \mathcal{A}$, there exist a diagonal positive definite matrix P and a positive scalar ϵ, such that $-(A_j^T P + PA_j + \epsilon I)$ is positive semidefinite.*

Then (iv) $\Rightarrow$ (iii) $\Rightarrow$ (ii) $\Rightarrow$ (i), but none of the converse implications is true. See, for example, [CL93, Hma94, KAS92, LPR98, MP89, RCDP93, WML94, Pol97] for proofs and further reading.

SLFs are also useful in the context of proving stability of switching systems. Motivated by this application, the following interesting facts are shown in [NB94], in both the Schur and Hurwitz cases: (i) a finite set of pairwise commuting asymptotically stable matrices admits a quadratic SLF; (ii) if $V(x) = x^T Px$ is an SLF for the matrices A_1 and A_2, then it is also an SLF for any positive linear combination of A_1 and A_2, i.e., for matrices of the form $\alpha A_1 + \beta A_2$, where $\alpha, \beta \geq 0$, and $\alpha + \beta > 0$; (iii) if $V(x) = x^T Px$ is an SLF for the matrices A_1 and A_2, then there exist neighborhoods $\mathcal{N}_1$ and $\mathcal{N}_2$ around A_1 and A_2, respectively, such that V remains a Liapunov function for any pair of matrices B_1 and B_2 belonging to the set $\mathcal{N}_1 \cup \mathcal{N}_2$. Note that fact (i) above was proved in [MMK98] independently in the Schur case.

Some additional criteria have been obtained in [OF97] for the simultaneous Schur diagonal stability of polytopes of special matrices: in particular, nonnegative and commutative. A linear map, called the Stein operator, $S_A : P \mapsto P - A^T PA$ is defined on the space of symmetric matrices, and several important properties of Schur diagonally stable matrices are enunciated and proved in a compact manner in terms of S_A. The results on convex invertible cones cited above are also relevant.

Proportional dominance used in [SŠ94] is a sufficient condition for the existence of a single diagonally-scaled infinity-norm in which all vertices of a polytope of nonnegative matrices have norm less than unity; see [BK96] for a discussion of this. Results for polytopes of nonnegative matrices can easily be extended to polytopes of M-matrices, since the latter are shifts of nonnegative matrices [SŠ97].

Diagonal and D-stability in applications
As mentioned in the text, the concept of D-stability arose in economics, but since then applications have surfaced in many other areas. The concept of D-stability plays an important role in the problem of decentralized stabilization of feedback control systems as well as that of decentralized integral controllability [MZ89, LNF91, YF90, CM94]. D-stability and total D-stability also arise in the area of mathematical population biology [SL83]. A variant of D-stability, called strong D-stability, occurs in the stability analysis of singularly perturbed systems [Abe86, AT86]. It is pointed out in [CM94] that strong D-stability of a matrix A is equivalent to the fact that A lies in the interior of the set of D-stable matrices, and the latter is characterized in [Har80].

An application of diagonal stability to balanced state space realizations of dynamical systems with a common Gramian matrix is discussed in [CL97b].

2.10 Appendix: Summary of the Theory of the Liapunov Equation

The linear matrix equation $AX + XB = C$
A linear matrix equation that is of great theoretical and practical importance is the following:

$$AX + XB = C. \tag{2.46}$$

Clearly the Liapunov equation, introduced in Chapter 1, is a special case of this equation with B replaced by A and A replaced by A^T. Equation (2.46) is to be solved for the unknown matrix X, given the matrices A, B, C. To do this, another definition of the product of two matrices is introduced. If $A = (a_{ij}) \in \mathbb{R}^{m \times m}$, $B \in \mathbb{R}^{n \times n}$, then the *Kronecker product* (sometimes also called the *direct* or *tensor* product) of A and B, written $A \otimes B$, is defined to be the partitioned matrix

$$A \otimes B = \begin{bmatrix} a_{11}B & a_{12}B & \cdots & a_{1m}B \\ a_{21}B & a_{22}B & \cdots & a_{2m}B \\ \vdots & \vdots & & \vdots \\ a_{m1}B & a_{m2}B & \cdots & a_{mm}B \end{bmatrix} = (a_{ij}B) \in \mathbb{R}^{mn \times mn}.$$

A fundamental property of the Kronecker product is the *mixed-product* property which relates the standard matrix product to the Kronecker product:

Lemma 2.10.1 *Let A be $m \times n$, B be $p \times q$, C be $n \times k$, and D be $q \times r$. Then $(A \otimes B)(C \otimes D) = AC \otimes BD$.*

Proof. Let $A = (a_{ih})$ and $C = (c_{hj})$. Partitioning according to the sizes of B and D, $A \otimes B = (a_{ih}B)$ and $C \otimes D = (c_{hj}D)$. The i, j block of $(A \otimes B)(C \otimes D)$

is

$$\sum_{h=1}^{n} a_{ih} B c_{hj} D = \left(\sum_{h=1}^{n} a_{ih} c_{hj} \right) B D.$$

This is exactly the i, j entry of AC times the block BD, which is the i, j block of $AC \otimes BD$. ∎

Other basic properties of the Kronecker product are given in [LT85, HJ91]: It is not commutative, but is distributive with respect to ordinary matrix addition. A vector-valued function that is associated with a matrix $A = (a_{ij}) \in \mathbb{R}^{m \times n}$ and also closely related to the Kronecker product is called the *vec-function* of A, denoted vec$A$ and defined as the vector formed by stacking the columns of A into one long vector in $\mathbb{R}^{mn}$. More formally,

$$\text{vec} A := \begin{pmatrix} a_{*1} \\ a_{*2} \\ \vdots \\ a_{*n} \end{pmatrix} \in \mathbb{R}^{m \times n},$$

where a_{*i} represents the ith column of A. An easily proved identity, useful in the study of the linear matrix equation (2.46), is the following:

$$\text{vec}(AX + XB) = ((I_n \otimes A) + (B^T \otimes I_m)) \text{vec} X. \tag{2.47}$$

The matrix $(I_n \otimes A) + (B \otimes I_m)$ is often called the *Kronecker sum* of the matrices A and B.

Simple relationships hold between eigenvalues of the matrices A and B and those of their Kronecker products and sums. Let $A = (a_{ij}) \in \mathbb{R}^{m \times m}$, $B = (b_{ij}) \in \mathbb{R}^{n \times n}$ have eigenvalues $\lambda_i, i = 1, \cdots, m$, and $\mu_j, j = 1, \cdots, n$ respectively.

Fact 2.10.2 *The eigenvalues of $A \otimes B$ are the mn numbers $\lambda_i \mu_j$, $i = 1, \cdots, m$, $j = 1, \cdots, n$. The eigenvalues of $(I_n \otimes A) + (B \otimes I_m)$ are the mn numbers $\lambda_i + \mu_j$, $i = 1, \cdots, m$, $j = 1, \cdots, n$.* □

The identity (2.47) and Fact 2.10.2 allow the characterization of situation in which equation (2.46) has a solution. This is done by using the vec-function to rewrite the matrix equation (2.46) as a system of simultaneous linear equations of the form $Lx = c$, where $L \in \mathbb{R}^{mn \times mn}$, $x, c \in \mathbb{R}^{mn}$. The result is as follows.

Theorem 2.10.3 *A matrix $X \in \mathbb{R}^{m \times n}$ is a solution of equation (2.46) if and only if the vector $x = \text{vec} X$ is a solution of the equation*

$$Lx = c,$$

where $L = (I_n \otimes A) + (B \otimes I_m)$ and $c = \text{vec} C$.

Proof. Immediate from identity (2.47) and Fact 2.10.2. ∎

The theory of linear systems of equations leads straightaway to the following corollary.

Corollary 2.10.4 *Equation (2.46) has a solution X if and only if rank$[L \ c] = $ rank$L$. It has a unique solution if and only if the matrix L is nonsingular.*

It is possible to give eigenvalue conditions for nonsingularity of L.

Theorem 2.10.5 *The equation (2.46) has a unique solution if and only if the matrices A and $-B$ have no eigenvalues in common.*

Proof. By Fact 2.10.2, the matrix L has eigenvalues $\lambda_i + \mu_j$, $i = 1, \cdots, m$, $j = 1, \cdots, n$. Applying the above corollary and recalling that a matrix is nonsingular if and only if its eigenvalues are nonzero, the proof is complete. ∎

In the special case that A and B are Hurwitz matrices, an explicit solution to (2.46) can be obtained.

Theorem 2.10.6 *If A and B are Hurwitz stable matrices, then the unique solution of equation (2.46) is given by*

$$X = -\int_0^\infty e^{At} C e^{Bt} dt. \tag{2.48}$$

□

Matrix stability and linear matrix equations

Stability can be defined in terms of the eigenvalues of a matrix A, as well in terms of the behavior of solutions to the linear differential equation $\dot{x} = Ax$. In fact, the relationship between the two definitions is given by the following proposition: *The matrix A is Hurwitz stable if and only if every solution vector $x(t)$ of $\dot{x} = Ax$ tends to the zero vector as $t \to \infty$.* In terms of inertia, a matrix $A \in \mathbb{R}^{n \times n}$ is Hurwitz stable if and only if its inertia is $(0, n, 0)$ and the Liapunov method, discussed below, expresses the inertia of any given real matrix in terms of the inertia of a related symmetric matrix. In particular, the Liapunov theory provides a criterion for Hurwitz stability of a matrix that avoids a direct calculation of its characteristic polynomial and the computation of the zeros of the latter.

The main Liapunov theorems follow (proofs can be found in [LT85, HJ91]).

Theorem 2.10.7 *Let $A, Q \in \mathbb{R}^{n \times n}$ and let Q be an arbitrary positive definite matrix.*

(i) If A is Hurwitz stable, then the equation

$$PA + A^T P = -Q \tag{2.49}$$

has a unique positive definite solution matrix P.

(ii) If there exists a positive definite matrix P satisfying equation (2.49), then A is Hurwitz stable.

A subtle point that emerges from the proof of the theorem above is that stability of A can be deduced if $x^T Q x < 0$ for just one eigenvector x associated with each eigenvalue of A^T. Thus if A has multiple eigenvalues, then Q need not be definite on the whole space $\mathbb{R}^n$.

The next theorem, which is a generalization of Liapunov's theorem, was proved independently in [DK74] and [OS62].

Theorem 2.10.8 *Let $A \in \mathbb{C}^{n \times n}$. If P is a Hermitian matrix such that*

$$PA + A^H P = -Q, \quad Q > 0, \tag{2.50}$$

then P is nonsingular and

$$n_+(A) = n_+(P), \quad n_-(A) = n_-(P), \quad n_0(A) = n_0(P). \tag{2.51}$$

Conversely, if $n_0(A) = 0$, then there exists a Hermitian matrix P such that equations (2.50) and (2.51) hold.   □

The next two theorems are concerned with the case in which the right-hand side of the Liapunov equation is semidefinite (instead of being definite). Complete proofs, together with the references to the original proofs, are in [LT85, p.448ff.]

Theorem 2.10.9 *If $A \in \mathbb{C}^{n \times n}$, $n_0(A) = 0$, and the Hermitian nonsingular matrix $P \in \mathbb{C}^{n \times n}$ satisfies the condition $PA + A^H P = -Q$, $Q \geq 0$, then $In\, A = In\, P$.*   □

In linear system theory, a commonly occurring sufficient condition that ensures the validity of the hypotheses P nonsingular and $n_0(A) = 0$ involves the idea of controllability of a matrix pair (A, B). Let A be $n \times n$ and B be $n \times m$. The pair (A, B) is *controllable* if and only if

$$\text{rank}\, [B \quad AB \quad \cdots \quad A^{n-1}B] = n.$$

Dually, the pair (A, C) is *observable* if the pair (A^T, C^T) is controllable.

Theorem 2.10.10 *If $PA + A^H P = -Q$, the matrix Q is positive semidefinite, and factorized as BB^H, and furthermore P is Hermitian and the pair (A, B) is assumed controllable, then $n_0(A) = n_0(P) = 0$ and $In\, A = In\, P$. Also, conclusions (i) and (ii) of Theorem 2.10.7 hold.*   □

The above results were also proved and stated in the following exhaustive form in [BL78] (Re $\lambda(A)$ denotes the real part of the eigenvalue λ of the matrix A):

Theorem 2.10.11 *For the equation $A^T P + PA = -C^T C$, the following assertions hold:*

(i) *If $P > 0$, then $\lambda(A) = 0$ or $Re\, \lambda(A) < 0$, for all C.*

(ii) *If $P > 0$ and (A, C) is an observable pair, then $Re\, \lambda(A) < 0$, for all C.*

(iii) *If $Re\, \lambda(A) < 0$ and (A, C) is an observable pair, then $P > 0$.*

(iv) *If $Re\, \lambda(A) < 0$ and $P > 0$, then the pair (A, C) is observable.*

(v) *If $Re\, \lambda(A) < 0$ and P is not positive definite, then the pair (A, C) is not observable.*

(vi) *If $Re\, \lambda(A) < 0$ and the pair (A, C) is not observable, then $P \geq 0$ and $\text{rank}\, P < n$.*   □

Proof. The proof of item (ii) is given as an illustration, since it involves the use of LaSalle's theorem (see the appendix to Chapter 3).
Proof of item (ii). Consider the Liapunov function $V(x) = x^T Px$ for the system $\dot{x} = Ax$. Clearly, $V(x) > 0$ and $\dot{V}(x) = -x^T C^T Cx \leq 0$. Let $M := \{x \in \mathbb{R}^n : \dot{V}(x) = 0, \forall t \geq o\}$. Then $\dot{V}(x) = -x^T C^T Cx = 0$, i.e. $\|Cx\|_2^2 = 0$, which is zero if and only if $y = Cx = 0$. Thus the set M can be rewritten as $C = \{x \in \mathbb{R}^n : Cx(t) = 0, \forall t \geq 0\}$. By observability of the pair (A, C), $y(t) = 0$, for all $t \geq 0$ implies that $x(t) = 0$, for all $t \geq 0$ as well, so that $M = \{0\}$. By LaSalle's theorem, global asymptotic stability of the zero solution of $\dot{x} = Ax$ follows and this implies the Hurwitz stability of A.   ∎

theorem, global asymptotic stability of the zero solution of $\dot{x} = Ax$ follows and
this implies the Hurwitz stability of A.   ∎

The Schwarz form of a matrix The $n \times n$ Schwarz form of a matrix below
plays an important role in stability theory (see [BS70]).

$$
S = \begin{bmatrix}
0 & 1 & 0 & \cdots & & 0 \\
-c_n & 0 & 1 & \ddots & & \vdots \\
0 & -c_{n-1} & & \ddots & & 0 \\
\vdots & \ddots & \ddots & & 0 & 1 \\
0 & & \cdots & 0 & -c_2 & -c_1
\end{bmatrix}
$$

satisfies the Liapunov equation $S^T P + PS = Q$, where

$$
P = \operatorname{diag}\left(c_1 c_2 \cdots c_n, c_2 c_3 \cdots c_n, \ldots, c_{n-1} c_n, c_n\right)
$$

and $Q = \operatorname{diag}(0, \ldots, 0, -2c_n^2)$. If $c_1, c_2, \ldots, c_n$ are nonzero real numbers, the pair
of matrices (Q, S^T) is controllable and hence

$$
n_+(S) = n - k, \quad n_-(S) = k, \quad n_0(S) = 0,
$$

where k is the number of positive terms in the sequence $c_1, c_1 c_2, c_1 c_2 c_3, \ldots,$
$c_1 c_2 \cdots c_n$.

3
Mathematical Models Admitting Diagonal-Type Liapunov Functions

The objective of this chapter is to introduce classes of dynamical systems that admit diagonal-type Liapunov functions and give the basic stability results, which are used in the remaining application-oriented chapters of the book. The word diagonal recalls that the prototypical function of this class is the quadratic form $x^T P x$, where P is a positive diagonal matrix and x a real vector. The results are stated in the most general form, without the additional hypotheses or assumptions that arise from the specifics of applications. Material on continuous- and discrete-time systems is classified in different, independent sections.

3.1 Classes of Nonlinearities and Diagonal-Type Liapunov Functions

The class of sector nonlinearities, which are used in many continuous- and discrete-time nonlinear systems in this book, is introduced below.

Definition 3.1.1 *The class $\mathcal{S}_c$ of functions is defined as follows:*

$$
\begin{aligned}
\mathcal{S}_c \quad := \quad & \{f(\cdot) | f : \mathbb{R} \to \mathbb{R}; \ f(\xi)\xi > 0; \ f(0) = 0; \\
& \text{and} \ \int_0^x f(\tau) \mathrm{d}\tau \to \infty \ \text{as} \ |x| \to \infty \}.
\end{aligned}
$$

This class is also referred to as the class of (*positive*) *infinite sector nonlinearities* . Some typical examples of infinite sector nonlinearities are $f(\xi) = \xi$, $f(\xi) = \xi^3$, $f(\xi) = \tanh \xi$, etc.

Another class of sector-bounded, time-varying nonlinearities is defined as follows.

Definition 3.1.2 *The class* $\mathcal{S}_d(k)$ *is defined as follows*

$$\mathcal{S}_d(k) := \left\{\phi(\cdot,\cdot)|\phi : \mathbb{R} \times \mathbb{N} \to \mathbb{R}; |\phi(\xi, k)| \leq |\xi|; \phi(0, k) = 0, \forall k \in \mathbb{N}\right\}$$

and also referred to as the class of sector *nonlinearities. If the notation* $\mathcal{S}_d$ *is used, the absence of* k *indicates that time-invariant nonlinearities are being considered.*

An example of a sector nonlinearity in $\mathcal{S}_d$ but not in $\mathcal{S}_c$ is $\phi(x) = \sin x$.
    The term diagonal nonlinearity is defined as follows.

Definition 3.1.3 *A function* $f : \mathbb{R}^n \to \mathbb{R}^n : x \mapsto f(x)$ *is called* diagonal *(sometimes* diagonal nonlinearity *or* separable nonlinearity*) if* $f_i(\cdot)$, *the ith component of* $f(\cdot)$, *is a function of* x_i *alone.*

A function is a *diagonal-type Liapunov function* for a given system if it proves the stability of the system through the use of matrix diagonal stability. This is an informal definition, but it is akin to the definition of a control Liapunov function [MADF98]. The reader will recall that several examples of this class of function were given in Chapter 1.
    The following positive definite functions are some of the most frequent candidates for *diagonal-type Liapunov functions.*

$$V_a = \sum_{i=1}^{n} d_i V_i^2. \tag{3.1}$$

$$V_b = \max_i\{d_i^{-1} \ V_i\}. \tag{3.2}$$

$$V_c = \sum_{i=1}^{n} d_i (V_i)^{1/2}. \tag{3.3}$$

$$V_d = \sum_{i=1}^{n} d_i V_i. \tag{3.4}$$

Note that the above functions are assumed to be positive definite; for instance, if $V_i = \int_0^{x_i} \psi_i(\tau)d\tau$, then $\psi_i(\cdot)$ must satisfy appropriate conditions so as to make V_i positive definite. One such condition that is crucial is that all the functions $\psi_i(\cdot)$ belong to the class $\mathcal{S}_c$, which guarantees radial unboundedness of the overall function V as well. The function V_d in (3.4) is also referred to in the literature as a Rosenbrock-type Liapunov function since it was introduced by Rosenbrock [Ros63].

3.2   Continuous-Time State-Space Models and Stability Results

The class of systems below was first considered by Persidskii [Per69] who also gave some basic stability results. The definition of absolute stability that is relevant for these Persidskii-type systems is also given below.

Definition 3.2.1 *A dynamical system* $\dot{x} = f(x)$, $x \in \mathbb{R}^n$, $f : \mathbb{R}^n \to \mathbb{R}^n$ *is said to be of* Persidskii-type *if it is of the form*

$$\dot{x}_i(t) = \sum_{j=1}^{n} a_{ij} f_j(x_j(t)), \quad i = 1, \ldots, n, \quad f_j \in \mathcal{S}_c, \forall j. \tag{3.5}$$

In addition to the assumption that $f_j \in \mathcal{S}_c, \forall j$, *it is also necessary to assume that the* f_js *satisfy some smoothness conditions in order to guarantee existence and uniqueness of solutions to (3.5). The dynamical system is also written compactly as follows:* $\dot{x} = Af(x)$, *where* $A = (a_{ij})$ *and* f *is a* diagonal nonlinearity *such that* $f \in \mathcal{S}_c^n$, *where* $\mathcal{S}_c^n$ *is the* n-fold Cartesian product of identical factors $\mathcal{S}_c$.

Definition 3.2.2 *The class of Persidskii-type systems defined by considering the dynamical system (3.5) for all nonlinearities* $f \in \mathcal{S}_c^n$ *is said to be* absolutely stable *(with respect to the nonlinearities* $f \in \mathcal{S}_c^n$) *if the zero solution of each system in the class is globally asymptotically stable.*

Persidskii's theorem is as follows.

Theorem 3.2.3 *The zero solution,* $x^* = 0$, *of the nonlinear system (3.5) is globally asymptotically stable for all* $f(\cdot) \in \mathcal{S}_c^\ell$ *if* $A \in \mathcal{D}_c(P)$. *Equivalently, for each member of the class of nonlinear systems* $\{\dot{x} = Af(x), \quad x(0) = x_0 : f \in \mathcal{S}_c^n\}$, *the positive definite function*

$$V(x) = 1/2 \sum_{i=1}^{n} p_i \int_0^{x_i} f_i(\tau) d\tau \tag{3.6}$$

is a diagonal-type Liapunov function establishing global asymptotic stability of the zero solution of the corresponding system.

Proof. Evaluating the time-derivative of the diagonal-type scalar Liapunov function $V(x)$ along the trajectories of (3.5) gives:

$$\dot{V}(x) = f^T(x)(PA + A^T P)f(x), \tag{3.7}$$

which, provided $A \in \mathcal{D}_c(P)$, is negative definite. In addition, for all i, the functions $f_i(\cdot)$ are in class $\mathcal{S}_c$, and this ensures radial unboundedness of V, so that $x^* = 0$ is globally asymptotically stable for all $f(\cdot)$ in $\mathcal{S}_c^\ell$. Equivalently, the class of systems defined by (3.5) as the f_js range over $\mathcal{S}_c$ is absolutely stable. ∎

Other classes of systems related to Persidskii-type dynamical systems, that arise in the applications in this book and the corresponding stability results are given below.

Theorem 3.2.4 *Consider the class of systems*

$$\dot{x} = Ax + D\Phi(x) \tag{3.8}$$

with D a nonpositive diagonal matrix and $\Phi(x) = (\phi_1(x_1), \cdots, \phi_n(x_n))^T \in \mathbb{R}^n$ such that, for all i, $\phi_i(\cdot)$ is in $\mathcal{S}_c$. The zero solution of this class is absolutely stable if the matrix A is in $\mathcal{D}_c$ and

$$V(x) = x^T P x, \ P = \mathrm{diag}(p_1, p_2, \ldots, p_n), \tag{3.9}$$

is a simultaneous quadratic diagonal Liapunov function for (3.8).   □

As defined earlier, the class of systems (3.8) for which this theorem is true is said to have *diagonally stable structure*.

The class of systems (3.8) is closely related to the well known class of *Lur'e systems* represented by the equations:

$$\begin{aligned} \dot{x} &= Ax + bf(y), \\ y &= cx, \end{aligned} \tag{3.10}$$

where $x \in \mathbb{R}^n$, $A \in \mathbb{R}^{n \times n}$, $b, c^T \in \mathbb{R}^n$ and the scalar function $f \in \mathcal{S}_c$. Here the stability question is the following: Under what conditions on the system matrices A, b, c can a diagonal-type Liapunov function ensure global stability of the equilibrium solution of the system (3.10). One way of finding such a condition is to observe that (i) the diagonal character of the nonlinearity f is preserved, if the vector c possesses only one nonzero element; and (ii) the positive infinite sector condition, satisfied by the nonlinearity, is preserved if the matrix $\Theta = (\theta_{ij}) := bc = (b_i c_j)$ satisfies the conditions $a_{ij} \theta_{ij} \geq 0$.

For example, suppose that $a_{ij} \theta_{ij} = 0$ for all $i \neq j$ and $\theta_{kk} \neq 0$ for some k with $a_{kk} \theta_{kk} \geq 0$. Then Theorem 3.2.4 can be applied with the diagonal matrix D in (3.8) defined as $D := \mathrm{diag}(d_1, \cdots, d_n)$, where $d_k < 0$ and $d_i = 0$, $i \neq k$, and the diagonal quadratic Liapunov function is given by (3.9). In this specific case, it may be said that the nonlinearity occurs in a "diagonal position." Observe that this is what happens in the example of Section 1.7, when the system is represented in the Lur'e form (1.46).

Another set of conditions on the matrices (A, b, c) that ensures global stability of the system (3.10) can be given when the matrix A is tridiagonal. For convenience, the notation $g(s)$ is used to denote the rational function $c(sI - A)^{-1}b$, also called the *transfer function*.

Definition 3.2.5 *The system (3.10) is said to belong to class $\mathcal{R}_1$ if: (i) This system has a minimal state space representation (A, b, c) such that $A = (a_{jk})$ is an $n \times n$ indecomposable tridiagonal matrix; $b^T = (\theta, 0, 0, \ldots, 0)$ and $c = (1, 0, 0, \ldots, 0)$; and (ii) $A \in \mathcal{D}_c$.*

If the system belongs to class $\mathcal{R}_1$, then the corresponding transfer function $g(s)$ is also said to belong to class $\mathcal{R}_1$. It should be noticed that if $g(s) \in \mathcal{R}_1$, then necessarily the relative degree $\delta g(s) = 1$. The two-step procedure to check if a given transfer function $g(s)$ belongs to the class $\mathcal{R}_1$ is as follows: (i) The given transfer function is subjected to the tridiagonal realization procedure in Section 3.3; (ii) the resulting tridiagonal system matrix is tested for diagonal stability by Theorem 2.2.7.

Definition 3.2.6 *The system of (3.10) is said to belong to the class $\mathcal{R}_2$ if: (i) for this system there exists a minimal state space representation (A, b, c) such that $A = (a_{jk})$ is an $n \times n$ indecomposable tridiagonal matrix; $b^T = (0, 1, 0, \ldots, 0)$; $c = (1, 0, \ldots, 0)$ and (ii) $A \in \mathcal{D}_c$.*

As in Definition 3.2.5, whenever the system is in class $\mathcal{R}_2$, then its transfer function $g(s)$ is also said to belong to class $\mathcal{R}_2$. If $g(s) \in \mathcal{R}_2$, then necessarily the relative degree $\delta g(s) = 2$ and $\theta = (a_{12})$.

With the above definitions stated, the following results hold.

Theorem 3.2.7 *If the system (3.10) is in class $\mathcal{R}_1$, then its zero solution is asymptotically stable for all nonlinearities in the positive infinite sector, and the corresponding diagonal Liapunov function that shows this is given by*

$$V(x) = x^T P x, \qquad (3.11)$$

where P is a positive diagonal solution of

$$PA + A^T P = -Q, \quad Q > 0, \qquad (3.12)$$

which is given by:

$$
\begin{aligned}
P &= \mathrm{diag}\,(p_1, p_2, \ldots, p_n) \\
p_1 &= |(a_{21})/(a_{12})|; \ \ p_2 = 1 \\
p_k &= \prod_{j=3}^{k} |(a_{j-1,j})/(a_{j,j-1})|; \quad k = 3, 4, \cdots, n
\end{aligned}
\qquad (3.13)
$$

and

$$\dot{V} = x^T (PA + A^T P)x \ - \ 2p_1\theta f(y)y < 0 . \qquad (3.14)$$

$\square$

The negative definiteness of $\dot{V}$ can be seen from the fact that $A \in \mathcal{D}_c(P)$ making the first term negative definite, while the sector condition guarantees that the second term is negative. Notice that the diagonal solution matrix P given by (3.13) for the tridiagonal matrix A is obtained from equation (2.10) in Lemma 2.2.4 for acyclic-3 matrices.

Theorem 3.2.8 *If the system (3.10) is in class $\mathcal{R}_2$, then its zero solution is absolutely stable for all nonlinearities in the positive infinite sector, and the corresponding diagonal-type Liapunov function that shows this is given* *by*

$$V(x) = x^T P x + (2/\theta) \int_o^y f(\tau)d\tau, \tag{3.15}$$

where P is also given by (3.13) and

$$\dot{V}(x) = x^T (PA + A^T P)x - 2(p_2|(a_{11})| /\theta)f(y)y < 0 . \tag{3.16}$$

□

Theorems 3.2.7 and 3.2.8 are also valid (with similar Liapunov functions) when classes $\mathcal{R}_1$ and $\mathcal{R}_2$ are defined for diagonally stable matrices, which are not necessarily tridiagonal.

A related class of systems that also has diagonally stable structure is obtained by interchanging the positions of the matrices A and D in (3.8). This gives the system

$$\dot{x} = A\Phi(x) + Dx. \tag{3.17}$$

This system is also absolutely stable under the condition $A \in \mathcal{D}_c(P)$, although the diagonal-type Liapunov function in this case is no longer quadratic and simultaneous, but involves integrals of the nonlinearities, as in (3.6), and is given by:

$$V(x) = \sum_{i=1}^{n} 2p_i \int_o^{x_i} \phi_i(\tau)d\tau.$$

This type of model describes Hopfield–Tank neural networks, which are the subject of Section 5.1.

Another class of models that admits diagonally stable structure is given below.

$$\dot{x}_i = \sum_{i=1 , i\neq j}^{n} a_{ij}f_j(x_j) + d_i\Phi_i(x_i), \tag{3.18}$$

for $i = 1, 2, \ldots, n$ or, in matrix notation,

$$\dot{x} = \tilde{A}f(x) + D\Phi(x), \tag{3.19}$$

where $x^T = (x_1, x_2, \ldots, x_n)$, $D = \text{diag}\,(d_1, \cdots, d_n)$ is a nonpositive diagonal matrix, $\tilde{A}$ is matrix $A = (a_{ij})$ with all diagonal entries set to zero; and $f(\cdot)$ and $\Phi(\cdot)$ belong to $\mathcal{S}_c^n$. For this class of systems, the stability result is as follows.

Theorem 3.2.9 *The zero solution of the class of systems (3.19) is absolutely stable if the matrix $C = \tilde{A} + D$ is such that the equation*

$$PC + C^T P = -Q \tag{3.20}$$

is verified for P and Q positive definite diagonal matrices (i.e., matrix C is in class $\mathcal{D}_{c2}$). The diagonal-type Liapunov function

$$V(x) = \sum_{i=1}^{n} 2p_i \int_0^{x_i} f_i(\tau) d\tau$$

ensures asymptotic stability.   □

Note that for the models (3.8), (3.17), and (3.18), it is possible to relax the condition of diagonal stability of matrix A to additive diagonal stability because of an additional negative definite term in the expression for the time derivative of the Liapunov function. Concrete examples of this are given in Sections 5.1.2 and 6.3.

A more general model related to the Persidskii model (3.5) is as follows.

$$\dot{x}_i(t) = \sum_{j=1}^{n} a_{ij} f_{ij}(x_j(t), t) \tag{3.21}$$

for $i = 1, \cdots, n$, and the functions f_{ij} belong to a suitable extension of the class $\mathcal{S}_c$ that includes time-varying nonlinearities $f_{ij}(x_j(t), t)$ and is denoted $\mathcal{S}_c(t)$.

The models described above were first introduced as models of perturbations of linear models in early papers on what is now known as robust stability theory. In this context, an unperturbed linear system (also referred to as *nominal*) is defined as

$$x_i^+ = \sum_{j=1}^{n} a_{ij} x_j, \quad i = 1, \cdots, n,$$

where $x = (x_1, \ldots, x_n)^T \in \mathbb{R}^n$, $A = (a_{ij}) \in \mathbb{R}^{n \times n}$, and x_i^+ denotes dx_i/dt in the continuous-time case, when $\mathbb{K} = \mathbb{R}^+$; or $x_i(t+1)$ in the discrete-time case, when $\mathbb{K} = \mathbb{N}$.

In order to state and prove the basic stability result for this class of systems, define the matrix $\overline{A} = (\overline{a}_{ij})$ with

$$\overline{a}_{ij} = \begin{cases} a_{ij}, & i = j \\ |a_{ij}|, & i \neq j, \end{cases} \tag{3.22}$$

which is called the *comparison matrix* associated to the matrix A (see Definition 2.1.1).

Theorem 3.2.10 *If the matrix A is such that $\overline{A}$ belongs to class $\mathcal{D}_c$, and there exist continuous functions $f_i(\cdot) \in \mathcal{S}_c$ such that the following entrywise function diagonal dominance type conditions hold, for all j:*

$$|f_{ij}(\xi, t)| \leq |f_j(\xi)| \leq |f_{jj}(\xi, t)|, \ \forall i \text{ and } \forall t \geq 0, \tag{3.23}$$

then the equilibrium $x = 0$ of all systems of the time-varying class (3.21) is globally asymptotically stable.

Proof. Using

$$V(x) = \sum_{i=1}^{n} p_i \int_{0}^{x_i} f_i(\tau)\mathrm{d}\tau; \qquad (3.24)$$

with $p_i > 0; i = 1, \cdots, n$, to be specified below. Computing $\dot{V}$ along the trajectories of (3.21) gives:

$$\dot{V}(x,t) = \sum_{i=1}^{n} \sum_{j=1}^{n} p_i a_{ij} f_i(x_i) f_{ij}(x_j,t).$$

From condition (3.23) and the fact that $\bar{a}_{ii} < 0$, for all i (which is a necessary condition for $\overline{A} \in \mathcal{D}_c$), it follows that

$$\dot{V}(x,t) \leq \sum_{i=1}^{n} \sum_{j=1}^{n} p_i \bar{a}_{ij} |f_i(x_i)| . |f_j(x_j)|. \qquad (3.25)$$

Using the vectorial norm defined in (2.7), inequality (3.25) can be rewritten in the form:

$$\dot{V}(x,t) \leq [\![f(x)]\!]^{\mathrm{T}} (\overline{A}^{\mathrm{T}} D + D\overline{A}) [\![f(x)]\!].$$

Since $\overline{A} \in \mathcal{D}_c(P)$ (with $P = (1/2)\mathrm{diag}\,(p_1, \cdots, p_n)$), it follows that $\dot{V}(x,t)$ is negative definite. ∎

In this theorem, given the hypothesis that $\overline{A} \in \mathcal{D}_c$, it must be assumed that the $a_{ii}, i = 1, \cdots, n$ are negative, since this is a necessary condition for $\overline{A} \in \mathcal{D}_c$. In the large scale systems approach (see Section 6.1), this corresponds to the standard assumption that the individual subsystems (in this case scalar) have to be asymptotically stable.

Although the Liapunov function (3.24) is apparently the same as the one in (3.6), there is a fundamental difference between the two. Equation (3.24) defines a *single time-invariant* Liapunov function (defined in terms of the *fixed* $f_i \in \mathcal{S}_c$) that *simultaneously* proves stability of the whole class of nonlinear time-varying systems (3.21), i.e., it works as a simultaneous Liapunov function for this class of systems.

Consider the time-varying version of the Persidskii system (3.5):

$$\dot{x}_i(t) = \sum_{j=1}^{n} a_{ij} f_j(x_j,t), \qquad (3.26)$$

for $i = 1, \cdots, n$. By analogy with the conditions (3.23), there must exist continuous functions $\phi_i(\xi)$, and positive numbers α_{im}, α_{iM} for $i = 1, \cdots, n$ such that

$$\alpha_{im} |f_i(\xi,t)| \leq |\phi_i(\xi)| \leq \alpha_{iM} |f_i(\xi,t)|.$$

The following corollary to Theorem 3.2.10 can now be stated.

Corollary 3.2.11 *The class of time-varying systems defined by (3.26) as the $f_{ij}s$ range over $S_c(t)$ is absolutely stable if the $n \times n$ matrix $B = (b_{ij})$ belongs to class $\mathcal{D}_c$, where $b_{ii} = a_{ii}/\alpha_{iM}$, and $b_{ij} = |a_{ij}|/\alpha_{jm}$.*   □

On the other hand, (3.6) actually defines a whole class of Liapunov functions, each one tailored to prove the stability of a particular system (namely the one defined by the f_is which appear in the model (3.5)), therefore (3.6) defines a Liapunov function that, although time-invariant, is *not* nonlinearity-invariant (i.e., is not simultaneous) and hence would not work without additional hypotheses for time-varying nonlinearities.

Note that this section dealt exclusively with the infinite sector nonlinearities. Some examples in which finite sector nonlinearities occur are treated in Chapter 6. The first applications of Persidskii's theorem to control system problems in the Western literature appeared in [KH79].

3.3 Seeking Hurwitz Diagonal Stability: State-Space Realization Procedures

It was shown in Section 1.7 that diagonal stability of a given matrix or system is not a coordinate free property. It was also pointed out in Section 2.3 that a nonderogatory Hurwitz matrix has an equivalent representation, via a change of coordinates, in Schwarz form (tridiagonal) and the latter form is diagonally semistable.

The objective in this section is to pursue this idea, seeking a special tridiagonal representation for an arbitrary system (A, b, c). As discussed in Section 2.3, tridiagonal matrices can be thought of as being in a canonical form for testing for diagonal stability, using Theorems 2.2.7 or 3.2.3, even when the linear system (A, b, c) is subject to nonlinear feedback, as in the case of Lur'e systems discussed in Section 3.2.

Given, without loss of generality, a single input, single output, linear system $(\tilde{A}, \tilde{b}, \tilde{c})$ in observable canonical form [Kai80], the associated transfer function $g(s) := \tilde{c}^T(sI - \tilde{A})^{-1}\tilde{b}$ is used below to give a procedure that finds another triple (A, b, c) with the same transfer function $g(s)$ but with matrix A in tridiagonal form, and with vectors b and c having some special structural properties. Representing the change of coordinates as the matrix T, the basic idea of the procedure is to set up a recurrence to find the entries of the triple (A, b, c) from the following "change of coordinate relations":

$$\begin{aligned} TA &= \tilde{A}T \\ Tb &= \tilde{b} \\ c &= \tilde{c}T. \end{aligned} \tag{3.27}$$

Consider a single input, single output linear system

$$\dot{x} = Ax + bu \tag{3.28}$$

$$y = cx, \tag{3.29}$$

where $x \in \mathbb{R}^n$; $A = (a_{ij}) \in \mathbb{R}^{n \times n}$ is a tridiagonal matrix; and the vectors $b^T = [b_1, b_2, \cdots, b_n]$ and $c = [c_1, c_2, \cdots, c_n]$ are in $\mathbb{R}^{1 \times n}$. Let $g(s) := c(sI - A)^{-1}b = n(s)/d(s)$ be the corresponding transfer function; assume also that $(\delta \leq 2)$, where $\delta := \text{degree } d(s) - \text{degree } n(s)$.

Now consider the following realization problems:

(i) (for $\delta = 1$) Find a state-space realization (3.28), (3.29) for $g(s)$ in which A is a tridiagonal matrix, $b_1 c_1 \neq 0$ and $b_j = c_j = 0$, for all $j \neq 1$.

(ii) (for $\delta = 2$) Find a state-space realization (3.28), (3.29) for $g(s)$ in which A is a tridiagonal matrix and b_2 and c_1 are the only nonzero elements of the vectors b and c.

Notice that in case (i)

$$g(s) = \theta_1 \det(sI_{n-1} - A_1)/\det(sI_n - A); \qquad \theta_1 = -b_1 c_1,$$

where A_1 is the $(n-1) \times (n-1)$ principal submatrix of matrix A, obtained by deleting the first row and column of A.

In case (ii),

$$g(s) = \theta_2 \det(sI_{n-2} - A_2)/\det(sI_n - A); \qquad \theta_2 = -a_{12}b_2 c_1,$$

where A_2 is the $(n-2) \times (n-2)$ principal submatrix obtained by deleting the first two rows and columns of the matrix A. Similarly, in the sequel $A_j (j = 0, 1, 2, \cdots, (n-1))$ denote the principal submatrices obtained by deleting the first j rows and columns of the matrix A ($A_0 := A$).

Let $A = (a_{ij})$ be a real $n \times n$ tridiagonal matrix. Then the following relationship between the elements of the matrix A and the coefficients of the characteristic polynomials of its lower principal submatrices A_i (as defined above) holds:

$$
\begin{aligned}
\alpha_i^k &= \bar{a}_{p,p} \alpha_{i-1}^{k-1} + \bar{a}_{p+1,p} \alpha_i^{k-2} \\
k &= n, n-1, \ldots, 1; \quad i = k-1, k-2, \ldots, 1, 0 \\
p &= (n-k) + 1,
\end{aligned}
\tag{3.30}
$$

where $\bar{a}_{p,p} = -a_{p,p}$; $\bar{a}_{p+1,p} = -a_{p+1,p}a_{p,p+1}$ and α_i^k are the coefficients of the characteristic polynomial $P_k(s) = s^k + \alpha_{k-1}^k s^{k-1} + \ldots + \alpha_0^k$ of the principal submatrix A_{n-k} defined as above. In addition, in (3.30), assume:

$$
\begin{aligned}
\bar{a}_{n+1,n} &= 0; \; \alpha_s^r = 1 (\text{if } r = s \geq 0); \\
\alpha_s^r &= 0 \; (\text{if } r < s \text{ or } r < 0 \text{ or } s < 0).
\end{aligned}
\tag{3.31}
$$

In case (i) the procedure derived from (3.30) and (3.31) leads to a unique realization in terms of $\bar{a}_{ij}$ and θ_1, whenever possible. In case (ii), however, the realization is not unique in terms of $\bar{a}_{ij}$ and θ_2.

Both realization procedures are obtained directly from the relationships (3.30) and (3.31), by choosing the indices i, p, k appropriately and solving the corresponding equations as shown in the examples that follow.

Example 3.3.1 *Relative degree,* $\delta = 1$. *Let*

$$g(s) = \frac{s^2 + \alpha_1^2 s + \alpha_0^2}{s^3 + \alpha_2^3 s^2 + \alpha_1^3 s + \alpha_0^3} = \frac{s^2 + s + 1}{s^3 + 3s^2 + 5s + 3}.$$

From (3.30) and (3.31) one has

$$
\begin{array}{lll}
& \alpha_2^3 = \bar{a}_{11} + \alpha_1^2 & \rightarrow \bar{a}_{11} = 2 \\
(\text{for } k = 3) & \alpha_1^3 = \bar{a}_{11}\alpha_1^2 + \alpha_0^2 + \bar{a}_{21} & \rightarrow \bar{a}_{21} = 2 \\
& \alpha_0^3 = \bar{a}_{11}\alpha_0^2 + \bar{a}_{21}\alpha_0^1 & \rightarrow \alpha_0^1 = \tfrac{1}{2}
\end{array}
$$

$$
\begin{array}{lll}
& \alpha_1^2 = \bar{a}_{22} + \alpha_0^1 & \rightarrow \bar{a}_{22} = \tfrac{1}{2} \\
(\text{for } k = 2) & & \\
& \alpha_0^2 = \bar{a}_{22}\alpha_0^1 + \bar{a}_{32} & \rightarrow \bar{a}_{32} = \tfrac{3}{4}
\end{array}
$$

$$
(\text{for } k = 1) \quad \alpha_0^1 = \bar{a}_{33} \qquad \rightarrow \bar{a}_{33} = \tfrac{1}{2}
$$
$$
(\text{unique solution})
$$

and the corresponding state-space realization is:

$$
A = \begin{bmatrix} -2 & 1 & 0 \\ -2 & -\tfrac{1}{2} & 1 \\ 0 & -\tfrac{3}{4} & -\tfrac{1}{2} \end{bmatrix} ; \ b = \begin{bmatrix} 1 \\ 0 \\ 0 \end{bmatrix} ; \ c^T = - \begin{bmatrix} 1 \\ 0 \\ 0 \end{bmatrix}.
$$

Example 3.3.2 *Relative degree* $\delta = 2$:

$$g(s) = \frac{s^2 + \alpha_1^2 s + \alpha_0^2}{s^4 + \alpha_3^4 s^3 + \alpha_2^4 s^2 + \alpha_1^4 s + \alpha_0^4} = \frac{s^2 + s + 2}{s^4 + 4s^3 + 8.5s^2 + 10s + 5.5}.$$

From (3.30) and (3.31), the following linear system is obtained:

$$
(\text{for } k = 4) \quad \begin{pmatrix} \alpha_3^4 - \bar{a}_{11} \\ \alpha_2^4 \\ \alpha_1^4 \\ \alpha_0^4 \end{pmatrix} = \begin{bmatrix} 1 & 0 & 0 & 0 \\ \bar{a}_{11} & 1 & 0 & 1 \\ 0 & \bar{a}_{11} & 1 & 1 \\ 0 & 0 & \bar{a}_{11} & 2 \end{bmatrix} \begin{pmatrix} \alpha_2^3 \\ \alpha_1^3 \\ \alpha_0^3 \\ \bar{a}_{21} \end{pmatrix}
$$

In this case, choosing $\bar{a}_{11} = 0.8$, *one determines* $\alpha_2^3, \alpha_1^3, \alpha_0^3,$ *and* $\bar{a}_{21}$, *and the realization depends upon this choice. To determine the remaining elements proceed as in the preceding example setting in (3.30) and (3.31),* $k = 3, 2, 1$ *and the following system matrices* (A, b, c) *are obtained.*

$$
A = \begin{bmatrix} -0.8 & -1 & 0 & 0 \\ 0.7 & -2.2 & 1 & 0 \\ 0 & -1.03 & -0.32 & 1 \\ 0 & 0 & -1.78 & -0.68 \end{bmatrix} ; \ b = \begin{bmatrix} 0 \\ 1 \\ 0 \\ 0 \end{bmatrix} , \ c^T = \begin{bmatrix} 1 \\ 0 \\ 0 \\ 0 \end{bmatrix}.
$$

In both examples, the transformed system is easily seen to have the property of diagonal stability and thus both example systems have the property of absolute stability with a nonlinear class $\mathcal{S}_c$ feedback term by Theorems 3.2.7 or 3.2.8.

It should be noted that there are, implicitly, similarity transformations given by a matrix T, associated to these realization procedures that provide convenient changes of coordinates; see [RKH86] where, in addition, block versions of the above state space realization schemes are also given.

3.4 Discrete-Time State-Space Models and Stability Results

This section is devoted to the discrete-time versions and analogs of the results presented in the previous section. In order to get a discrete-time analog of Theorem 3.2.3, which is the basic Persidskii theorem, the lemma below is needed.

Lemma 3.4.1 *Consider the functions ϕ_i in $\mathcal{S}_d(k)$ for all i and let Φ : $\mathbb{R}^n \times \mathbb{N} \to \mathbb{R}^n : \big((x_1,\ldots,x_n)^T,k\big) \mapsto (\phi_1(x_1,k),\ldots,\phi_n(x_n,k))^T$, denoted as $\Phi \in \mathcal{S}_d(k)^n$ for brevity. Let $P = \mathrm{diag}\,(p_1,\ldots,p_n) \in \mathbb{R}^{n\times n}$, $p_i > 0, \forall i$ be a positive diagonal matrix. Then*

(i) $\forall\, x \in \mathbb{R}^n, \quad \Phi(x(k),k)^T\, P\, \Phi(x(k),k) \leq x^T P x;$

(ii) $\forall\, x \in \mathbb{R}^n, \forall A \in \mathbb{R}^{n\times n}, \quad \Phi(Ax(k),k)^T\, P\, \Phi(Ax(k),k) \leq x^T A^T P A\, x.$

Proof. (i) $\forall P \in \mathbb{R}^{n\times n}$, P positive diagonal, $\forall \Phi \in \mathcal{S}_d(k)^n$,

$$
\begin{aligned}
\Phi(x(k),k)^T P\, \Phi(x(k),k) &= \sum_{i=1}^{n} p_i \phi_i(x_i(k),k)^2 = \sum_{i=1}^{n} p_i |\phi_i(x_i(k),k)|^2 \\
&\leq \sum_{i=1}^{n} p_i x_i^2 = x^T P x.
\end{aligned}
$$

(ii) Substitute x by Ax in (i) above. ∎

The following theorem is a discrete-time version of Theorem 3.2.3.

Theorem 3.4.2 *Let $A = (a_{ij}) \in \mathbb{R}^{n\times n}$, $x = (x_1,\ldots,x_n)^T \in \mathbb{R}^n$. Then, for all functions $\Phi \in \mathcal{S}_d(k)^n$, the zero solution of the difference equation*

$$x(k+1) = A\Phi(x(k),k)) \tag{3.32}$$

is globally asymptotically stable if the matrix A is diagonally stable.

Proof. Since $A \in \mathcal{D}_d$, there exists a positive diagonal matrix P such that $A^T P A - P$ is negative definite. Define the candidate Liapunov function $V(x) := x^T P x$ and let $\Delta V(x(k),k) := V(x(k+1)) - V(x(k))$. Then:

$$\Delta V(x(k),k) = \Phi^T(x(k),k)A^T P A\Phi(x(k),k) - x^T(k)P x(k).$$

There are two cases to consider.

- $x(k) \neq 0$ and k such that $\Phi(x(k), k) = 0$, then:

$$\Delta V(x(k), k) = -x^T(k)Px(k) < 0.$$

- $x(k)$ and k such that $\Phi(x(k), k) \neq 0$, then:

$$
\begin{aligned}
\Delta V(x(k), k) &= \Phi^T(x(k), k)A^T PA\Phi(x(k), k) - x^T(k)Px(k) \\
&\leq \Phi^T(x(k), k)A^T PA\Phi(x(k), k) - \Phi^T(x(k), k)P\Phi(x(k), k) \\
&= \Phi^T(x(k), k)(A^T PA - P)\Phi(x(k), k) < 0,
\end{aligned}
$$

where the first inequality follows from Lemma 3.4.1(i) and the second inequality from the fact that $A \in \mathcal{D}_d$.   ∎

A version of Theorem 3.4.2 for time-invariant nonlinearities appeared in [KH84a], where a diagonal-type Liapunov function, $V(x) = f(x)^T Pf(x)$, that is nonlinearity dependent was used.

As in the case of Theorem 3.4.4, Theorem 3.4.2 is also valid for the system

$$x(k+1) = \Phi(Ax(k), k) \tag{3.33}$$

and to show this, one uses the same quadratic diagonal Liapunov function and Lemma 3.4.1(ii), i.e., the result below is valid.

Theorem 3.4.3 *If $A \in \mathcal{D}_d$ then the zero solution of system (3.33) is globally asymptotically stable, for all $\Phi \in \mathcal{S}_d(k)^n$.*   □

Note that these results are valid for time-varying nonlinearities, which is not the case in continuous-time since different conditions must be used for the time-invariant and time-varying cases.

A generalized discrete-time version of the Persidskii-type system, corresponding to the continuous-time system (3.26), is given by the equations below.

$$x_i(k+1) = \sum_{j=1}^{n} a_{ij}\phi_{ij}(x_j(k), k); \tag{3.34}$$

for $i = 1, \cdots, n$, and the functions $\phi_{ij}(\cdot, \cdot)$ belong to the class $\mathcal{S}_d(k)$ (Definition 3.1.2) for all i, j.

Theorem 3.4.3 and the model (3.34) are used, for example, in digital filtering (see Section 5.4).

In the context of robust stability, this type of system, as in the continuous-time case, accommodates general classes of parameter perturbations and time-varying nonlinearities that act simultaneously on the nominal linear model $x(k+1) = Ax(k)$. Notice, once again, that parameter-

and time-dependent perturbations on the entries of matrix A, of the type: $(a_{ij} + \delta a_{ij}(t))$ (additive), and $(a_{ij}\delta a_{ij}(t))$ (multiplicative), are special cases of model (3.34). For the class of discrete-time systems (3.34) the basic stability result based on diagonal stability is given below.

Theorem 3.4.4 *The equilibrium $x = 0$ of all systems in the class (3.34) is globally asymptotically stable if the matrix $|A|$ belongs to class $\mathcal{D}_d$.*

Proof. Using the diagonal Liapunov function $V(x) = \sum_{i=1}^{n} p_i x_i^2$, the difference $\Delta V(x(k), k)$ has the form:

$$\Delta V(x(k), k) = \sum_{i=1}^{n} p_i [x_i^2(k+1) - x_i^2(k)],$$

and since $\phi_{ij}(\cdot, \cdot) \in \mathcal{S}_d(k)$ one has:

$$x_i^2(k+1) \leq \left(\sum_{j=1}^{n} |a_{ij}\phi_{ij}(x_j(k), k)| \right)^2 \leq \left(\sum_{j=1}^{n} |a_{ij}||x_j(k)| \right)^2,$$

and using the vectorial norm defined in (2.7), gives

$$\Delta V(x(k), k) \leq [\![x(k)]\!]^{\mathrm{T}} \left(|A^{\mathrm{T}}||P||A| - P \right) [\![x(k)]\!].$$

Since $|A| \in \mathcal{D}_d(P)$, it follows that $\Delta V(x(k), k) < 0$. ∎

The above theorem is the discrete-time version of Theorem 3.2.10. Note also that it holds for the model below.

$$x_i(k+1) = \sum_{j=1}^{n} \phi_{ij}(a_{ij}x_j(k), k). \tag{3.35}$$

A related result holds for the following modification of model (3.34) in which different time-varying delays $d_{ij}(k)$ are present in the state variables:

$$x_i(k+1) = \sum_{j=1}^{n} a_{ij}x_j(d_{ij}(k)), \tag{3.36}$$

where $k \leq d_{ij}(k) \leq k - d$. Once again, as discussed in Section 1.3, diagonal stability of matrix $|A|$ ensures the asymptotic stability of (3.36). In other words, given $|A| \in \mathcal{D}_d$, the linear system $x(k+1) = Ax(k)$ has robust properties even in the presence of varying delays between the different state variables, and this is discussed in more detail in Section 3.6.

In Theorems 3.2.10 and 3.4.4, note that the corresponding Liapunov functions $V(x)$ satisfy $V(x) \to \infty$ with $\|x\| \to \infty$ and this fact guarantees global asymptotic stability. Furthermore, for Theorems 3.2.10 and 3.4.4, stated in the time-varying case, it is easy to see that V, $-\dot{V}$, and $-\Delta V$ are all positive and bounded away from zero, which is a requirement for asymptotic stability in the time-varying case.

An important feature that emerges from this analysis is that, in Theorems 3.2.4, 3.2.10, 3.4.2, and 3.4.4, a simultaneous Liapunov function suffices to prove the asymptotic stability of an entire class of perturbed systems. However, in Theorem 3.2.3, although matrix diagonal stability guarantees asymptotic stability via the existence of a diagonal-type Liapunov function, the difference is that the corresponding Liapunov function is constructed from the nonlinearities or perturbations. This means that for a given nominal diagonally stable system, a Liapunov function which testifies to the stability for a given state-dependent perturbation does not do so for a different perturbation in the same sector, i.e., this Liapunov function is not simultaneous.

Notice also that Theorems 3.2.10 and 3.4.4 are directly related to the results obtained in the large scale systems literature, which utilize properties of M-matrices that constitute a subset of the set of diagonally stable matrices. In the context of large scale systems, the robust characteristics provided by M-matrices, in conjunction with simultaneous Liapunov functions, are known [Šil78], and here they appear explicitly (also see Section 6.1) Also see Section 2.4, where the question of persistence of diagonal stability under perturbations is discussed.

Thus the use of diagonal-type Liapunov functions makes possible simple derivations of time-varying generalizations of some known stability criteria for nonlinear systems.

In addition, in some cases, the sufficient condition of diagonality becomes necessary. For example, if it is assumed that the zero solution of (3.35) is globally asymptotically stable for all $\phi_{ij} \in \mathcal{S}_d(k)$, then it is possible to prove the converse of Theorem 3.4.4. In other words, if stability is to be maintained for the whole class of sector-bounded nonlinearities, then the sufficient condition of Theorem 3.4.4 also becomes necessary, based on the necessary and sufficient version of Fact 2.5.7 for nonlinearities in $\mathcal{S}_d(k)$ which is stated here as follows:

Lemma 3.4.5 *For all integers k, for all vectors $x \in \mathbb{R}^n$, and for all functions $\Phi \in \mathcal{S}_d(k)^n$, $\Phi(x,k)^T P \Phi(x,k) \leq x^T P x$ if and only if P is diagonal.*

Proof.   Follows immediately from Fact 2.5.7   ∎

Lemma 3.4.5 appears without proof in [MMR78], where Φ is a nonlinearity describing the overflow characteristic of a digital filter and is central to the proof of a sufficient condition (diagonal stability of a nominal system matrix) for the absence of overflow oscillations in a digital filter (see Section 5.4).

The results of this section are extremely general in the sense that nothing needs to be known about the nonlinearities, other than the fact that they are sector-bounded. This tends to make the results conservative if they are applied to special cases in which more information is available on the types of nonlinear perturbations acting on the systems in question. For example, when the nonlinear models describe digital filters, the nonlinearities (e.g., saturation, quantization, etc.) are completely known, and this knowledge can be exploited to obtain less conservative results (see the discussion in Section 5.4).

An important class of matrices related to the class of diagonally stable matrices was introduced in [LM94], and developed further in [CG99a, CG99b, SVD97] motivated by the particular case of saturation nonlinearities.

Definition 3.4.6 *A matrix P is said to be* positive diagonally dominant *if it is positive definite and row diagonally dominant, i.e.,*

$$\forall i \,, \ |p_{ii}| \geq \sum_{j \neq i} |p_{ij}| \tag{3.37}$$

Note that, since a positive definite matrix must have positive diagonal entries, the absolute value signs on the diagonal entries p_{ii} may be removed. A useful lemma on positive diagonally dominant matrices is the following.

Lemma 3.4.7 [CG99a] *If the matrix* $P = (p_{ij})$ *is positive diagonally dominant and there exist real numbers α_{ij} such that for all i, j:*

$$|\alpha_{ij} + \alpha_{ji}| \leq -(\alpha_{ii} + \alpha_{jj}), \tag{3.38}$$

then

$$\sum_{i,j} p_{ij}\alpha_{ij} \leq 0.$$

Proof. Note that $p_{ij} = p_{ji}$. Thus

$$
\begin{aligned}
2\sum_{i \neq j} p_{ij}\alpha_{ij} &= \sum_{i \neq j} p_{ij}(\alpha_{ij} + \alpha_{ji}) \\
&\leq \sum_{i \neq j} |p_{ij}||\alpha_{ij} + \alpha_{ji}| \\
&\leq -\sum_{i \neq j} |p_{ij}|(\alpha_{ii} + \alpha_{jj}) \\
&= -2\sum_{i \neq j} |p_{ij}|\alpha_{ii} \\
&\leq -2\sum_{i} p_{ii}\alpha_{ii},
\end{aligned}
$$

where the last inequality follows from $-\alpha_{ii} \geq 0$ (from (3.38)) and the positive diagonal dominance of P.   ∎

The class $\mathcal{OL}$ of odd and 1-Lipschitz nonlinearities is defined as follows:

$$\mathcal{OL} := \{\phi : \mathbb{R} \to \mathbb{R} \text{ such that } \forall s, t \in \mathbb{R}, |\phi(s) + \phi(t)| \leq |s + t|\} \,.$$

The reason for the somewhat strange name is that a function in $\mathcal{OL}$ is odd and has a Lipschitz constant of one. To see this, set $t = -s$ and $t = -t$ respectively. This means that the standard saturation function, the hyperbolic tangent, the sigmoid, the sine function, and the sector functions defined by $|\phi(s)| \leq |s|, \forall s \in \mathbb{R}$, all belong to $\mathcal{OL}$, although the sign function does not.

Lemma 3.4.8 [CG99a] *The matrix P is positive diagonally dominant if and only if for all $\phi \in \mathcal{OL}$, for all $\zeta \in \mathbb{R}^n - \{0\}$,*

$$0 < \phi(\zeta)^T P \phi(\zeta) \leq \zeta^T P \zeta.$$

Note that the inequality above implies that P is positive definite.

Proof. (*only if*) $\phi \in \mathcal{OL}$ means that for all $s, t \in \mathbb{R}$,

$$|\phi(s) + \phi(t)| \leq |s + t|$$
$$|\phi(s) - \phi(t)| \leq |s - t|.$$

By multiplication of the respective sides of this inequality:

$$(\phi(s) \pm \phi(t))^2 \leq (s \pm t)^2,$$

which implies

$$2|\phi(s)\phi(t) - st| \leq ((\phi(s)^2 - s^2) + (\phi(t)^2 - t^2)). \qquad (3.39)$$

Note that

$$\phi(\zeta)^T P \phi(\zeta) - \zeta^T P \zeta = \sum_{ij} \phi(\zeta_i) p_{ij} \phi(\zeta_j) - \sum_{ij} \zeta_i p_{ij} \zeta_j$$
$$= \sum_{ij} p_{ij}(\phi(\zeta_i)\phi(\zeta_j) - \zeta_i\zeta_j)$$

Now let $\alpha_{ij} := \phi(\zeta_i)\phi(\zeta_j) - \zeta_i\zeta_j$. Then (3.39) implies (3.38) and hence Lemma 3.4.7 applies.

(*if*) It will be shown that if $P > 0$ is not diagonally dominant, then there exists $\phi \in \mathcal{OL}$ and $\zeta \in \mathbb{R}^n$ such that

$$\zeta^T P \zeta - \phi(\zeta)^T P \phi(\zeta) < 0,$$

or, equivalently,

$$(\zeta - \phi(\zeta))^T P(\zeta + \phi(\zeta)) < 0.$$

In fact, it is always possible to choose $\phi = \text{sat}$. Since P is not diagonally dominant, there exists an index i and a positive real number $\epsilon > 0$ such that

$$(2 + \epsilon)p_{ii} < \sum_{j \neq i} 2|p_{ij}|.$$

Thus, it suffices to choose

$$\zeta_j + \phi(\zeta_j) = \begin{cases} -2\,\text{sgn}(p_{ij}) & \text{if } j \neq i \\ 2 + \epsilon & \text{if } j = i \end{cases}$$

$$\zeta_j - \phi(\zeta_j) = \begin{cases} 0 & \text{if } j \neq i \\ \epsilon & \text{if } j = i \end{cases}$$

or, equivalently,

$$\zeta_j = \begin{cases} -\text{sgn}(p_{ij}) & \text{if } j \neq i \\ 1 + \epsilon & \text{if } j = i. \end{cases}$$

This concludes the proof of the lemma. ∎

The next proposition is the analog of Theorem 3.4.2 in the case when the nonlinearities belong to $\mathcal{OL}$ instead of $\mathcal{S}_d$.

Proposition 3.4.9 [CG99a] *If there exists a positive diagonally dominant P such that $A^T P A - P < 0$, then for all $\phi \in \mathcal{OL}$, the zero solution of the difference equation $x(k + 1) = A\Phi(x(k))$ is globally asymptotically stable.*

Proof. Note that, without loss of generality, scaling of the matrix P allows the Stein linear matrix inequality to be written as:

$$A^T P A - P \leq I. \tag{3.40}$$

Let $V = \zeta^T P \zeta$. Then (3.40) implies that

$$
\begin{aligned}
\Delta V \; :=\; & V(x(k+1)) - V(x(k)) \\
=\; & x(k+1)^T P x(k+1) - x(k)^T P x(k) \\
=\; & \phi(x(k))^T A^T P A \phi(x(k)) - x(k)^T P x(k) \\
=\; & (\phi(x(k))^T A^T P A \phi(x(k)) - \phi(x(k))^T P \phi(x(k))) - \\
& (x(k)^T P x(k) - \phi(x(k))^T P \phi(x(k))) \\
\leq\; & -\phi(x(k))^T \phi(x(k)) - (x(k)^T P x(k) - \phi(x(k))^T P \phi(x(k)))
\end{aligned}
$$

which, together with Lemma 3.4.8, implies that V is a Liapunov function such that its decrement ΔV is bounded away from zero unless the function $\phi(x(k))^T \phi(x(k)) \to 0$ and $x(k)^T P x(k) - \phi(x(k))^T P \phi(x(k)) \to 0$ (since both terms are nonnegative). Thus $x(k) \to 0$ as $k \to \infty$. ∎

This proposition motivated the introduction, in [CG99a, CG99b], of a class of matrices referred to as dd-stable.

Definition 3.4.10 *A matrix A is said to be dd-stable if there exists a positive diagonally dominant matrix P such that $P - A^T P A$ is positive definite.*

The following example shows that the class of diagonally stable matrices is strictly contained in the class of dd-stable matrices.

Example 3.4.11 *In [LM94], the following matrix is given:*

$$A = \begin{bmatrix} 0.6 & -0.2 \\ 0.3 & 1.1 \end{bmatrix}.$$

Since $a_{22} > 1$, by Proposition 2.8.7 (i), $A \notin \mathcal{D}_d$. However, A is dd-stable, since the positive diagonally dominant matrix $P = \begin{bmatrix} 1 & 0.5 \\ 0.5 & 0.8 \end{bmatrix}$ is such that $P - A^T P A$ is positive definite.

Class $\mathcal{OL}$ is more restricted than class $\mathcal{S}_d(k)$, and this explains why Lemma 3.4.8 above allows the larger class of dd-stable matrices than the class of diagonal matrices in Lemma 3.4.5.

No theoretical characterization of the class of dd-stable matrices is known, although some conditions that guarantee that the inverse of a positive diagonally dominant matrix has the same property are known [CG99b]. The following lemma is also of interest, since it gives a computable (LMI) characterization of positive diagonally dominant matrices.

Lemma 3.4.12 [CG99a] *A symmetric matrix P is positive diagonally dominant if and only if P is positive definite and there exists a symmetric matrix R such that*

$$\forall i \neq j, \qquad r_{ij} \geq 0, \; p_{ij} + r_{ij} \geq 0$$
$$\forall i, \qquad p_{ii} \geq \sum_{j \neq i}(p_{ij} + 2r_{ij})$$

which involves $n(n-1)/2$ variables r_{ij} in addition to p_{ij} and n^2 inequalities in addition to $P > 0$.

Proof. Sufficiency is obvious since

$$
\begin{aligned}
p_{ii} &\geq \sum_{j \neq i}(p_{ij} + 2r_{ij}) \\
&= \sum_{j \neq i}(|p_{ij} + r_{ij}| + |-r_{ij}|) \\
&\geq \sum_{j \neq i}|p_{ij}|.
\end{aligned}
$$

Necessity follows from choosing

$$
r_{ij} = \begin{cases} -p_{ij} & \text{if } \; p_{ij} < 0 \\ 0 & \text{if } \; p_{ij} \geq 0. \end{cases}
$$

■

3.4.1   Discrete-Time Interval Systems

The concept of matrix diagonal stability arises in the analysis of a special kind of matrix polytope known as an interval matrix. A real interval matrix arises from an ordinary real matrix by supposing that each entry, instead of being exactly specified, is only known to belong to a certain interval of the real line. Thus, the term interval matrix, although widely used and also adopted here, is really a misnomer, because it actually refers to a set of matrices. This is reflected in the notation by using a calligraphic letter such as $\mathcal{A}$, $\mathcal{B}$, etc., to denote an interval matrix since, throughout the book, this represents a set of matrices. More specifically, given a pair of matrices $\underline{A} = (\underline{a}_{ij})$ and $\overline{A} = (\overline{a}_{ij})$ of the same dimension, an *interval matrix* is defined as follows.

Definition 3.4.13 *The interval matrix* $\mathcal{A}(\underline{A}, \overline{A})$ *is defined as the set of matrices*

$$\{A = (a_{ij}) \text{ such that } \underline{a}_{ij} \leq a_{ij} \leq \overline{a}_{ij}\}.$$

When the lower and upper bounds are clear from the context, they are suppressed, and interval matrix is denoted as just $\mathcal{A}$. *If all matrices contained in* $\mathcal{A}(\underline{A}, \overline{A})$ *are Schur (Hurwitz) stable, then* $\mathcal{A}(\underline{A}, \overline{A})$ *is said to be Schur (respectively Hurwitz) stable.*

Given a set of k real matrices denoted $\{A_i\}_{i=1}^k$, their *convex hull*, $\mathcal{G}$, is the set defined as

$$\mathcal{G} := \left\{ \sum_{i=1}^k \alpha_i A_i, \quad \forall \alpha_i \geq 0 \text{ such that } \sum_{i=1}^k \alpha_i = 1 \right\}.$$

The set $\mathcal{G}$ is called a *matrix polytope* and the matrices A_i are called its vertex matrices. It is easy to show that every $n \times n$ interval matrix can be expressed as a polytope (with at most 2^{n^2} rank-one matrices as vertices). Thus, theoretically, the stability of an interval matrix is equivalent to the stability of a rank-one matrix polytope, using the terminology of [Cox94].

Interval matrices have been proposed to model parametric uncertainties in state space dynamical systems and much research has gone into characterizing stability. Interest peaked with the publication of Kharitonov's result [Kha78] characterizing the Hurwitz stability of interval polynomials in terms of four extreme polynomials, since it was shown that the stability of interval matrices cannot, in general, be decided by the stability of the extreme or vertex matrices alone, in addition to being a computationally hard problem [PR93].

In the context of uncertain state space models, with the uncertainty being modeled by interval matrices, an adequate description of the model as well as the type of stability being referred to is necessary. In order to avoid some of the confusing terminology that arises in the literature on interval systems, this section will adopt the more precise notion of difference inclusion.

Given an interval matrix $\mathcal{A}(\underline{A}, \overline{A})$ of size $n \times n$, the vector valued function $x(k)$ is said to be a *solution of the difference inclusion*

$$x(k+1) \in \mathcal{A}(\underline{A}, \overline{A})x(k) \tag{3.41}$$

if there exists a time-dependent selection of matrices $A(k) \in \mathcal{A}(\underline{A}, \overline{A})$, $k \geq k_0$, and an initial value $x(k_0) = x_0$, such that $x(k)$ is a solution of the time-varying system associated to this selection:

$$x(k+1) = A(k)x(k), \quad x(k_0) = x_0. \tag{3.42}$$

The zero solution of the difference inclusion (3.41) is called *exponentially stable* if there exist constants $\mu > 0$ and $\eta \in (0, 1)$ such that the estimate

$$\|x(k)\| \leq \mu \eta^k \|x_o\|$$

holds for every solution of (3.42), every selection $A(k) \in \mathcal{A}(\underline{A}, \overline{A})$, and every choice of k_0. In this case, by abuse of language, the difference inclusion is said to be exponentially stable.

In general, it is difficult to decide the exponential stability of an inclusion [Bar88a, TB97a, TB97b]. However, certain special cases can be treated within the framework of matrix diagonal stability and simultaneous diagonal Liapunov functions, as is shown below. First, it is necessary to introduce the notion of quadratic stability of an inclusion.

Definition 3.4.14 *If there exist a single positive definite matrix P and $\epsilon > 0$ such that, for all $A \in \mathcal{A}(\underline{A}, \overline{A})$ and for all $x \in \mathbb{R}^n$:*

$$x^T(A^T P + PA)x \leq -\epsilon \|x\|_2^2,$$

then the difference inclusion (3.41) is said to be quadratically stable.

The following fact is easy to show:

Fact 3.4.15 *If the difference inclusion (3.41) is quadratically stable, then it is exponentially stable.*   □

Clearly, if either quadratic or exponential stability of (3.41) hold, then the interval matrix $\mathcal{A}(\underline{A}, \overline{A})$ is Schur stable, and, equally clearly, the converse is not true.

The quadratic stability analysis of an inclusion is conveniently carried out in terms of a majorant matrix.

Definition 3.4.16 *Given an interval matrix $\mathcal{A}(\underline{A}, \overline{A})$, the nonnegative constant majorant matrix is defined as follows:*

$$U_{\mathcal{A}} = (u_{ij}), \quad where \ u_{ij} := \max\{|\underline{a}_{ij}|, |\overline{a}_{ij}|\}. \tag{3.43}$$

Whenever the interval matrix being majorized is clear from the context, the subscript $\mathcal{A}$ is dropped.

Using the above definitions, the following theorem can be stated.

Theorem 3.4.17 *The difference inclusion (3.41) is exponentially stable if the nonnegative majorant matrix $U_{\mathcal{A}}$ is in class $\mathcal{D}_d$.*

Proof. Let $U_{\mathcal{A}}$ be abbreviated as U for the duration of this proof. Consider the diagonal quadratic function:

$$V(x) = |x|^T P |x|. \tag{3.44}$$

Define $\Delta V(x(k), k) := V(x(k+1)) - V(x(k))$. Since $V(\cdot)$ is a scalar function, and P is a positive diagonal matrix, for all $A(k) \in \mathcal{A}$ one has:

$$\begin{aligned} V(x(k+1)) &\leq |x|^T |A(k)|^T P |A(k)| |x| \\ &\leq |x|^T U^T P U |x|, \text{ by the definition of } U. \end{aligned}$$

Now,

$$\begin{aligned}
\Delta V(x(k), k) &= V(x(k+1)) - V(x(k)) \\
&\leq |x|^T U^T P U |x| - |x|^T P |x| \\
&= |x|^T (U^T P U - P) |x| \\
&< 0 \text{ , since } U \in \mathcal{D}_d(P).
\end{aligned}$$

To complete the proof, observe that $\lambda_{min}(P)\|x\|^2 \leq V(x) \leq \lambda_{max}(P)\|x\|^2$.

∎

A comparison theorem was used in [BP90] to show that the above sufficient condition, which they stated equivalently as $\rho(U) < 1$, actually guarantees bounded-input, bounded-output stability of the corresponding difference inclusion system with inputs and outputs. The alternative short proof given above serves to place Theorem 3.4.17 in the more general context of robustness enjoyed by systems which admit simultaneous diagonal Liapunov functions and to suggest an extension to the nonlinear case, which is also made below. In addition to this, it is easier to design controllers based on this Stein LMI condition, as examples in Section 6.4 show. Another diagonal-type Liapunov function used in the analysis of interval systems appears in [MM97].

For some classes of interval matrices, it is possible to prove a converse to Theorem 3.4.17 in the following sense. If the difference inclusion system associated to the class is exponentially stable, then the majorant matrix of the class must be diagonally stable, as the following lemma shows.

Lemma 3.4.18 *The difference inclusions associated to the following classes of interval matrices $\mathcal{A}(\underline{A}, \overline{A})$ are exponentially stable if and only if the corresponding nonnegative majorant matrices U_A are Schur diagonally stable.*

(i) *Symmetric-interval matrices: $\mathcal{A} := \{A : -\overline{A} \preceq A \preceq \overline{A}\}$.*

(ii) *Nonnegative-interval matrices: $\mathcal{A} := \{A : 0 \preceq \underline{A} \preceq A \preceq \overline{A}\}$.*

(iii) *Nonpositive-interval matrices: $\mathcal{A} := \{A : \underline{A} \preceq A \preceq \overline{A} \preceq 0\}$.*

(iv) *Right- or left-shifted symmetric interval matrices; i.e., the interval matrix $\mathcal{A}(\underline{A}, \overline{A})$ is such that either $|\underline{A}| \succeq \overline{A}$ or $|\underline{A}| \preceq \overline{A}$.*

Proof. (*if*) In all cases, the proof of sufficiency follows from Theorem 3.4.17.

(*only if*) For necessity the general strategy is to first notice that exponential stability of the inclusion implies Schur stability of every matrix in the interval set. If, in particular, it can be shown that the *nonnegative* majorant matrix (or its negative) is actually contained in the interval, then it must be Schur stable and hence, by Lemma 2.7.25, in class $\mathcal{D}_c$. The proof is completed by exhibiting the expression for U in each of the four cases:

(i) $U = \overline{A}$; (ii) $U = \overline{A}$; (iii) $U = |\underline{A}| = -\underline{A}$; (iv) (a) $|\underline{A}| \preceq \overline{A}$ implies $U = \overline{A}$; (b) $|\underline{A}| \succeq \overline{A}$ implies $U = -\underline{A}$.   ∎

Stability of interval polynomials

The necessary and sufficient condition for the Schur stability of an interval polynomial given in [MK86] is rederived here as a corollary to Lemma 3.4.18(i).

Corollary 3.4.19 *The family of polynomials*

$$\mathcal{F} := \left\{ p(z) = z^n + \sum_{j=0}^{n-1} a_j z^j \text{ such that } \forall j, \ |a_j| \le l_j \right\} \tag{3.45}$$

is Schur stable (i.e., all roots of all polynomials lie within the unit circle in the complex plane) if and only if $\sum_{i=0}^{n-1} l_i < 1$.

Proof. The Schur stability of $\mathcal{F}$ is implied by the exponential stability of the difference inclusion system

$$x(k+1) \in \mathcal{C}(\underline{C}, \overline{C}) x(k), \tag{3.46}$$

where $\mathcal{C}(\underline{C}, \overline{C})$ is the interval top companion form matrix associated to the class of polynomials $p(z)$ in (3.45) and where $\overline{C}$ is the companion matrix which has the vector $(l_0, \ldots, l_{n-1})$ as its first row; and where $\underline{C} = -\overline{C}$. From Lemma 3.4.18(i), the difference inclusion system (3.46) is exponentially stable if and only if $\overline{C} \in \mathcal{D}_d$. Note that $\overline{C}$ is a nonnegative companion form matrix with all row sums equal to one except for the first which is $\sum_{i=0}^{n-1} l_i$ and therefore, by Theorem 2.7.23 and Lemma 2.7.25, $\overline{C} \in \mathcal{D}_d$ if and only if $\sum_{i=0}^{n-1} l_i < 1$.   ∎

Schur diagonal stability of the majorant matrix is also necessary and sufficient for the exponential stability of the associated difference inclusion for the class of triangular matrices.

Lemma 3.4.20 *The difference inclusion corresponding to a interval triangular matrix $\mathcal{A}(\underline{A}, \overline{A})$ ($\underline{A}$ and $\overline{A}$ are triangular) is exponentially stable if and only if the nonnegative majorant matrix U is Schur diagonally stable.*

Proof. (*if*) Follows from Theorem 3.4.17.

(*only if*) Since the difference inclusion is exponentially stable, this implies that every matrix in $\mathcal{A}(\underline{A}, \overline{A})$ is Schur stable, which implies that all the diagonal entries of each matrix are strictly less than unity in absolute value. This, in turn, implies that the triangular majorant matrix U also has diagonal entries strictly less than unity in absolute value and is therefore Schur stable. To complete the proof, recall that Schur stable triangular matrices are in $\mathcal{D}_d$ (Corollary 2.7.12).   ∎

The continuous-time counterpart of this result is in [CLR97].

Stability of interval Morishima matrices

Another important class of interval systems is related to the Morishima matrices defined below.

Definition 3.4.21 *A matrix A is called a* Morishima matrix *if there exists a signature matrix $K \in \mathbb{K}_{\mathrm{sgn}}$ such that $KAK = |A|$.*

The class of Morishima matrices is a subclass of the class of *checkerboard* matrices (see Definition 2.7.14). It was shown (Proposition 2.7.15) that a Schur stable checkerboard matrix admits a diagonal solution to its Stein equation. The following fact is an immediate consequence.

Fact 3.4.22 *A Schur-stable Morishima matrix is diagonally stable.* $\square$

Fact 3.4.23 *If A is a Morishima matrix and $A \in \mathcal{D}_d(P)$, then $|A| \in \mathcal{D}_d(P)$.*

Proof. Since A is a Schur-stable Morishima matrix, there exists $P > 0$, diagonal, such that $A^T P A - P < 0$ and consequently, for the corresponding nonsingular signature matrix S

$$0 > S^{-1} A^T SPSAS^{-1} - S^{-1} PS^{-1} = |A|^T P |A| - P.$$

$\blacksquare$

Let $\underline{A}, \overline{A}$ be Morishima matrices. Then $\mathcal{A}(\underline{A}, \overline{A})$ is called an *interval Morishima matrix*.

In addition to the usual majorant matrix $U_{\mathcal{A}}$ (abbreviated below to U) and linear time-varying systems associated to an interval matrix, the following additional items are associated to an interval Morishima matrix.

$$x(k+1) \quad = \quad |A(k)| x(k), \quad A(k) \in \mathcal{A}(\underline{A}, \overline{A}), \tag{3.47}$$

$$\widetilde{U} \quad := \quad S^{-1} U S, \text{ where } S \text{ is the appropriate signature matrix.}$$

Let a solution of the *modulus difference inclusion* be defined as a time-dependent selection $A(k) \in \mathcal{A}(\underline{A}, \overline{A})$, $k \geq k_0$, and an initial value $x(k_0) = x_0$, such that $x(k)$ is a solution of the system (3.47).

With these definitions, the following result holds.

Lemma 3.4.24 [BK96] *The following statements are equivalent for interval Morishima systems:*

(i) U is Schur diagonally stable, i.e., $U \in \mathcal{D}_d(P)$.

(ii) $V(x) = |x|^T P |x|$ is a diagonal simultaneous Liapunov function for the entire class (3.47) obtained by all possible choices of time-dependent selection $A(k) \in \mathcal{A}(\underline{A}, \overline{A})$.

(iii) $\widetilde{U}$ is Schur diagonally stable, i.e., $\widetilde{U} \in \mathcal{D}_d(P)$.

(iv) $V(x) = |x|^T P |x|$ is a diagonal simultaneous Liapunov function that proves the quadratic stability of the difference inclusion system (3.41).

Proof. The essential part of the proof is (iii) implies (iv) and (i) implies (ii), the other parts being straightforward. Consider again the diagonal and quadratic function (3.44) $(V(x) = |x|^T P |x|)$ for the difference inclusion system (3.41). Note the following equality:

$$V(x) = |x|^T P |x| = |x|^T SPS|x|,$$

where the diagonal matrix S is the signature matrix corresponding to the transformation $A \to |A|$. Now for the system (3.47):

$$
\begin{aligned}
V(x(k+1)) &= |A(k)x(k)|^T SPS|A(k)x(k)| \\
&\leq |x(k)|^T |A(k)|^T SPS|A(k)||x(k)| \\
&\leq |x(k)|^T U^T SPSU|x(k)| \\
&= |x(k)|^T S\widetilde{U}^T P\widetilde{U}S|x(k)| \, ;
\end{aligned}
$$

thus $\Delta V(x(k)) \leq |x(k)|^T S(\widetilde{U}^T P\widetilde{U} - P)S|x(k)|,$

and since $\widetilde{U} \in \mathcal{D}_d(P)$ it follows that $\Delta V(x(k))$ is negative definite. This proves $(iii) \Rightarrow (iv)$. A similar development is used to show $(i) \Rightarrow (ii)$, and the equivalence $(iii) \Leftrightarrow (i)$, is consequence of Fact 3.4.23. ■

Note that, from Fact 3.4.22, Schur stability and Schur diagonal stability of U and $\widetilde{U}$ are equivalent.

From the above lemma, it follows that, if it is known that the interval matrix to be tested for stability is Morishima, then the majorant matrix $\widetilde{U}$, which may have both positive and negative entries, is the correct majorant matrix to use, rather than the nonnegative majorant matrix U.

Nonlinear time-varying systems and the matrix Hadamard product

Here the results obtained in the previous section are extended to the class of nonlinear systems defined below.

Let ϕ_{ij} belong to the set $\mathcal{S}_d(k)$ for $i = 1, \ldots, n$; $j = 1, \ldots, n$ and define a nonlinear time-varying system as follows:

$$x_i(k+1) = \sum_{j=1}^{n} a_{ij}\phi_{ij}(x_j(k), k), \tag{3.48}$$

for $i = 1, \ldots, n$. Consider the whole class $\mathcal{N}$ of nonlinear time varying systems of the type (3.48) obtained by letting the ϕ_{ij} range over *all* functions in $\mathcal{S}_d(k)$. For this class, the following proposition is easily proven.

Proposition 3.4.25 [BK91] $\{x(k) : k = 0, 1, 2, \ldots\}$ *is a trajectory or solution of the nonlinear time-varying system (3.48) if and only if it is a* *trajectory of the linear time-varying system below*

$$x_i(k+1) = \sum_{j=1}^{n} a_{ij} h_{ij}(k) x_j(k), \tag{3.49}$$

where for all $k \in \mathbb{N}$, $|h_{ij}(k)| \leq 1$ and the $h_{ij}(\cdot)$ are suitably chosen.   □

Using the Hadamard product and defining the matrix $H(k) = (h_{ij}(k))$, the system (3.49) can be rewritten as follows:

$$x(k+1) = [A \circ H(k)] x(k) + Bu(k), \tag{3.50}$$

where for all $k \in \mathbb{N}$, $|H(k)| \preceq \mathbb{1}$, where $\mathbb{1}$ is defined to be a square matrix of appropriate dimension with all entries equal to 1.

Since it is clear that

$$-|A| \preceq A \circ H(k) \preceq |A|, \tag{3.51}$$

a stability result for the nonlinear time varying system (3.48) can now be formulated as follows:

Theorem 3.4.26 *The difference inclusion associated to the interval matrix $\mathcal{A}(-|A|, |A|)$ or, equivalently, the class of nonlinear time varying systems $\mathcal{N}$ is globally exponentially stable if and only if $|A| \in \mathcal{D}_d$.*

Proof. From (3.51), it is clear that the difference inclusion in question is associated to an interval symmetric system. Thus the result is immediate from Lemma 3.4.18(i).   ∎

From the perspective of Hadamard-stable matrices characterized in Chapter 2, the above theorem can also be restated as follows:

Theorem 3.4.26a *The difference inclusion system associated to the interval matrix $\mathcal{A}(-|A|, |A|)$ or, equivalently, the class of nonlinear time varying systems $\mathcal{N}$ is globally exponentially stable if and only if matrix A is Hadamard stable.*   □

Related results on feedback stabilization of interval systems can be found in Section 6.4.

3.5 Models for Asynchronous Systems

This section presents classes of state-space models of discrete-time systems subject to time-varying delays in the state. These classes of systems also model iterative parallel or distributed computations that are subject to desynchronization—such systems are known as asynchronous systems or desynchronized systems and are discussed in detail in Chapter 4. It is shown that the stability analysis of this class of systems can be carried out using a diagonal-type Liapunov function and classical results on asynchronous systems can be easily rederived with this function.

First, consider the nonlinear state-space model described by the block iteration:

$$x_i(k+1) = \sum_{j=1}^{n} A_{ij}(x(k),k)x_j(k), \quad k = 0,1,2,\cdots, \qquad (3.52)$$

where $x_i(k) \in \mathbb{R}^{n_i}$, $N := \sum_{i=1}^{n} n_i$, $x(k)^T := \left(x_1(k)^T,\ldots,x_m(k)^T\right) \in \mathbb{R}^{1 \times N}$ and $A_{ij}(x(k),k) \in \mathbb{R}^{n_i \times n_j}$. Of course, some restrictions on the functions $A_{ij}(\cdot,\cdot)$ are required in order to avoid unbounded solutions to the difference equations (3.52). In fact, in order to ensure that the zero solution $x_i(k) = 0$, for all i,k, is indeed a solution of (1.16), the following is needed.

Assumption 3.5.1 (Well posedness)   *The matrix valued functions $H_{ij}(0,\ldots,0,k)$ are well-defined for all k (i.e., assume finite values).*

Now, introducing time-varying delays in the state variables of the subsystems in (3.52), the following model is obtained.

$$x_i(k+1) = \sum_{j=1}^{n} A_{ij}\left(x_1(d_{i1}(k)),\ldots,x_n(d_{in}(k)),k\right)x_j(d_{ij}(k)), \qquad (3.53)$$

for $i = 1,2,\cdots,n$, $k = 0,1,2,\cdots$, and where for all k, the positive integer-valued functions $d_{ij}(k)$ are subject to the following assumption.

Assumption 3.5.2 (Bounded delays)

$$\exists\, d \in \mathbb{N}, \forall k \in \mathbb{N}, \forall i,j \in \{1,\ldots,m\}, \; d_{ij}(k) \in \{k,k-1,\ldots,k-d\}. \qquad (3.54)$$

Note that the $d_{ij}(k)$s are functions of three variables: update or iteration number k, sending subsystem j, and receiving subsystem i.

Observe that the nonlinear time-varying discrete-time system (3.53) admits the zero solution $x_i(k) = 0$, for all i,k. The model (3.53) can be written in standard state-space form by stacking the variables $x_i(k)$ and their delayed versions $x_i(k-\ell)$, $\ell = 1,\ldots,d$, in an expanded state vector, which is denoted $\mathbf{x}$ (see Section 4.2.1 for further details).

The problem of stability of a linear discrete-time system subject to *time-varying, possibly unbounded* delays was studied by Chazan and Miranker [CM69] in the context of what they termed "chaotic" iterative computation. The chaos referred to is exactly the arbitrary variation in the time delays that are usually assumed to be either absent or else constant and bounded.

The main result obtained in [CM69] was generalized to the case of non-linear systems in [ET82]. For the case of partial asynchronism, a proof via Liapunov functions, including an estimate for the rate of exponential convergence, was given in the sequence of papers [KBŠ90, BKM91], and is discussed in detail in Chapter 4. Abstract and very general versions of the Liapunov function approach applicable to both partial and total asynchronism may also be found in [BT89, KKKK83, KKKK84a, KKKK84b, AKKK92]. In summary, these results may be stated in the form of the following general theorem, which combines the results of [ET82, KBŠ90, BKM91].

In order to state the generalization of the main result of Chazan and Miranker using a diagonal-type Liapunov function, an assumption on the nonlinearities is needed.

Assumption 3.5.3 *Let $g(x) = (g_1(x), \ldots, g_n(x))^T$ and let each g_i satisfy the following "block-Lipschitz" condition:*

$$\forall x, y \in \text{Dom}(g_i), \quad \|g_i(x) - g_i(y)\| \leq \sum_{j=1}^{n} l_{ij} \|x_j - y_j\|. \tag{3.55}$$

Theorem 3.5.4 *Let D denote the domain of g and assume that $g(D) \subset D$. Consider the system below:*

$$x_i(k+1) = g_i(x^i(k)), \qquad i = 1, \ldots, n, \tag{3.56}$$

where $x_i(k+1) \in \mathbb{R}^n$ is the value of the ith component after updating, and $x^i(k) := (x_1(d_{i1}(k))^T, \ldots, x_m(d_{im}(k))^T)^T$. The unique equilibrium x^* is exponentially stable under the class of all partial asynchronisms if $L \in \mathcal{D}_d$, where $L = (l_{ij})$. □

The proof of Theorem 3.5.4 depends on the following theorem, which is essentially the result of [KBŠ90] for the system (3.53). Define the aggregate matrix $H = (h_{ij})$, where h_{ij} is the supremum of $\|\overline{A}_{ij}(k)\|_\infty$ over all possible values of $x_j(d_{ij}(k))$, for all j, where

$$\overline{A}_{ij}(k) := A_{ij}(x_1(d_{i1}(k)), \ldots, x_n(d_{in}(k)), k).$$

Thus,

$$H = (h_{ij}) := \sup_{\mathbf{x}, k} \|\overline{A}_{ij}(k)\|_\infty. \tag{3.57}$$

Theorem 3.5.5 *The zero solution of (3.53) is globally exponentially stable if $H \in \mathcal{D}_d$.*

Proof. See Section 4.2.1. □

By Lemma 2.7.25, the diagonal stability condition on H is equivalent to requiring the spectral radius $\rho(H) < 1$, which is easy to verify numerically. The proof of this theorem is based on a weighted infinity norm which is a diagonal type Liapunov function. Details of the proof are in Section 4.2.1.

A time-varying discrete version of the well-known Bellman lemma [GŠ73] is useful for a proof of Theorem 3.5.4 based on a comparison principle.

Lemma 3.5.6 *Let $z(k; k_0, z_0) \in \mathbb{R}^n$, $\forall k$ be a solution of the vector difference inequality*

$$z(k+1) \preceq H(k)z(k), \tag{3.58}$$

where $\forall k, H(k) \in \mathbb{R}^{n \times n}$ and the initial condition is given by

$$z_0 = z(k_0; k_0, z_0). \tag{3.59}$$

Let $y(k; k_0, y_0)$ be a solution of the associated comparison vector difference equation

$$y(k + 1) = H(k)y(k) \qquad (3.60)$$

with initial condition given by $y_0 = y(k_0; k_0, y_0)$.

If $z_0 = y_0$, and for all $k \geq k_0, H(k) = (h_{ij}(k)) \succeq 0$, then for all $k \geq k_0$, one has

$$z(k; k_0, z_0) \preceq y(k; k_0, z_0).$$

Proof. Make the following induction hypothesis:

$$z(k; k_0, z_0) \preceq y(k; k_0, z_0), \qquad (3.61)$$

for all $k \in \{k_0, k_0 + 1, \ldots, n\}$, $k = k_o, k_o + 1 \ldots$. Multiplying both sides of the inequality (3.61) by the nonnegative matrix $H(k)$ gives by (3.58) and (3.60)

$$z(k + 1; k_0, z_0) \preceq y(k + 1; k_0, z_0), \qquad (3.62)$$

which is the desired inequality.   ∎

The above proof hinges on the observation that if $v^T := (v_1, \ldots, v_n)$ and $w^T := (w_1, \ldots, w_n)$ are any two vectors in $\mathbb{R}^n$ (not necessarily nonnegative) that satisfy $v \preceq w$, and if $A \succeq 0$ is any nonnegative matrix, then $Av \preceq Aw$.

Lemma 3.5.7 *Under the bounded delays and well-posedness assumptions, given a nonnegative matrix $H = (h_{ij}) \in \mathbb{R}^{n \times n}$ and a n-dimensional nonnegative vector $z(k) \succeq 0$ that satisfies the asynchronous difference inequality:*

$$z_i(k + 1) \preceq \sum_{j=1}^{n} h_{ij} \, z_j(d_{ij}(k)) \quad \text{for } i = 1, \ldots, n, \qquad (3.63)$$

$H \in \mathcal{D}_d$ *is a sufficient condition for $z(k)$ to tend exponentially to zero when $k \to \infty$, or equivalently,*

$$\lim_{k \to \infty} z(k) = 0, \qquad \text{if} \qquad H \in \mathcal{D}_d.$$

Proof. By the bounded delays assumption, it is possible to define the state vector $Z \in \mathbb{R}^{n(d+1)}$ as:

$$Z(k) = \begin{pmatrix} Z_1(k) \\ \vdots \\ Z_n(k) \end{pmatrix} \qquad \text{where} \quad Z_j(k) := \begin{pmatrix} z_j(k) \\ z_j(k - 1) \\ \vdots \\ z_j(k - d) \end{pmatrix} \in \mathbb{R}^{d+1}.$$

$$(3.64)$$

From (3.63) it follows that:

$$\forall k \in \mathbb{N}, \; Z(k + 1) \preceq \mathcal{H}(k) \; Z(k), \qquad (3.65)$$

where $\mathcal{H}(k) \in \mathbb{R}^{n(d+1) \times n(d+1)}$ is a nonnegative, time-varying matrix the elements of which are ones, zeros, or the elements of H (see Example 1.4.2 in Section 1.4). Let $Y(k)$ be a solution of the following difference equation:

$$\forall k \in \mathbb{N}, \qquad Y(k+1) \quad = \quad \mathcal{H}(k)Y(k). \qquad (3.66)$$

Then, by Theorem 3.5.5, $H \in \mathcal{D}_d$ is a sufficient condition to ensure that $Y(k)$ tends exponentially to zero. Consequently, given that $\mathcal{H}(k)$ is nonnegative and that the initial conditions are the same ($Z(0) = Y(0)$), Lemma 3.5.6 implies that:

$$\lim_{k \to \infty} z(k) = 0$$

thus completing the proof.   ∎

The proof of the above lemma is similar in spirit to a standard type of proof in the area of large scale systems. The strategy is to start from a description of the system dynamics, apply norm majorizations to arrive at a difference/differential inequality, and to finally conclude stability of the original system through a comparison principle. The reader will find many examples of such reasoning in [MM77, Šil78, Šil91]. More details on difference inequalities can be found in [Aga92], whereas differential inequalities are the subject of [LL69].

3.6   Discrete-Time Systems with Delays

The study of discrete-time systems consisting of interconnected subsystems and subjected to delays in the states of the subsystems has a long history (the earliest papers appear to be [Kra57, KB59]). The first basic results were derived and stated in the context of control theory problems and iterative computation and more specifically, in the area of parallel computation, where the delays model communication, software, and other delays in information exchange between processors. The stability result of the previous section, Theorem 3.5.5, is interpreted in this section in the context of discrete-time systems consisting of interconnected subsystems and subjected to delays in the states of the subsystems. The results in this section were first presented in [KB93a].

For ease of reference the model of an interconnected system with delays described by the equation (3.53) is repeated below:

$$x_i(k+1) = \sum_{j=1}^{n} A_{ij}\left(x_1(d_{i1}(k)), \ldots, x_n(d_{in}(k)), k\right) x_j(d_{ij}(k)). \qquad (3.67)$$

This model was guaranteed to be stable under the Schur diagonal stability condition of a reference matrix in Section 3.5. It implies that $\{\overline{A}_{ij}(k)\}$ is a sequence bounded in norm for all k. Furthermore, it also implies that the zero solution is the global equilibrium. From Lemma 2.7.25, the diagonal stability condition can be equivalently stated as a quasidominance condition. For scalar blocks (i.e. $n_i = 1, \forall i$), and constant coefficients, the system (3.67) becomes

$$x_i(k+1) = \sum_{j=1}^{n} a_{ij} x_j(d_{ij}(k)). \tag{3.68}$$

Defining the matrix $A := (a_{ij})$, the Schur diagonal stability of $|A|$, equivalent to its quasidominance, is also the condition rediscovered in [WH94, Theorem 1].

It is interesting to compare Theorem 3.5.5 for (3.68) to the continuous-time results presented in [LA80], and, in order to do this, the continuous-time model is presented below.

$$\dot{x}_i(t) = a_{ii} x_i(t) + \sum_{i \neq j} a_{ij} x_j(t - T_{ij}), \tag{3.69}$$

where $t \in [0, \infty)$, $i = 1, \cdots, n$. The vector $x = (x_1(t), \cdots, x_n(t))^T$ is called the instantaneous state vector, the scalars $T_{ij} > 0$ are referred to as the time delays, while $A = (a_{ij})$ is defined as the system matrix. The notion of an ϵ-perturbation of the system (3.69) is defined as the one obtained from it by replacing the elements of the system matrix $A = (a_{ij})$ by ϵ-perturbations $a_{ij}^{\epsilon} = a_{ij} + \delta_{ij}$, where $|\delta_{ij}| \leq \epsilon$. The main result in [LA80] is stated below.

Theorem 3.6.1   *(i) If A is quasidiagonal dominant, then the system (3.69) is stable for any set of (fixed) time-invariant delays; and*

(ii) if there exists $\epsilon > 0$ such that all ϵ-perturbations of the nominal system (3.69) with delays are stable, for all choices of time-invariant delays, then the system matrix A is quasidiagonal dominant.

Discrete-time analogs of these results are as follows.

Theorem 3.6.2   *(i) If $I - |A|$ is quasidiagonal dominant, then the system with delays is stable for all uniformly bounded time-varying delays;*

(ii) If $I - |A|$ is not quasidiagonal dominant and nonsingular, then there exists a regular sequence of delays, such that the zero solution of the resulting system with delays is unstable.

Theorem 3.6.2(i) was proved above—it is merely a statement of Theorem 3.5.5 in words for the system (3.68). The result in Theorem 3.6.2(ii) is essentially contained in Chazan and Miranker's classic paper [CM69]; more accessible proofs can be found in [BT89, SBKK98] together with much additional discussion.

An interesting point in this connection is that delays as small as one are sufficient to cause divergence when $\rho(|A|) > 1$, even if $\rho(A) < 1$. If

$\rho(|A|) = 1$, then it is possible to construct a system with bounded delays that does not have a stable zero solution (for proofs of both facts, see [SBKK98]).

A comparison of the discrete- and continuous-time results may be made in the following terms.

Theorem 3.6.2(i) is similar to its counterpart Theorem 3.6.1(i), but in the discrete-time case, more is true. Namely, the delays are allowed to be time-varying (thus there is no "nominal delayed system").

Theorem 3.6.2(ii), like its counterpart 3.6.1(ii), is a necessity result, except that it is stated as a contrapositive and ensures "robustness with respect to delays," as opposed to the continuous-time "robustness with respect to parametric ϵ-perturbations." Actually, this fact motivates the terminology r-absolute and p-absolute stability *with respect to the delays*, used in the Russian literature [KKKK83].

Note also that, from the result [Har80] that the set of diagonally stable matrices (equivalently, by Lemma 2.7.25, quasidiagonal dominant in the case of nonnegative matrices) is an open set in any matrix-norm induced topology, it follows that Theorem 3.6.2(i) also guarantees robustness with respect to perturbation in the matrix A (norm-bounded or ϵ-perturbations).

It should be noted that the results above all pertain to the "delay independent" case: In other words, the results are valid for arbitrary bounded delays. Another class of "delay-dependent" results is available. The stability conditions depend on the value of the delay(s), assumed to be fixed. The latter class of results is evidently less conservative. Both stability and stabilization results of this type, applicable to uncertain systems as well, may be found in [LdS97] and the references therein.

The stability problem of systems subject to state feedback with delay has been studied by several authors using the following model.

$$x(k+1) \;=\; A(k)x(k) + B(k)u(k), \tag{3.70}$$
$$u(k) \;=\; Kx(k-T). \tag{3.71}$$

The stability condition obtained in [WH94] is:

$$\|A(k)\| + \|B(k)K\| \le \delta < 1, \quad \forall k \ge M. \tag{3.72}$$

Notice that this is a very restrictive condition, since it implies that $\|A(k)\| \le \delta < 1$, which, in turn, implies that $\rho(|A(k)|) < 1$ (by Corollary 2.7.28) and hence the robust exponential stability of the zero solution of the system $x(k+1) = A(k)x(k)$, using arguments similar to the interval matrix case (see Section 3.4.1 and Theorem 3.4.4). By the same token it is also clear that the condition (3.72) is easier to meet *in the absence of feedback*. Thus a reasonable interpretation of this result is as follows: A nominally stable system tolerates a small amount of delayed state feedback (which is destabilizing), where small is to be understood in the sense of satisfying (3.72).

3.7   Notes and References

Robust stability regions in parameter space

Robust stability regions for the classes of systems discussed in this chapter can be derived on the basis of characterizations of diagonal stability of the matrices $A, \overline{A}$ and $|A|$.

In case the nominal system matrix is not diagonally stable, it is possible, by means of state feedback, to determine a new nominal system matrix $(A + BF)$ such that $(A + BF) \in \mathcal{D}_c$ (or $\mathcal{D}_d$), using, for example, the procedure developed in [GS86, CGL97] (also see [PT81, MKB92]). This procedure is a form of *robustification* of the nominal system and is discussed in Section 6.4. Another possibility is to use a similiarity transformation to look for a new representation of the system in which the transformed matrix belongs to $\mathcal{D}_c$, in the spirit of [Yed85].

Notice that the emphasis in this chapter is on the determination of robust stability conditions without any additional requirements on the dynamics, such as optimality with respect to some index. These requirements can, however, be incorporated into the analysis, based on, for example, [GC87, PTS77, SŠ88] (see also Section 6.4).

Systems with saturation nonlinearities

Consider the stability problem of a system with saturation nonlinearities $\sigma(\cdot)$ $(\sigma(x) := \mathrm{sgn}(x)$, if $|x| \geq 1$, else $\sigma(x) := x)$, namely

$$\dot{x}(t) = A\sigma(x(t)). \tag{3.73}$$

It is clear that system (3.73) as well as the system $\dot{x}(t) = \sigma(Ax(t))$ are Persidskii-type systems, so that $A \in \mathcal{D}_c$ is a sufficient condition for global asymptotic stability of the zero solution. However, given the specific nature of the saturation nonlinearity, less restrictive conditions can be obtained and this is done for two-dimensional or planar systems in [AD96, MSV96] and, for system of any dimension, using the concept of dd-stable matrices in [LM94, HM98] and, from a more general perspective, in [CG99a, CG99b].

There is also a connection between the variable structure systems of the type (1.36) studied in Section 1.6 and a system with saturation nonlinearities (3.73), and it can be seen through a change of scale in the time variable, $t = \epsilon^{-1}\tau$. This results in the equivalent system (from the stability viewpoint)

$$\frac{ds}{d\tau} = A\,\sigma(\epsilon^{-1}s). \tag{3.74}$$

Clearly as $\epsilon \to 0$, system (3.74) tends to system (1.36). Therefore the stability of (3.73) implies the stability of (1.36), but not conversely.

Stabilizing systems with delays in the state feedback

In the time-invariant case, an exhaustive study of the stabilizability of systems in controllable canonical form with delays in the state feedback not exceeding two was carried out in [YA88]. Furthermore, their results show that, for systems with unit delay in the state feedback, it is possible to stabilize *unstable* nominal systems. Thus it seems likely that, with the necessary changes, some analogous results hold true in the time-varying case.

3.8   Appendix: Summary of Liapunov Stability Theory

Basic stability definitions

Consider the vector difference equation

$$x(k + 1) = f(x(k), k), \quad x(0) = x_0, \tag{3.75}$$

where $x(k) \in \mathbb{R}^n$ and $f : \mathbb{R}^n \times \mathbb{N} \to \mathbb{R}^n$. It is assumed that $f(x, k)$ is continuous in x. Equation (3.75) is said to be *autonomous* or *time-invariant* if the variable k does not appear explicitly in the right-hand side of the equation, i.e., $f(x(k), k) = f(x(k))$. A vector $x^* \in \mathbb{R}^n$ is called an *equilibrium point* of equation (3.75) if $f(x^*, k) = x^*$ for all $k > 0$. In most of the literature it is assumed that a convenient change of coordinates allows x^* to be taken as the origin (zero vector); this equilibrium is called the *zero solution*. The notation $x(x_0, k, 0)$ is used to denote a sequence of vectors (alternatively written $\{x(k)\}$) that starts from the initial condition $(x(0) = x_0)$ and satisfies (3.75). Such a sequence is called a *solution* of (3.75).

The stability definitions used in this book are collected below

Definition 3.8.1 *The equilibrium point of (3.75) is said to be:*

(i) stable *if given ε and $0 \geq 0$, there exists $\delta = \delta(\varepsilon, 0)$ such that $\|x_0 - x^*\| < \delta$ implies $\|x(x_0, k, 0) - x^*\| < \varepsilon$ for all $k \geq 0$,* uniformly stable *if δ may be chosen independent of 0 and* unstable *if it is not stable.*

(ii) attractive *if there exists $\mu = \mu(0)$ such that $\|x_0 - x^*\| < \mu$ implies*

$$\lim_{k \to \infty} x(x_0, k, 0) = x^*,$$

uniformly attractive *if the choice of μ is independent of 0.*

(iii) asymptotically stable *if it is stable and attractive, and* uniformly asymptotically stable *if it is uniformly stable and uniformly attractive.*

(iv) exponentially stable *if there exist $\delta > 0$, $M > 0$ and $\eta \in (0, 1)$ such that $\|x(x_0, k, 0) - x^*\| \leq M\|x_0 - x^*\|\eta^k$, whenever $\|x_0 - x^*\| < \delta$.*

(v) *a solution $x(x_0, k, 0)$ is* bounded *if for some positive constant B,*

$$\|x(x_0, k, 0)\| \leq B,$$

for all $k \geq 0$, and where B may depend on each solution.
If, in item (iii), $\mu = \infty$ or in item (iv) $\delta = \infty$, the corresponding stability property is said to be global.

In general, exponential stability implies all the other types. For linear systems, exponential stability and uniform asymptotic stability are equivalent. For time-invariant systems each stability property is equivalent to its uniform counterpart and this is also true for periodic systems. Note that attractivity does not imply stability. A good discussion of this can be found in [Hur67, Ort73, Ela96].

Liapunov stability theorems

The major tool in the stability analysis of nonlinear difference and differential equations was introduced by Liapunov in his famous memoir published in 1892 [Lia49].

Consider the time-invariant equation

$$x(k+1) = f(x(k)), \tag{3.76}$$

where $f : G \to \mathbb{R}^n, G \subset \mathbb{R}^n$, is continuous. Assume that x^* is an equilibrium point of the difference equation, i.e., $f(x^*) = x^*$.

Let $V : \mathbb{R}^n \to \mathbb{R}$ be defined as a real valued function. The *decrement* or *variation* of V relative to (3.76) is defined as:

$$\Delta V(x(k)) := V(x(k+1)) - V(x(k)) = V(f(x(k))) - V(x(k)).$$

Note that if $\Delta V(x) \leq 0$, then V is nonincreasing along solutions of (3.76).

Definition 3.8.2 *The function V is said to be a* Liapunov function *on a subset H of $\mathbb{R}^n$ if: (a) V is continuous on H; and (b) the decrement $\Delta V \leq 0$, whenever x and $f(x)$ are in H.*

Let $B(x, \rho)$ denote the open ball of radius ρ and center x defined by $B(x, \rho) := \{y \in \mathbb{R}^n : \|y - x\| < \rho\}$.

Definition 3.8.3 *The real-valued function V is said to be* positive definite at x^* *if (i) $V(x^*) = 0$ and (ii) $V(x) > 0$ for all $x \in B(x, \rho)$, for some $\rho > 0$.*

The first Liapunov stability theorem is now stated.

Theorem 3.8.4 *If V is a Liapunov function for (3.76) on a neighborhood H of the equilibrium point x^*, and V is positive definite with respect to x^*, then x^* is stable. If, in addition, $\Delta V(x) < 0$, whenever x and $f(x)$ are in H and $x \neq x^*$, then x^* is asymptotically stable. Moreover, if $G = H = \mathbb{R}^n$ and*

$$V(x) \to \infty \quad as \quad \|x\| \to \infty$$

then x^* is globally asymptotically stable.

Definition 3.8.5 *(i) For a subset $G \subset \mathbb{R}^n$, x is a* limit point *of G if there exists a sequence $\{x_i\}$ in G with $x_i \to x$ as $i \to \infty$.*
(ii) The closure *$\overline{G}$ of G is defined to be the union of G and all its limit points.*
(iii) Considering equation (3.76), the positive orbit *$O^+(x_0)$ is defined as $O^+(x) := \{x(x_0, k, 0) : k \in \mathbb{N}\}$. $O^+(x_0)$ is denoted $O(x_0)$ for brevity.*
(iv) The limit set *$\Omega(x_0)$, also referred to as the* positive limit set*, of x_0 is the set of all possible limit points of x_0. Thus*

$$\Omega(x_0) := \{y \in \mathbb{R}^n : x(k_i) \to y \text{ as } k_i \to \infty \text{ for some subsequence } \{k_i\} \text{ of } \mathbb{N}\}.$$

(v) A set A is positively invariant *if $O(x_0) \subset A$ for every $x_0 \in A$.*

It is easily shown that both $O(x_0)$ and $\Omega(x_0)$ are positively invariant.

Let V be a positive Liapunov function on a subset G of $\mathbb{R}^n$. Define

$$E := \{x \in \overline{G} : \Delta V(x) = 0\}.$$

Let M be the *maximal invariant subset* of E, i.e., M is defined as the union of all invariant subsets of E.

With this terminology, the important LaSalle theorem (also called the Krasovskii–LaSalle theorem) is stated as follows:

Theorem 3.8.6 *If there exists a positive definite Liapunov function V for equation (3.76) on $G \subset \mathbb{R}^n$, then each solution of (3.76) which remains in G for all $k \geq 0$ is either unbounded or approaches the maximal invariant subset M that contains the set $E := \{x \in \overline{G} : \Delta V(x) = 0\}$.*

A common use of this theorem is to conclude stability of the origin in the case when $M = \{0\}$.

Continuous-time analogs of these theorems are as follows. The reader who wishes greater rigor and details should consult [Hah67, HS74, Vid93].

$$\dot{x}(t) = f(x(t), t), \quad x(t_0) = x_0, \tag{3.77}$$

where $t_0 \geq 0$, $x(t) \in \mathbb{R}^n$ and $f : \mathbb{R}^n \times \mathbb{R}_+ \to \mathbb{R}^n$ is continuous. The equation (3.77) is called *time-invariant* or *autonomous* when the right-hand side does not depend on t

$$\dot{x} = f(x). \tag{3.78}$$

It is further assumed throughout below that the equations (3.77), (3.78) have unique solutions corresponding to each initial condition x_0. In the time-invariant case, this happens, for example, if f satisfies a Lipschitz condition

$$\forall x, y \in B \subset \mathbb{R}^n, \|f(x) - f(y)\| \leq \ell \|x - y\|.$$

The constant ℓ is known as a *Lipschitz constant* for f and f is sometimes referred to as *Lipschitz-continuous*. The terms *locally Lipschitz* and *globally Lipschitz* are used in the obvious manner to refer to the domain over which the Lipschitz condition holds. The Lipschitz property is stronger than continuity (implies uniform continuity), but weaker than continuous differentiability. A simple global existence and uniqueness theorem in the time-invariant case is as follows.

Theorem 3.8.7 *Let $f(x)$ be locally Lipschitz on a domain $D \subset \mathbb{R}^n$, and let W be a compact subset of D. Let the initial condition $x_0 \in W$ and suppose it is known that every solution of (3.78) lies entirely in W. Then there is a unique solution that is defined for all $t \geq 0$.*   □

Consider the system (3.77) where f is locally Lipschitz and suppose that $x^* \in \mathbb{R}^n$ is an equilibrium point of (3.77); that is,

$$f(x^*, t) \equiv 0.$$

Since a change of variables can shift the equilibrium point to the origin, all definitions and theorems below are stated for this case. It is also common to speak of the properties of the *zero solution*, i.e., $x(t) \equiv 0, \forall t$.

Definition 3.8.8 *The equilibrium point $x = 0$ (equivalently, the zero solution) of (3.77) is said to be*

(i) *stable, if for arbitrary t_0 and each $\epsilon > 0$, there is a $\delta = \delta(\epsilon, t_0) > 0$ such that $\|x(t_0)\| < \delta$ implies $\|x(t)\| < \epsilon$, for all $t \geq t_0$.*
The idea is that the entire trajectory stays close to zero, if the initial condition is close enough to zero.

(ii) *unstable, if not stable.*

(iii) asymptotically stable, *if it is stable and if a convergence condition holds: for arbitrary t_0, there exists $\delta_1(t_0)$ such that $\|x(0)\| < \delta_1(t_0)$ implies*

$$\lim_{t \to \infty} x(t) = 0.$$

(v) uniformly stable *and* uniformly asymptotically stable *if δ in (i) and δ_1 in (ii) can be chosen independently of t_0.*

(vi) globally asymptotically stable *when δ_1 in (iii) can be taken arbitrarily large.*

(vii) exponentially stable *when, in addition to stability, $\|x(t_0)\| < \delta_1(t_0)$ implies*

$$\|x(t)\| \le K\|x(t_0)\| \exp[-\alpha(t - t_0)]$$

for some positive α and K.

Liapunov theorems for time-invariant systems

Let $V(x)$ be a real scalar function of $x \in \mathbb{R}^n$ and let D be a closed bounded region in $\mathbb{R}^n$ containing the origin.

Definition 3.8.9 $V(x)$ *is* positive definite *(semidefinite) in D, denoted $V > 0$ ($V \ge 0$), if $V(0) = 0$, $V(x) > 0$ ($V(x) \ge 0$) for all $x \ne 0$ in D. $W(x)$ is negative definite (negative semidefinite) if and only if $-W(x)$ is positive definite (positive semidefinite).*

Theorem 3.8.10 *Consider the equation (3.78) and let $V(x)$ be a positive definite real valued function defined on D a closed bounded region containing the origin of $\mathbb{R}^n$. The zero solution of (3.78) is:*

(i) stable *if $\dot{V} = (grad\,V)f(x) \le 0$ (the derivative of V along the trajectories of (3.78)).*

(ii) asymptotically stable *if $\dot{V}$ (see item (i)) is negative definite, or, alternately $\dot{V}(x) \le 0$, but $\dot{V}$ not identically zero along any trajectory except $x \equiv 0$.*

(iii) globally asymptotically stable *if in item (ii), $D = \mathbb{R}^n$ and $V(x) \to \infty$ as $\|x\| \to \infty$.*

(iv) exponentially stable *if in item (ii) there holds $\alpha_1\|x\|^2 \le V(x) \le \alpha_2\|x\|^2$ and $-\alpha_3\|x\|^2 \le \dot{V}(x) \le -\alpha_4\|x\|^2$, for some positive α_i.* □

A function $V(x)$ which allows a proof of a stability result using one of the items of this theorem is called a *Liapunov function*.

For the time-varying case (3.77), some modifications are needed. Consider real scalar functions $V(x,t)$ of the vector $x \in \mathbb{R}^n$ and time $t \in \mathbb{R}_+$, defined on a closed bounded region D containing the origin.

Definition 3.8.11 $V(x,t)$ *is* positive definite *in D, denoted $V > 0$, if $V(0,t) = 0$ and there exists $W(x)$ with $V(x,t) \ge W(x)$ for all x,t and $W > 0$. $V(x,t)$ is nonnegative definite in D if $V(0,t) = 0$ and $V(x,t) \ge 0$ for all x,t.*

Observe that the derivative of V along the trajectories of (3.77) is given by

$$\dot{V} = (grad\,V)^T f(x,t) + (\partial V/\partial t).$$

With these changes, items (i) and the first part of item (ii) of Theorem 3.8.10 hold. If $V(x,t) \le W_1(x)$ for all t and some positive definite W_1, then uniformity holds in both cases. In item (iii), if $W(x) \leq V(x,t) \leq W_1(x)$ with $W(x) \to \infty$ as $\|x\| \to \infty$, then uniform global asymptotic stability holds. Item (iv) is valid as stated, without change.

LaSalle's invariance principle says, roughly speaking, that if one can find a function V such that $\dot{V}$ is negative definite, and, in addition, it can be shown that no system trajectory stays forever at points where $\dot{V} = 0$, then the origin is asymptotically stable. To formalize this, the following definition is needed.

Definition 3.8.12 *A set M is said to be an* invariant set *with respect to (3.78) if $x(0)$ in M implies that $x(t)$ in M for all t in $\mathbb{R}$.*

LaSalle's theorem is now stated.

Theorem 3.8.13 *Let D be a closed and bounded (compact) set with the property that every solution of (3.78) which starts in D remains for all future time in D. Let $V : D \to \mathbb{R}$ be a continuously differentiable function such that $\dot{V}(x) \leq 0$ in D. Let E be the set of all points in D where $\dot{V}(x) = 0$. Let M be the largest invariant set in E. Then every solution starting in D approaches M as $t \to \infty$.*   □

It is clear that the second part of item(ii) of Theorem 3.8.10 is a special case of LaSalle's theorem—in fact, a very important special case known as the Barbashin–Krasovskii theorem. LaSalle's theorem extends Liapunov's theorem in at least three important ways: First, the negative definite requirement of the latter is relaxed to negative semidefiniteness; second, it can be used when the system has an equilibrium set rather than an isolated equilibrium point; third, the function $V(x)$ does not have to be positive definite.

4

Convergence of Asynchronous Iterative Methods

The objective of this chapter is to show how Jacobi-type iterative methods for the solution of linear and nonlinear equations and, more specifically asynchronous versions of these methods, can be analyzed in the framework introduced in this book. By regarding these methods as discrete-time interconnected systems with varying delays in the interconnections, the analysis based on a diagonal-type Liapunov function introduced in Chapter 3 leads to computable conditions for convergence based on nonnegative stable (and hence diagonally stable) matrices, as well as estimates for the speed of convergence of different iterative methods. A perspective on the analysis of interconnected systems using diagonal-type Liapunov functions that complements the approach in this chapter is to be found in Chapter 6.

4.1 What Are Asynchronous Iterative Methods?

A computer that has only one processor is referred to as *sequential*. The term *parallel computer* is used to denote a computing machine that has more than one processor or computer, working in parallel, and devoted to the solution of a single problem. The term computer is used to mean either an independent machine or a processor within a larger machine. This usually presupposes that: (i) the problem has been decomposed into as many subproblems as there are computers; and (ii) that there exists some method of putting together the solutions to the subproblems in order to get the solution to the original undecomposed problem. A parallel iterative method is one that realizes both these tasks in a straightforward manner. The use of vector, array, and parallel computers with different high-speed low-cost alternative architectures that appeared in the last decade, has provoked significant changes in languages, programming techniques, and algorithms in general. In scientific computing in particular, the different ways that numerical methods are conceived or adapted to these machines and architectures have been stimulating theoretical and applied research in different areas.

Complexity, convergence, and performance are examples of such areas. Many old ideas and techniques that were abandoned in the past, because of the lack of appropriate computing power, have been resuscitated. Examples of this process are the ideas of decomposition or partitioning and hierarchical as well as iterative methods for the solution of large sets of algebraic equations.

More specifically, in order to solve the classical linear equation $Ax = b$ when the dimension of A is of the order of many thousands of variables, Gaussian elimination combined with sparsity techniques is considered to be one of the most efficient methods for sequential computers. However, iterative methods like Jacobi, Gauss–Seidel, SOR, conjugate–gradient, etc., which exhibit slow convergence in sequential machines, have demonstrated good performance in parallel computers in many applications (see, for example, [DFK92]).

Most direct solution techniques do not have an inherently parallel structure and require a considerable amount of data communication among computers when implemented in parallel. Iterative methods, on the other hand, require less data communication and lend themselves to concurrent implementation, this being partly due to the inherent parallelism frequently associated with such methods (e.g., the Jacobi iterative method is essentially parallel). These facts have justified the reevaluation of the classical iterative methods implemented in parallel computers particularly with regard to convergence conditions as well as speed of convergence.

A parallel (or distributed) implementation of an iterative algorithm is one in which the computational load is shared by several computers while coordination is maintained by information exchange via communication links. Essentially, there are two modes of operation of such implementations: synchronous and asynchronous (for a more detailed discussion of synchronism versus asynchronism, see [BT89, ÜD92]). In the synchronous mode, the point of departure is some iterative algorithm which is guaranteed to converge to the correct solution under the usual circumstances of sequential computation in a single computer. The computational load of the typical iteration is then divided in some way (for example, blockwise) among the available computers, and it is assumed that the computers exchange all necessary information regarding the outcomes of the current iteration before a new iteration can begin. Such synchronous parallel iterative methods have an obvious disadvantage: the need of synchronization of the iterations, which is a nontrivial task for a large-scale computation; and the fact that the speed of computation is limited to that of the slowest computer, so that the faster computers spend considerable amounts of time in an idle or wait status.

The second mode of operation is asynchronous: computation and communication is carried out at the various computers completely independent of the progress in other computers. This extreme model of complete independence of each computer was introduced by Chazan and Miranker [CM69] (for point iterative methods) as a generalization of the free steering iterative method introduced by Ostrowski [Ost55]. Subsequently many researchers proposed various hybrid and nonlinear methods (see [Bau78, TPM83] and references therein). Extensions of synchronous algorithms to the asynchronous case and the corresponding convergence conditions have also been much investigated recently (see the surveys [SBKK98, Szy98, SB98]).

The usual spectral radius convergence condition for sequential linear stationary or parallel synchronous methods does not guarantee convergence when asynchronous parallel computation methods are used. A well-known and stronger convergence condition for classical block-iterative methods such as the Jacobi and Gauss–Seidel methods to solve $Ax = b$ is the diagonal dominance of the matrix A. It is shown in this chapter, using a weighted infinity norm diagonal-type Liapunov function, that even when the block-iterative method is nonlinear, nonstationary, and implemented in parallel using asynchronous communication, a related condition, namely quasidominance of a nonnegative aggregate matrix, which is equivalent to diagonal stability, still guarantees the convergence of the method.

4.2   A Mathematical Model for Asynchronous Block-Iterations

A mathematical state-space model for a class of nonlinear time-varying parallel iterative methods is presented in this section. Using this model, which generalizes several models of the Chazan–Miranker type, together with Liapunov techniques, it is shown that the diagonal stability of a certain matrix guarantees exponential convergence of this class of iterative methods.

Traditionally, in order to treat convergence problems of iterative and block-iterative methods such as Jacobi, Gauss–Seidel, and SOR, that are linear and time-invariant, techniques based on the Perron–Frobenius and Gerschgorin theorems ([Var62, You71]) and H-matrices ([Rob69]) have been used. When disturbances caused by finite arithmetic and variable delays introduced by asynchronisms and memory access times, etc., are considered, the models become nonlinear and time-varying. It is shown that the above-mentioned techniques, combined with Liapunov techniques, can still be used to find convergence conditions for different classes of nonlinearities. In the synchronous–sequential context, the generalized distance functions used in [Pol71, Pol82] etc. can be regarded as Liapunov functions, as has been pointed out in [Ber83].

It is well known [OR70] that the problem of solving the equation

$$f(x) = 0, \qquad x \in \mathbb{R}^n, \qquad f : D \to \mathbb{R}^n, \tag{4.1}$$

where $f : \mathbb{R}^n \to \mathbb{R}^n$ is sufficiently smooth, and D is a closed subset of $\mathbb{R}^n$ can be transformed into the problem of finding the fixed points of an iteration that is written as follows:

$$x(k+1) = g(x(k), k), \qquad k = 0, 1, 2, \cdots, \tag{4.2}$$

with $x(0)$ given, and where $g : \mathbb{R}^n \times \mathbb{N} \to \mathbb{R}^n$. Finding fixed points x^* that satisfy $x^* = g(x^*, k)$, $k = 0, 1, 2, \cdots$ becomes equivalent to solving (4.1). Note also that the classical iterative methods to solve the linear equation $Ax = b$ such as Jacobi, Gauss–Seidel, successive overrelaxation, etc., correspond to the choice of $f(x) = Ax - b$ in (4.1) and appropriate choice of g in (4.2).

In the context of asynchronous parallel computation of solutions of large systems of equations, it is natural to consider block-partitioned versions of (4.2). Finally, equation (4.2) is viewed here as a discrete-time dynamic system. Thus convergence of the iterative method to the solution of (4.1) is implied by the asymptotic stability of this discrete-time system. In order to be able to use the powerful arsenal of Liapunov tecniques to solve the latter problem, a state-space model for a general asynchronous block-iterative method is given in the sequel.

Consider the general synchronous block-iterative method described by the equation:

$$z_i(k+1) = \sum_{j=1}^n H_{ij}(z(k), k) z_j(k), \tag{4.3}$$

for $i = 1, 2, \cdots, n$; $k = 0, 1, 2, \cdots$, and where $z_i(k) \in \mathbb{R}^{n_i}$ for all k, $N := \sum_{i=1}^n n_i$, $z(k)^T := \left(z_1(k)^T, \cdots, z_m(k)^T \right) \in \mathbb{R}^{1 \times N}$ and $H_{ij}(z(k), k) \in \mathbb{R}^{n_i \times n_j}$, for all k.

Clearly, if $z(k) := x(k) - x^*$ (where x^* is the fixed point of (4.2) and the solution of (4.1)), then (4.3) may be thought of as the error equation of a time-varying, block-iterative method, and convergence of this method is implied by the asymptotic stability of the zero solution of the discrete-time system described by (4.2). The term synchronous is used in the following sense: The evaluation of each of the n subvectors $z_i(k+1)$ is assigned, at time k, to one of n computers and the exchange of subvectors between computers is synchronized to occur at the same time instant (clock pulse), before a new iteration can begin.

To arrive at a general state-space description of a class of asynchronous versions of models of the type (4.3), delayed vectors are introduced into the right-hand side of (4.3) (returning to the notation x instead of z) to get:

$$x_i(k+1) = \sum_{j=1}^{n} H_{ij}\left(x_1(d_{i1}(k)), \cdots, x_n(d_{in}(k)), k\right) x_j(d_{ij}(k)), \qquad (4.4)$$

for $i = 1, 2, \cdots, n$, $k = 0, 1, 2, \cdots$. For $j = 1, \cdots, n$ and for all k, the integers $d_{ij}(k)$ belong to the set $\{k, k-1, \cdots, k-d\}$ for some integer $d > 0$ which represents a uniform upper bound on the delays. In other words, no iteration of the asynchronous iterative method being modeled uses any information that was generated more than d iterations ago. The values of the individual delay terms $d_{ij}(k)$ are determined by individual computation times, communication delays, memory access delays, software access delays, etc., all of which are, in practice, time-varying and bounded. The following notation is used in this section: $\mathbf{n} := \{1, \cdots, n\}$, $\mathbf{d} := \{0, 1, ..., d\}$ and $(\mathbf{d} - \mathbf{1}) := \{0, 1, ..., d-1\}$, and the bounded delays assumption (3.5.2) holds here.

Note that the $d_{ij}(k)$s are functions of three variables: iteration number k, sending computer index j, and receiving computer index i. The uniformly bounded delay asynchronism is referred to in [BT89] as *partial asynchronism* to distinguish it from the case of possibly unbounded delays. Note that this assumption is realistic in the sense that it is understood that none of the computers involved in the computation remains idle for a period larger than d time units.

In order to ensure that the zero solution $x_i(k) = 0$, for all i, k, is indeed a solution of (4.4), the well posedness assumption (3.5.1) is also needed.

Note that the equation (4.4) describes a discrete-time dynamical system that is nonlinear and time-varying. By the well posedness assumption, it admits the zero solution $x_i(k) = 0$, for all i, k. As pointed out above, in this book, the convergence of an asynchronous iterative method really refers to asymptotic (or exponential) stability of the zero solution of (4.4).

The model (4.4) can be written in a standard state-space representation by stacking the variables $x_i(k)$ and their delayed versions $x_i(k - \ell)$, $\ell = 1, \cdots, d$, in a state-vector. To do this, the following notation is introduced:

$$\begin{aligned}
\overline{H}_{ij}(k) &:= H_{ij}\left(x_1(d_{i1}(k)), x_2(d_{i2}(k)), \cdots, x_n(d_{in}(k)), k\right) & (4.5) \\
x_{i,\ell}(k) &:= x_i(k-\ell), \quad \ell = 0, \cdots, d, \quad i = 1, \cdots, n & (4.6) \\
\overline{x}_i(k)^T &:= \left(x_i(k)^T, x_{i,1}(k)^T, \cdots, x_{i,d-1}(k)^T, x_{i,d}(k)^T\right) & (4.7) \\
\mathbf{x}(k)^T &:= \left(\overline{x}_1(k)^T, \overline{x}_2(k)^T, \cdots, \overline{x}_n(k)^T\right), & (4.8)
\end{aligned}$$

and, finally, switching functions $\Psi_{ij}(\overline{x}_j(k), k)$ that have the following property: For each triple i, j, k, $\Psi_{ij}(\overline{x}_j(k), k)$ takes the value of exactly one of the variables in the set

$$\{x_j(k-d), x_j(k-d+1), \cdots, x_j(k)\},$$

which is—and this is the crucial observation—alternatively written as

$$\{x_{j,d}(k), x_{j,d-1}(k), \cdots, x_j(k)\}.$$

This is a form appropriate for state-space representation. Thus equation (4.4) can be written in an expanded state-space form, where $\mathbf{x}(k)$ defined in (4.8) is the state-vector, as follows:

$$
\begin{aligned}
x_i(k+1) &= \sum_{j=1}^{n} \overline{H}_{ij}(k)\Psi_{ij}\left(\overline{x}_j(k), k\right), \\
x_{i,1}(k+1) &= x_i(k), \\
x_{i,2}(k+1) &= x_{i,1}(k), \\
&\vdots \\
x_{i,d}(k+1) &= x_{i,d-1}(k) \qquad i = 1, 2, \cdots, n,
\end{aligned}
\tag{4.9}
$$

with initial conditions $x_{i,\ell}(0)$, $\ell = 0, 1, \cdots, d$, appropriately chosen.

The use of delays to model asynchronisms can also be viewed from the perspective that the reference or global clock is always set or synchronized with the fastest computer so that the states of the remaining computers are understood to be delayed with respect to it.

4.2.1 Convergence Conditions for the Asynchronous Case

In this subsection Liapunov techniques are used to derive sufficient conditions for stability of the zero solution $x_{i,\ell}(k) = 0$, for all i, ℓ, k of (4.9). As pointed out in the previous section, convergence is being used loosely and synonymously with stability. Note that the asynchronous model (4.9) includes the synchronous model (4.3) as a special case (the switching functions $\Psi_{ij}(\overline{x}_j(k), k)$ simply choose the variables $x_j(k)$ for all i, j, k). As a result, the only general necessary conditions that can be obtained are those necessary for the convergence of the corresponding synchronous iterative computation.

Defining an aggregate matrix $H = (h_{ij})$, using the notation introduced in (4.5)–(4.8):

$$h_{ij} := \sup_{\mathbf{x},k} \|\overline{H}_{ij}(k)\|_\infty, \tag{4.10}$$

the main result of this section can be stated.

Theorem 4.2.1 *Under the well posedness and bounded delays assumptions (3.5.2) and (3.5.1), the zero solution of (4.9) is globally exponentially stable if $H \in \mathcal{D}_d$.*

Proof. Consider the diagonal-type Liapunov function candidate:

$$V(k) := \max_{i \in \mathbf{n}, \ell \in \mathbf{d}} \left\{ w_i^{-1}\|x_i(k)\|_\infty, w_i^{-1}\|x_{i,\ell}(k)\|_\infty \right\}, \tag{4.11}$$

where the w_i are positive numbers (weights). Then, from (4.9),

$$V(k+1) =$$

$$\max_{i \in \mathbf{n}, \ell \in (\mathbf{d}-1)} \left\{ w_i^{-1} \left\| \sum_{j=1}^{n} \overline{H}_{ij}(k) \psi_{ij}(\overline{x}_j(k), k) \right\|_{\infty}, w_i^{-1} \|x_{i,\ell}(k)\|_{\infty} \right\}$$

$$\leq \max_{i \in \mathbf{n}, \ell \in \mathbf{d}} \left\{ \sum_{j=1}^{n} w_j w_i^{-1} \|\overline{H}_{ij}(k)\|_{\infty} w_j^{-1} \|\psi_{ij}(\overline{x}_j(k), k)\|_{\infty}, w_i^{-1} \|x_{i,\ell}(k)\|_{\infty} \right\}$$

$$\leq \max_{i \in \mathbf{n}, \ell \in \mathbf{d}} \left\{ \max_{i \in \mathbf{n}} \left\{ w_i^{-1} \sum_{j=1}^{n} w_j \|\overline{H}_{ij}(k)\|_{\infty} \right\} \times \max_{j \in \mathbf{n}} \left\{ w_j^{-1} \|\psi_{ij}(\overline{x}_j(k), k)\|_{\infty} \right\}, \right.$$

$$\left. w_i^{-1} \|x_{i,\ell}(k)\|_{\infty} \right\}$$

$$\leq \max_{i \in \mathbf{n}, \ell \in \mathbf{d}} \left\{ \max_{j \in \mathbf{n}} \left\{ w_j^{-1} \|\psi_{ij}(\overline{x}_j(k), k)\|_{\infty} \right\}, w_i^{-1} \|x_{i,\ell}(k)\|_{\infty} \right\}$$

$$\leq \max_{i \in \mathbf{n}, \ell \in \mathbf{d}} \left\{ w_i^{-1} \|x_i(k)\|_{\infty}, w_i^{-1} \|x_{i,\ell}(k)\|_{\infty} \right\} = V(k),$$

i.e., $V(k+1) \leq V(k)$.

Note that the penultimate majorization follows from (4.10) and from the fact that by Lemma 2.7.25, $H \in \mathcal{D}_d$ if and only if there exist n positive real numbers w_i, $i = 1, \cdots, n$ such that

$$\max_{1 \leq i \leq n} \left\{ w_i^{-1} \sum_{j=1}^{n} w_j h_{ij} \right\} < 1. \tag{4.12}$$

The last majorization is valid because the range of $\Psi_{ij}(\overline{x}_j(k), k)$ is the set $\{x_{j,d}(k), x_{j,d-1}(k), \cdots, x_j(k)\}$.

From the state-space equation (4.9), using the bounded delays assumption, it follows that the function $V(k)$ defined in (4.11) can stay constant for at most d steps, after which it must decrease, since after d steps all the subvectors have been updated: i.e., subjected to a weighted infinity norm contraction (4.12). In other words, there exists $\delta \in (0,1)$ such that

$$V(k) - V(k+d+1) \geq \delta V(k). \tag{4.13}$$

Equation (4.13) implies that there is a subsequence $\{V(k_i)\}$ of $\{V(k)\}$ with

$$k_i \leq i(d+1) \tag{4.14}$$

which decays exponentially fast; i.e., for some $a > 0$, $b \in (0,1)$

$$V(k_i) \leq ab^i. \tag{4.15}$$

From definition 4.11, if $\epsilon := \min_i\{w_i^{-1}\} = (\max_i\{w_i\})^{-1} > 0$, then

$$\forall k, \qquad \epsilon\|\mathbf{x}(k)\|_\infty \leq V(k), \tag{4.16}$$

Equation (4.15) now implies that the subsequence $\{\|\mathbf{x}(k_i)\|_\infty\}$ of $\{\|\mathbf{x}(k)\|_\infty\}$ decays at least at the same rate:

$$\|\mathbf{x}(k_i)\|_\infty \leq \tilde{a}b^i \quad \tilde{a} > 0. \tag{4.17}$$

It remains to show that, as a result, $\{\|\mathbf{x}(k)\|_\infty\}$ also decays exponentially fast. For arbitrary k, there exists a greatest k_i with

$$k_i \leq k < k_i + d + 1. \tag{4.18}$$

Note that, as explained in the example 1.4.2, equation (4.9) can be written in matrix form as follows:

$$\mathbf{x}(k+1) = H_a(\mathbf{x}(k), k)\mathbf{x}(k), \tag{4.19}$$

where $H_a(\mathbf{x}(k), k)$ is a state-dependent time-varying matrix defined by the $\overline{H}_{ij}(k)$s and Ψ_{ij}s of (4.9). From (4.18) and (4.19) it follows that for arbitrary k, $\mathbf{x}(k)$ can be written as

$$\mathbf{x}(k) = H_a(\mathbf{x}(k-1), k-1)\cdots H_a(\mathbf{x}(k_i), k_i)\mathbf{x}(k_i) \tag{4.20}$$

with $k \geq k_i \geq k - d$.

Since, by hypothesis, the $\overline{H}_{ij}(k)$s are bounded for all k, it follows from (4.17) and (4.20) that there exists a constant $a' > 0$, independent of $\mathbf{x}(k_i)$, such that

$$
\begin{aligned}
\|\mathbf{x}(k)\|_\infty \;\leq\;& a'b^i \\[4pt]
\leq\;& a'b^{k_i(d+1)^{-1}} && \text{[by (4.14), since } 0 < b < 1] \\[4pt]
=\;& \frac{a'}{(b')^d}(b^{(d+1)^{-1}})^d b^{k_i(d+1)^{-1}} && \text{[where } b' = b^{(d+1)^{-1}} < 1] \\[4pt]
=\;& a^* b^{(d+1)^{-1}(d+k_i)} && \text{[where } a^* = a'/(b')^d] \\[4pt]
\leq\;& a^*(b')^k && \text{[by (4.18)]}
\end{aligned}
\tag{4.21}
$$

and this completes the proof. ∎

Condition(4.12) implies that $\{\overline{H}_{ij}(k)\}$ is a (norm-) bounded sequence for all k and, furthermore, that the zero solution is the only global equilibrium of (4.9). From (4.21) it follows that as $d \to \infty$, $b' \to 1$, and $a^* \to a'$, so that convergence could be slower with larger delays. The proof of this theorem mimics a technique in [AM81] except that a diagonal-type weighted infinity norm function is used instead of a quadratic Liapunov function.

Some additional observations on this theorem are as follows. H-matrices, introduced in [Ost55], are diagonally similar to a strictly diagonally dominant matrix (see [FP67]). Free-steering iterative methods, in which the order of performing the iterations is arbitrary, converge when the iteration matrix is an H-matrix. Since the model (4.9) permits free-steering, as well as delays and a class of nonlinearities, Theorem 4.2.1 is a generalization to a larger class of iterative methods, provided that a certain aggregate matrix (in this case $H = (h_{ij})$) is an H-matrix.

From the perspective of interconnected systems, Theorem 4.2.1 can be viewed as a generalization of the result [Šil78, Theorem 4.12]. First, it is a block version of that result; second, and more important, condition (4.12) assures exponential stability of system (4.3) even when time-varying delays are present in the interconnections between the subsystems represented by $x_i(k + 1) = \overline{H}_{ii}(k)x_i(k)$. In fact, since Theorem 4.2.1 ensures robust stability of the system (4.4) under parameter disturbances as well as changes in the delays (see Section 3.6), it is natural that the stability condition is based on nonnegative matrices (another perspective on this is in Section 6.1).

Theorem 4.2.1 also generalizes the classical result of Chazan and Miranker [CM69], which is derived for stationary linear pointwise iterative methods, to nonstationary block-iterative methods and, moreover, a class of nonlinear iterative methods is also permitted.

The model (4.3) is sufficiently general to encompass many models proposed earlier and is suited to the derivation of convergence conditions using a diagonal-type Liapunov function. Advantages of a Liapunov function approach are the following. First, the convergence conditions can be expressed in terms of the aggregate matrix that is associated to the block partition of the matrix. This simplifies the testing of the condition for large scale systems because no matrix inversions are required and the infinity norm is easy to evaluate. Second, explicit derivation of estimates for the rates of convergence is possible in the synchronous case (see the next section) and a clear relationship is obtained between the delays and the convergence rate in the asynchronous case.

It is straightforward to modify the proof technique above to handle iterative methods with overlapping blocks, which is the subject of Section 4.3.

4.2.2   Convergence Conditions for the Synchronous Case

The specialization of Theorem 4.2.1 to the case of synchronous iterative methods represented by model (4.3) can be derived using a diagonal quadratic Liapunov function. This quadratic function does not seem to be suitable to handle the delays that appear in model (4.9) (see [BT89, ÜD92] for examples and explanations). Note also that the vector and matrix norms are not specified because the results in this section are valid for any p-norm, $p \in [1, \infty]$.

Consider system (4.3) (replacing z by x) and let

$$h_{ij} = \sup_{x,k} \|H_{ij}(x(k), k)\|, \quad H := (h_{ij}), \quad \overline{H}(k) := (H_{ij}(x(k), k)). \quad (4.22)$$

Using (4.22) and a vectorial norm defined in (2.7), the following inequality holds:

$$[\![\overline{H}(k)x(k)]\!] \preceq H[\![x(k)]\!], \tag{4.23}$$

and the following result can be proved.

Lemma 4.2.2 *The zero solution of (4.3) is globally exponentially stable if $H \in \mathcal{D}_d$.*

Proof. Let $D = \mathrm{diag}\,(d_1, \ldots, d_n)$ and the diagonal Liapunov function be defined as

$$V(x(k)) \quad := \quad \sum_{i=1}^{n} d_i \|x_i(k)\|^2 \tag{4.24}$$

$$= \quad \langle [\![x(k)]\!], D[\![x(k)]\!] \rangle. \tag{4.25}$$

For system (4.3), the following majorizations hold.

$$V(x(k+1)) \quad = \quad \langle [\![\overline{H}(k)x(k)]\!], D[\![\overline{H}(k)x(k)]\!] \rangle \tag{4.26}$$

$$\leq \quad \langle H[\![x(k)]\!], DH[\![x(k)]\!] \rangle \quad \text{[by (4.23)]} \tag{4.27}$$

$$\leq \quad \rho(H)^2 V(x(k)). \tag{4.28}$$

Note that (4.28) follows from the Perron–Frobenius theorem, since H is a nonnegative matrix and $[\![x(k)]\!]$ is a nonnegative vector for all k. Let $\Delta V(x(k)) := V(x(k+1)) - V(x(k))$. From (4.27) it is concluded that

$$\Delta V(x(k)) \leq \langle [\![x(k)]\!], (H^T D H - D)[\![x(k)]\!] \rangle. \tag{4.29}$$

The right-hand side of (4.29) is always negative (for $x(\cdot) \neq 0$) if and only if $H \in \mathcal{D}_d$. Finally, since $V(\cdot)$ is radially unbounded, by (4.28) the proof is completed: i.e., the system is globally exponentially stable with decay rate or rate of convergence given by $\rho(H)^2 < 1$ (by Lemma 2.7.25). ∎

From (4.28), it can also be concluded that

$$\Delta V(x(k)) \leq - \left[1 - \rho(H)^2\right] V(x(k)). \tag{4.30}$$

The right-hand side of (4.30) is always negative because $V(\cdot)$ is always positive (for $x(\cdot) \neq 0$) and, from Lemma 2.7.25, H is in $\mathcal{D}_d$ if and only if $\rho(H) < 1$ which implies that $(1 - \rho(H)^2) > 0$. This constitutes an alternate proof of Lemma 4.2.2.

Consider, for example, the classical Block–Jacobi method to solve $Ax = b$ with a conformal partition of $A = (A_{ij})$, the matrix that appears in the error equation has the form:

$$\begin{aligned} H_{ij}(x(k), k) &= A_{ii}^{-1} A_{ij}; \\ H_{ii}(x(k), k) &= 0. \end{aligned} \tag{4.31}$$

Consequently, for this case, using (i) the definition of block quasidominance [Oku78, Definition 1] and (ii) the equivalence of block quasidominance of $A = (A_{ij})$ and the diagonal stability of H, Lemma 4.2.2 implies that:

$$\sum_{\substack{j \neq i}}^{n} \|A_{ii}^{-1} A_{ij}\| \frac{w_j}{w_i} < 1, \tag{4.32}$$

for $i = 1, 2, \ldots, n$ and this ensures the convergence of the Block–Jacobi method. Furthermore, if in (4.31) it is assumed that $h_{ii} = 0$ and $h_{ij} = \|A_{ii}^{-1}\| \|A_{ij}\|$, then

$$\sum_{\substack{j \neq i}}^{n} \|A_{ii}^{-1}\| \cdot \|A_{ij}\| \frac{w_j}{w_i} < 1, \tag{4.33}$$

for $i = 1, 2, \ldots, n$ also ensures the convergence of the Block-Jacobi method (by Lemma 4.2.2).

4.3   Asynchronous Iterations to Solve Almost Linear Equations

This section presents convergence conditions for asynchronous block-iterative methods for the solution of the so-called almost linear equation described below. For this commonly occurring class of fixed point problems it is also natural to consider the synchronous and asynchronous versions of the classical block-iterative methods (such as Jacobi) discussed in the previous section.

Consider an equation of the form

$$Ax = f(x), \ A \in \mathbb{R}^{n \times n}, \ f : \mathbb{R}^n \to \mathbb{R}^n, \tag{4.34}$$

in the special case where f is a block-diagonal Lipschitz-continuous function. Such an equation is known as an *almost linear* equation [OR70]. The objective in this section is to find conditions on a block-splitting of the matrix A and conditions on f such that the associated asynchronous block-iterative method converges to the desired solution.

The special case of a block-diagonal, Lipschitz-continuous nonlinear operator f is quite common in practical applications such as the solution of two point boundary value problems for ordinary differential equations [Ort72] and the load-flow problem for electrical networks [BKM91]. The main ideas of the analysis provided here are to manipulate the asynchronous error equation in order to represent it in the form of a standard vector difference inequality, to use Theorem 4.2.1 to prove exponential stability of the associated comparison vector difference equation; and to use Lemma 3.5.7 to show that a suitable norm of the error tends to zero. Although results similar to those obtained here can be derived differently, the simplicity of the diagonal-stability-based Liapunov approach justifies its presentation.

Let a Cartesian product decomposition of $\mathbb{R}^n$ be given:

$$\mathbb{R}^n = \mathbb{R}^{n_1} \times \cdots \times \mathbb{R}^{n_m}, \; n_1 + \cdots + n_m = n. \tag{4.35}$$

A block-splitting of the matrix A, conformal with the decomposition (4.35) is given:

$$A = M - N, \tag{4.36}$$

where M is block-diagonal, $\mathbf{m} := \{1, 2, \ldots, m\}$ and

$$M = \operatorname{diag}(M_1, \cdots, M_m), \; M_i \in \mathbb{R}^{n_i \times n_i}, \; \det M_i \neq 0, \forall i \in \mathbf{m} \tag{4.37}$$

$$N_{ij} \in \mathbb{R}^{n_i \times n_j}, N := (N_{ij}). \tag{4.38}$$

Note that (4.37) implies that M is nonsingular. In what follows, to avoid an excess of notation, a sort of Einstein convention is used, by which all statements in which an index i or j appears is assumed to hold for all $i, j \in \mathbf{m}$.

Let $D \subset \mathbb{R}^n$ and $D = D_1 \times \cdots \times D_m$ with $D_i \subset \mathbb{R}^{n_i}$. It is assumed that $f : D \to f(D)$ is block-diagonal, i.e., if $x = (x_1^T, \ldots, x_m^T)^T \in D, x_i \in D_i$, and if $f = (f_1, \ldots, f_m)^T$, then f_i depends only on x_i. More precisely, the following assumption holds.

Assumption 4.3.1 $D \subset \mathbb{R}^n, D = D_1 \times \cdots \times D_m, D_i \subset \mathbb{R}^{n_i}$.

$$f_i : D_i \to f_i(D_i) : x_i \mapsto f_i(x_i),$$
$$f : (x_1^T, \ldots, x_m^T)^T \mapsto (f_1(x_1)^T, \ldots, f_m(x_m)^T)^T. \tag{4.39}$$

It is also assumed that a single norm $\| \cdot \|$ is used on the space $\mathbb{R}^{n_i}, i = 1, \ldots, m$ and on $\mathbb{R}^n$ as well, and that each f_i is Lipschitz-continuous with constant l_{f_i} with respect to this norm:

Assumption 4.3.2

$$\forall x_i, y_i \in D_i, \; \|f_i(x_i) - f_i(y_i)\| \leq l_{f_i} \|x_i - y_i\|. \tag{4.40}$$

Substituting the splitting (4.36) in (4.34) yields:

$$x = M^{-1} N x + M^{-1} f(x) =: g(x), \tag{4.41}$$

where $g = M^{-1} N + M^{-1} f$. Equation (4.34) has a solution in a set D, if and only if there exists a fixed point $x^* \in D$ for g, i.e., $x^* = g(x^*)$. Throughout this section, the following assumption is also made.

Let D admit a Cartesian decomposition as in Assumption 4.3.1, then the following assumption also holds:

Assumption 4.3.3 *There exists a closed set $D \subset \mathbb{R}^n$ such that $g(D) \subset D)$, where g is defined in (4.41) above, which contains exactly one fixed point x^* of the operator g.*

Then, denoting the fixed point in the set D by $x^* = (x_1^{*T}, \ldots x_m^{*T})^T$, and using (4.37)—(4.39) one obtains

$$x_i^* = \Sigma_{j=1}^n M_i^{-1} N_{ij} x_j^* + M_i^{-1} f_i(x_i^*), \tag{4.42}$$

which is the fixed point equation that must be satisfied by solutions of (4.34).

The assumption that the time-varying delays that model asynchronism are uniformly bounded in time (over all processors) by a positive integer d is also made here. Thus the general bounded-delay asynchronous block-iterative method based on (4.41) is written as:

$$x_i(k+1) = \sum_{j=1}^m M_i^{-1} N_{ij} x_j(d_{ij}(k)) + M_i^{-1} f_i(x_i(d_{ii}(k))). \tag{4.43}$$

Subtracting (4.42) from (4.43) yields the following asynchronous error equation for the ith component:

$$x_i(k+1) - x_i^* = \sum_{j=1}^m M_i^{-1} N_{ij}(x_j(d_{ij}(k)) - x_j^*) + M_i^{-1}[f_i(x_i(d_{ii}(k))) - f_i(x_i^*)]. \tag{4.44}$$

Two error vectors are defined below.

$$e_i(k+1) := x_i(k+1) - x_i^*; \quad e_j(d_{ij}(k)) := x_j(d_{ij}(k)) - x_j^*. \tag{4.45}$$

Then (4.44) can be rewritten as:

$$e_i(k+1) = \sum_{j=1}^m M_i^{-1} N_{ij} e_j(d_{ij}(k)) + M_i^{-1}[f_i(x_i(d_{ii}(k))) - f_i(x_i^*)]. \tag{4.46}$$

Taking norms in (4.46) and using (4.40) gives:

$$\|e_i(k+1)\| \le \sum_{i=1}^m \|M_i^{-1} N_{ij}\| \cdot \|e_j(d_{ij}(k))\| + M_i^{-1} l_{f_i} \|e_i(d_{ii}(k))\|. \tag{4.47}$$

The use of assumption 4.3.2 above (Lipschitz-continuity of the f_is in the domains D_i) implicitly uses the further hypothesis that for all $k \in \mathbb{N}$, $x_i(d_{ii}(k)) \in D_i$ and $x_i^* \in D_i$. This is guaranteed by assumption 4.3.3. The following notation is needed in the sequel.

$$z_i(k+1) := \|e_i(k+1)\|; \quad z_j(d_{ij}(k)) := \|e_j(d_{ij}(k))\|; \tag{4.48}$$

$$z_{i,p}(k) := \|e_i(k-p)\|, \quad p = 0, 1, \ldots, d \tag{4.49}$$

$$h_{ij} := \begin{cases} \|M_i^{-1} N_{ii}\| + \|M_i^{-1}\| l_{f_i} & i = j \\ \\ \|M_i^{-1} N_{ij}\|, & i \ne j \end{cases} \tag{4.50}$$

$$\text{and} \quad H := (h_{ij}). \tag{4.51}$$

All z variables are nonnegative scalar variables and the variables $z_{i,p}(k)$ introduced in (4.49) are norms of delayed error variables so that, by the bounded delays assumption, it is only necessary to consider $p = 0, 1, \ldots, d$.

Denoting $[\![C]\!] = ([\![C_{ij}]\!]) \in \mathbb{R}^{m \times m}$ where $C = (C_{ij}) \in \mathbb{R}^{n \times n}$ and $C_{ij} \in \mathbb{R}^{n_i \times n_j}$, the nonnegative $m \times m$ matrix H can be expressed as:

$$H = [\![M^{-1}N]\!] + [\![M^{-1}]\!]L_f, \tag{4.52}$$

where,

$$L_f := \operatorname{diag}(l_{f_1}, \ldots, l_{f_n}). \tag{4.53}$$

Inequality (4.47) is now rewritten as the asynchronous difference inequality

$$z_i(k+1) \le \sum_{j=1}^{m} h_{ij} z_j(d_{ij}(k)). \tag{4.54}$$

As in (4.8) using the expanded state vector:

$$z_a(k)^T = (z_1(k), z_{1,1}(k), \cdots, z_{1,d}(k), \cdots, z_m(k), \cdots, z_{m,d}(k)), \tag{4.55}$$

one can rewrite the m scalar asynchronous difference inequalities (4.54) as a synchronous vector difference inequality in $\mathbb{R}_+^{md}$ as follows.

$$\forall k \in \mathbb{N}, \ z_a(k+1) \preceq H_a(k) z_a(k), \tag{4.56}$$

where, for all $k, H_a(k)$ is an $md \times md$ time-varying nonnegative matrix whose elements are the (nonnegative) elements of H, ones, and zeros (for details of its definition and interpretation, see Example 1.4.2).

Let $y_a(k)$ be a solution of the associated comparison vector difference equation, i.e.,

$$\forall k \in \mathbb{N}, \ y_a(k+1) = H_a(k) y_a(k). \tag{4.57}$$

Then, by Theorem 4.2.1, $H \in \mathcal{D}_d$ is a sufficient condition for $y_a(k)$ to tend to zero exponentially as $k \to \infty$. Since, for all $k, H_a(k)$ is a nonnegative matrix, Lemma 3.5.6 now implies that $z_a(k) \to 0$, provided that (4.56) and (4.57) start from the same initial condition. This means that the asynchronous block-iterative method defined in (4.43) is locally convergent for all initial conditions in D. In other words, the following result has been proved.

Theorem 4.3.4 *Consider a matrix A with a block-splitting as in equation (4.36), and a block-diagonal nonlinear operator f with Lipschitz-continuous components f_i (assumptions 4.3.1 and 4.3.2). If the nonnegative matrix H (4.50) is Schur diagonally stable and the operator $g = M^{-1}N + M^{-1}f$ is subject to assumption 4.3.3, then the asynchronous block-iterative method (4.43) to solve the fixed-point problem $Ax = f(x)$ is locally convergent to the fixed-point $x^* \in D$, for all initial conditions in D.* $\square$

For a nonnegative matrix H, Lemma 2.7.25 shows that there are many conditions equivalent to $H \in \mathcal{D}_d$, some of which may be easier to check—for example, the condition that $\rho(H) < 1$. An obvious sufficient condition is that $\|H\|$ be strictly less than unity in any matrix norm, the infinity norm being one which lends itself to easy computation. Note also that, by Theorem 4.2.1, local exponential stability of the asynchronous block-iterative method is actually ensured, and this is stronger than proving local convergence or attractivity.

Some generalizations of Theorem 4.3.4 fall out of this Liapunov approach in a natural manner.

Theorem 4.3.5 *If a nonstationary splitting of the matrix A is given, i.e., in equations (4.37)—(4.39), M_i, N_{ij} are replaced by $M_i(k), N_{ij}(k)$; the l_{f_i}'s are Lipschitz constants of the nonstationary f_i's and the reference matrix entries h_{ij} are redefined as:*

$$h_{ij} := \begin{cases} \sup_k\{\|M_i^{-1}(k)N_{ii}(k)\| + \|M_i^{-1}(k)\|l_{f_i}\}, & i = j \\ \sup_k\{\|M_i^{-1}(k)N_{ij}(k)\|\}, & i \neq j \end{cases} \qquad (4.58)$$

$$H := (h_{ij}) \qquad (4.59)$$

then, under assumptions 4.3.1 to 4.3.3, the nonstationary version of Theorem 4.3.4 holds, i.e., $H \in \mathcal{D}_d$ is a sufficient condition for the convergence of the nonstationary asynchronous block-iterative method.   □

In the numerical analysis literature, the word *nonstationary* is commonly used in the same sense that *time-varying* is used in the control and systems literature. Thus the above theorem is a generalization of Theorem 4.3.4 to the time-varying case.

A generalization of Theorem 4.3.4 in which a different relaxation factor ω_i may be used for each processor is also easy to obtain. It is useful in situations in which different subproblems have different values of optimal relaxation factors.

Theorem 4.3.6 *If relaxation factors $\omega_i > 0$, are introduced into the asynchronous iterative method (4.43) as follows:*

$$x_i(k+1) = x_i(d_{ii}(k)) + \omega_i(y_i - x_i(d_{ii}(k))), \qquad (4.60)$$

where

$$y_i := \sum_{j=1}^{m} M_i^{-1} N_{ij} x_j(d_{ij}(k)) + M_i^{-1} f_i(x_i(d_{ii}(k))) \qquad (4.61)$$

and if $H_\omega \in \mathcal{D}_d$ where $H_\omega = (h_{ij}^\omega), \omega = (\omega_1, \ldots, \omega_m)$ and

$$h_{ij}^\omega := \begin{cases} |1 - \omega_i| + \omega_i(\|M_i^{-1}\|l_{f_i} + \|M_i^{-1}N_{ii}\|), & i = j \\ \omega_i\|M_i^{-1}N_{ij}\| & i \neq j, \end{cases} \qquad (4.62)$$

then the relaxed asynchronous iterative method defined by (4.60) and (4.61) converges.

Proof.   *Mutatis mutandis,* the proof of Theorem 4.3.4.   □

Corollary 4.3.7 *If $\forall i \in \mathbf{m}$, $\omega_i = \omega$, then,*

$$H_\omega = \omega H + |1 - \omega|I, \tag{4.63}$$

so that setting $\omega = 1$ recovers H of (4.51). Let $\rho(H) < 1$. Since $H \succeq 0$,

$$\rho(H_\omega) = \omega\rho(H) + |1 - \omega|. \tag{4.64}$$

Thus, given the classical condition [CM69],

$$0 < \omega < 2/(1 + \rho(H)), \tag{4.65}$$

the spectral radius condition $\rho(H_\omega) < 1$ guarantees convergence.   □

Using the techniques of Theorem 4.3.4, it is possible to define convergent asynchronous versions of the parallel multisplitting methods introduced in [Whi86] for the nonlinear algebraic equation $Ax + f(x) = b$ where A is an M-matrix, f a diagonal function with continuous nondecreasing components, and b a vector in $\mathbb{R}^n$. It is also possible to consider overlapping block-decompositions of A, as discussed in the next section. Various sufficient conditions for the convergence of asynchronous iterative methods have been derived in the literature under slightly different hypotheses on the classes of allowable asynchronisms. For instance, assume, as in [Mie74, ET82], that the delays are functions of j (sending processor) and k (time) *only.* In other words, for each time k, the same set of delayed variables is used by all the processors which are updating. Under this slightly more restrictive hypothesis, Theorem 4.3.4 becomes equivalent to the result that, if g is a contraction with respect to a vectorial norm, then a class of asynchronous iterative methods to solve the fixed point equation $u = g(u)$ converges [Mie74]. It is also observed in the latter paper that a contraction with respect to a vectorial norm is always a contraction in an appropriately chosen weighted infinity norm. In fact, it was proved in [ET82] that if g is a weighted infinity norm contraction, then a class of asynchronous iterative methods to solve $u = g(u)$ converges. In addition to the hypothesis on the delays mentioned above, the class of totally asynchronous iterative methods is considered in [Mie74, ET82].

The hybrid contraction mapping/difference inequality approach followed above leads to results similar to those of other approaches; it allows greater flexibility (time and space variation) in the choice of delays which model asynchronisms, but restricts them to be uniformly bounded in time.

4.4   Parallel Asynchronous Team Algorithms

A new context for asynchronism arose with the introduction of the so-called *team algorithms*, which are hybrids of different algorithms. Combinations of different algorithms to solve a specific problem were first proposed, in the context of power system applications, in [DTS71] for sequential computers and later again in [TPM83], where they were first called team algorithms, in a parallel asynchronous computing environment, in which such methods have a natural implementation. These "hybrid" methods aim to retain the desirable properties of the constituent methods and to generate a synergetic effect that results in several advantages as well as considerable speedup.

Contemplating an implementation on a parallel computer with a shared-memory architecture available at the time, a *blackboard* model was used in [TPM83] which also presented several numerical examples showing that a combination of the Newton method with the Steepest Descent method, i.e., a team algorithm, is able to solve problems in a given domain where the Newton method fails because the Jacobian becomes singular, and the Steepest Descent method needs a very large number of iterations to converge.

The objective in this section is to formalize the notion of team algorithms, in order to present sufficient convergence conditions for different versions (in a parallel asynchronous computing environment), based on the results of the previous sections.

There are several advantages in combining different algorithms, depending on the characteristics of the problem to be solved, as in the case of an application to the so-called load flow problem of electrical networks, reported on in [BKB96].

This section also considers the solution of the equation (4.1), and as in the previous sections of this chapter, the main idea is to use a system of p computers in such a way that each computer solves only a part of the whole system and communicates its partial result to the other computers to finally solve the global problem (4.1). As in the previous sections, the term computer denotes either a processor within a parallel machine or a an independent computing unit.

In what follows, to avoid an excess of notation, once again, a sort of Einstein convention is used, by which all statements in which an index i or j appears is assumed to hold for all $i, j \in \mathbf{m}$.

Let a vector $x \in D \subset \mathbb{R}^n$ be partitioned as:

$$x = \left(x_1^T, x_2^T, \ldots, x_m^T \right)^T, \qquad x_i \in D_i, \tag{4.66}$$

and the function $f(x)$, no longer diagonal, conformally partitioned as:

$$f(x) = \left(f_1^T(x), f_2^T(x), \ldots, f_m^T(x) \right)^T, \qquad f_i : D \to \mathbb{R}^{n_i}. \tag{4.67}$$

Consequently equation (4.1) may be rewritten in the form of m *subproblems*:

$$f_i(x) = 0. \tag{4.68}$$

and each subproblem $f_i(x) = 0$ is to be solved by a (possibly different) iterative algorithm represented by the maps g_i, each of which updates x_i, i.e.,

$$x_i \leftarrow g_i(x). \tag{4.69}$$

An iterative algorithm is conveniently chosen such that the fixed point $x^* = g(x^*)$ of the map

$$g(x) = \left(g_1(x)^T, \ldots, g_m(x)^T\right)^T, \qquad g_i(x) : D \to D_i, \tag{4.70}$$

is a solution of (4.1) in D, i.e., $f(x^*) = 0$.

The mathematical expression for $g_i(x)$ depends on the specific algorithm chosen to solve the corresponding subproblem. In general, no single algorithm can solve all subproblems equally well, and it is clearly better to choose, for each subproblem, the algorithm that is best suited for it. In case two or more different algorithms are chosen and combined, this hybrid of algorithms is called a *team algorithm.*

In this class of algorithms, each computer i attempts to solve its local subproblem (4.68) using a map g_i that updates x_i, transmitting the updated value to the other computers. If the communication between computers is governed by a global clock, then the *synchronous* version of a team algorithm can be written as:

$$x_i(k+1) = g_i(x(k)), \quad k = 0, 1, 2, \cdots. \tag{4.71}$$

More interesting is an asynchronous implementation of (4.71) that allows computation of the subvectors x_i in a desynchronized manner.

The version of the vector x, available to computer i at iteration k, is denoted as $x^i(k)$ and is given by:

$$x^i(k) := \left(x_1\big(d_{i1}(k)\big)^T, \ldots, x_m\big(d_{im}(k)\big)^T\right)^T. \tag{4.72}$$

Using this notation, the general asynchronous block-iterative team algorithm, under the bounded delay assumption, based on the update laws (4.69) may be written as:

$$x_i(k+1) = g_i(x^i(k)), \tag{4.73}$$

and it represents a non-overlapped *asynchronous* block team algorithm in which each computer i tries to solve its local subproblem (4.68) by updating x_i, i.e., applying map g_i and using the most recently received values of x_j for all $j \neq i$. After computing the new value of x_i, computer i communicates its results to the remaining computers, without blocking or interruptions of the latter, and initiates a new iteration. The process continues until a given stopping criterion is met.

Throughout this section the assumption of uniqueness of fixed points of the map g is also made.

In the notation introduced above, $x^* = \left(x_1^{*T}, x_2^{*T}, \ldots, x_m^{*T}\right)^T$, $x_i^* \in D_i$, for all $i \in \mathbf{m}$, $x_i^* = g_i(x^*)$, and, since it is being assumed that g represents an iterative method to solve for the fixed point, then $f(x^*) = 0$.

The concept of *block-Lipschitz continuity* is needed for what follows.

Definition 4.4.1 *A function $g(x)$ satisfying (4.70) is called* block-Lipschitz continuous *in a given norm with respect to a Cartesian decomposition $D_1 \times \cdots \times D_m$ defined in (4.35) if $\forall x, y \in D$ satisfying (4.66) the following inequality holds:*

$$\|g(x) - g(y)\| \leq \sum_{i=1}^{m} \ell_i \|x_i - y_i\|. \tag{4.74}$$

The constants ℓ_i are called block-Lipschitz constants.

Block-Lipschitz continuity is equivalent to Lipschitz continuity, but often more convenient to use.

Lemma 4.4.2 *A function $g(x)$ is block-Lipschitz continuous if and only if it is Lipschitz continuous.*

Proof. Given the equivalence of norms, the proof can be carried out, without loss of generality, using the standard definition of the 1-norm.

(*only if*) Let g be Lipschitz continuous in the 1-norm with Lipschitz constant ℓ:

$$\|g(x) - g(y)\|_1 \leq \ell \|x - y\|_1 = \ell \sum_{i=1}^{m} \|x_i - y_i\|_1,$$

which shows that g is block-Lipschitz constant with constants $\ell_i = \ell$ in 1-norm.

(*if*) Let g be block-Lipschitz continuous in 1-norm with constants ℓ_i, and let $\ell_{\max} := \max_i \ell_i$.

$$\|g(x) - g(y)\|_1 \leq \sum_{i=1}^{m} \ell_i \|x_i - y_i\|_1 \leq \ell_{\max} \sum_{i=1}^{m} \|x_i - y_i\|_1 = \ell_{\max} \|x - y\|_1,$$

thus completing the proof.   ∎

It is assumed, for simplicity, that a single norm $\|\cdot\|$ is used in all spaces $\mathbb{R}^{n_i}$, $i = 1, \ldots, m$ and on $\mathbb{R}^n$ as well. However, this could be easily generalized along the lines of [OŠ85]. The following assumption is also made:

Assumption 4.4.3 (Block-Lipschitz continuity): *Each map $g_i(x)$ of equation (4.73) is block-Lipschitz continuous, i.e.,*

$$\forall x, y \in D, \qquad \|g_i(x) - g_i(y)\| \leq \sum_{j=1}^{m} l_{ij} \|x_j - y_j\|. \qquad (4.75)$$

The *error vector* $e = (e_1^T, \cdots, e_m^T)^T \in \mathbb{R}^n$ is defined as in (4.45) and, also as defined in equation (4.48), a *reduced error vector* $z \in \mathbb{R}^n$ is used. Within this framework, the theorem below is an immediate generalization of Theorem 4.3.4 and provides convergence conditions for different classes of team algorithms.

Theorem 4.4.4 *Under assumptions (3.5.2) and (4.4.3), the asynchronous block-iterative algorithm given by (4.73) converges to the unique fixed point x^* in D if*

$$L \in \mathcal{D}_d,$$

where $L := (l_{ij}) \in \mathbb{R}^{m \times m}$, and where the l_{ij}s are the block-Lipschitz constants of the g_is. $\square$

The matrix L, as defined above, is referred to as a *reference matrix*.

This theorem gives a general convergence condition for asynchronous fixed-point iterations and can be considered as a more general formulation of Theorems 4.2.1 and 4.3.4. It will be shown below that, in the present formulation, it is easily applicable to many variants of team algorithms.

As mentioned above, the basic idea of a team algorithm was suggested in [TPM83], where it was proposed to solve the algebraic system (4.1), using two (or more) iterative methods to compute different versions of the vector x in different computers and combine these versions in a process (or computer) called the *administrator*. An asynchronous implementation of this particular team algorithm, using the notation in (4.72), can be mathematically represented as follows.

$$\begin{aligned}
\chi_1(k+1) &= g_1(x^1(k)) \\
\chi_2(k+1) &= g_2(x^2(k)) \\
x(k+1) &= c\,x(k) + w_1\,\chi_1^3(k) + w_2\,\chi_2^3(k).
\end{aligned} \qquad (4.76)$$

This team algorithm is also called a *fully overlapped team algorithm* and works as follows. Each computer i, $(i = 1, 2)$ evaluates a different update of the complete vector x using a different algorithm represented by the map g_i, $(i = 1, 2)$. These different versions of the vector x are denoted by χ_i, $(i = 1, 2)$ respectively and are used by the third computer (the so-called administrator) to evaluate a new update of the vector x, using properly chosen positive weights $(w_1, w_2 > 0)$ and the scalar c is chosen to ensure that, at the solution x^*, the following equality holds:

$$x = \chi_1 = \chi_2 = x^* \in \mathbb{R}^n,$$

which, in turn, implies that:

$$c = 1 - w_1 - w_2. \tag{4.77}$$

Since this algorithm is asynchronous, the vectors χ_1, χ_2 and x, in accordance with the notation (4.72) appear with a superscript; e.g., $x^1(k)$, $x^2(k), \chi_1^3(k), \chi_2^3(k)$, etc., meaning that these vectors, at iteration k, correspond to the respective delayed versions of the vector x available in computers 1 and 2; and the delayed versions of χ_1 and χ_2 available to computer 3 (the administrator).

To derive a sufficient condition for the convergence of this team algorithm, assuming that each map g_i has a Lipschitz constant l_i in the given domain D, and using an expanded state vector $z \in \mathbb{R}^{3 \times n}$ defined as:

$$z = \begin{pmatrix} z_1 \\ z_2 \\ z_3 \end{pmatrix} := \begin{pmatrix} \chi_1 \\ \chi_2 \\ x \end{pmatrix}, \qquad z_i \in \mathbb{R}^n. \tag{4.78}$$

Equations (4.76) may be rewritten as:

$$\begin{pmatrix} z_1(k+1) \\ z_2(k+1) \\ z_3(k+1) \end{pmatrix} = \begin{pmatrix} g_1(z_3^1(k)) \\ g_2(z_3^2(k)) \\ c\, z_3^3(k) + w_1\, z_1^3(k) + w_2\, z_2^3(k) \end{pmatrix}, \tag{4.79}$$

which may be viewed as a nonoverlapped block team algorithm of the form (4.73) with the reference matrix H given by:

$$H = \begin{bmatrix} 0 & 0 & l_1 \\ 0 & 0 & l_2 \\ w_1 & w_2 & |c| \end{bmatrix}, \tag{4.80}$$

and the following corollary of Theorem 4.4.4 may now be stated.

Corollary 4.4.5 *Under the assumptions of bounded delays, well-posedness and uniqueness in D, the completely overlapped asynchronous team algorithm given by (4.76), converges to the unique solution x^* in D, if $H \in \mathcal{D}_d$, where H is given by (4.80).* □

The team algorithm given by (4.76) is referred to as *fully overlapped* because all computers evaluate updates of the complete vector x, and these updates are then combined in some prespecified way to yield a global update which is sent to all computers, this being the task of the administrator.

In order to appreciate the advantage of solving the same problem in parallel using a fully overlapped team algorithm, consider a problem that may be solved using a fast algorithm (such as the Newton method) represented by the map g_1, but which is not able to ensure convergence in the whole region D, or alternatively may be solved using a slow algorithm (such as a gradient method) represented by a map g_2 with Lipschitz constant $l_2 < 1$ which, however, converges in the whole domain D.

With a fully overlapped team algorithm, it is possible to ensure convergence to the solution using Corollary 4.4.5, by choosing the following weights:

$$w_1 = \epsilon; \qquad w_2 = 1 - \epsilon; \qquad \text{where} \quad 0 < \epsilon < \frac{1 - l_2}{l_1 - l_2}. \qquad (4.81)$$

Such a combination takes advantage of the speed of the algorithm represented by g_1 ensuring, at the same time, that it will converge in the whole region D. In fact, several examples presented in [TPM83] show that the fully overlapped team algorithm combining Newton and Steepest Descent methods can solve, in a few iterations, problems that the Newton method alone (g_1) fails to solve, while the Steepest Descent method (g_2) needs hundreds of iterations to get close to the solution.

The use of asynchronous block team algorithms opens up the possibility of solving large scale problems efficiently using distributed- or shared-memory parallel computers, and the use of fully overlapped team algorithms makes possible the solution of problems that none of the individual algorithms is able to solve independently. . For obvious reasons, the fully overlapped team algorithm does not lead to an appreciable speedup, and this motivates the use of partially overlapped team algorithms, to be discussed below.

Nonoverlapped, fully, and partially overlapped team algorithms are special cases of a *generalized team algorithm*, which admits all possible combinations of overlapping, i.e., one or more computers may calculate the value of one or more variables using different algorithms, and a properly specified *administrator* calculates a weighted sum of the overlapped variables to be used in subsequent calculations. The basic ingredients of this type of algorithm are listed below.

(i) a partition of the problem $f(x) = 0$ into m subproblems $f_i(x) = 0$, $i = 1, \ldots, m$;

(ii) a specification of p different algorithms represented by the functions $g_1, \ldots, g_p$;

(iii) a specification of an $m \times p$ matrix $\Psi = (\psi_{ij})$, called the *assignment matrix* associated to the team algorithm, where $\psi_{ij} = 1$ (respectively 0) if the subvector x_i is (respectively is not) updated by the map g_j in computer j, $j = 1, 2, \cdots, p$, and the rows of Ψ are ordered such that the first r rows are associated to subvectors that are updated by only one computer;

(iv) a specification of an *administrator*, i.e., indication of how and in which computer(s) the overlapped variables are combined for use in subsequent calculations.

It is clear that the fully overlapped team algorithm is a special case of a generalized team algorithm, where: $r = 0$, $m = 1$, $p = 2$, and $\Psi = [\, 1 \quad 1 \,]$

and that a nonoverlapped team algorithm is also a special case, where the assignment matrix is an identity matrix; i.e., $\Psi = I$, and $p = m = r$.

If overlapping is used, then a subvector x_i may be updated by more than one computer; thus there may exist several different *versions* of the vector x_i depending on the evaluating computer (or map). In this context, the following notation is adopted:

χ_{ij} denotes the version of subvector x_i calculated using map g_j;

x_i denotes the subvector calculated by the administrator from the different subvectors χ_{ij}s.

The artifice of overlapping some equations helps to get better convergence, for example, in cases when there exist strongly coupled blocks. The first step in solving a problem using a team algorithm is the partitioning of the problem into an appropriate number of subproblems. This step may be done using any existing partitioning algorithm: see, for example, [CT89, Van83, SŠ91, VFK92] and references therein. Here, it was assumed that the partition is given.

Example 4.4.6 *A TA combines two different algorithms* $(p = 2)$ *that are represented by their corresponding maps* g_1 *and* g_2*. This TA is used to solve a problem partitioned in three blocks* $(m = 3)$*, in such a way that* x_1 *is updated only by* g_1*,* x_2 *is updated only by* g_2*, and* x_3 *is updated by both* g_1 *and* g_2*.*

The assignment matrix associated to this TA is the following:

$$\Psi = \begin{bmatrix} 1 & 0 \\ 0 & 1 \\ 1 & 1 \end{bmatrix}. \tag{4.82}$$

Column 1 indicates that g_1 updates x_1 and x_3, while row 3 indicates that x_3 is updated by both g_1 and g_2. Note that $r = 2$ because only x_1 and x_2 are updated in separate computers (x_1 by g_1 and x_2 by g_2).

The partially overlapped team algorithm with the associated assignment matrix (4.82) has a parallel asynchronous implementation that can be expressed by the following equations.

$$\begin{aligned}
\text{Computer 1: } x_1(k+1) &= g_1(x^1(k)) & (4.83) \\
\chi_{31}(k+1) &= g_1(x^1(k)) & (4.84) \\
\text{Computer 2: } x_2(k+1) &= g_2(x^2(k)) & (4.85) \\
\chi_{32}(k+1) &= g_2(x^2(k)) & (4.86) \\
\text{Computer 3: } x_3(k+1) &= c\,x_3^3(k) + w_1\,\chi_{31}^3(k) + w_2\,\chi_{32}^3(k). & (4.87)
\end{aligned}$$

Note that the administrator calculates x_3 using a weighted sum of different versions of subvector x_3, where $c = 1 - w_1 - w_2$. It should be noted that $x_3^3, \chi_{31}^3, \chi_{32}^3 \in \mathbb{R}^{n_3}$; and they are different versions of the same subvector.

The partially overlapped team algorithm represented by the equations above dedicates a computer to the exclusive task of evaluation of a weighted sum of vectors (administrator). In principle, this is not an efficient implementation and was used here only for illustrative purposes. More efficient implementations and a more complete discussion on different implementations can be found in [BKB96].

The simplicity of deriving the reference matrix of a team algorithm is illustrated by the following example.

Example 4.4.7 *An asynchronous iterative team algorithm implemented in two computers ($p = 2$) is used to solve the following nonlinear system of equations:*

$$\begin{aligned} x_1^2 + a_{11}x_1 + a_{12}x_2 &= 0, \\ x_2^2 + a_{21}x_1 + a_{22}x_2 &= 0, \end{aligned} \tag{4.88}$$

where for $i, j \in \{1,2\}$, $a_{ii} > 2$, and for $i \neq j$, $0 \leq a_{ij} < 1$. The domain of interest is $D = \{x \in \mathbb{R}^2 : \|x\|_\infty \leq 1\}$.

For the maps $g_i(\cdot)$, the corresponding block-Lipschitz constants, can be evaluated using the following estimate:

$$\forall i, j \in \{1,2\}, \qquad l_{ij} \geq \max_{x \in D} \left| \frac{\partial g_i}{\partial x_j} \right|. \tag{4.89}$$

If for example, a block team algorithm uses a Newton–Raphson method (NR) in computer 1 and a Component Solution method (CS, as in Bertsekas and Tsitsiklis, [BT89]) in computer 2, then, the reference matrix H of the block team algorithm can be easily derived by using the appropriate rows of the reference matrices of the individual algorithms of the team. Therefore, in this case, H is given by:

$$H = \begin{bmatrix} \dfrac{2(a_{11} + a_{12} - 1)}{(a_{11} - 2)^2} & \dfrac{a_{12}}{a_{11} - 2} \\[3ex] \dfrac{a_{21}}{\sqrt{a_{22}^2 - 4a_{21}}} & 0 \end{bmatrix},$$

which was formed with the first row of the reference matrix H_{NR} followed by the second row of the reference matrix H_{CS}.

From a practical viewpoint, asynchronous algorithms outperform synchronous algorithms for most partitions and most initial conditions, thus indicating their usefulness for computation of the solutions to large scale problems. However an adequate choice of partition is crucial to ensure good performance of an asynchronous parallel algorithm. The partitioning problem is difficult [VFK92, SŠ91] and, in this context, the theoretical convergence conditions derived in this chapter provide some useful guidelines for the choice of a partition. For example, a partition that leads to a reference matrix that has a higher degree of block-dominance than another, generally leads to iterative methods that converge faster [BM93, ZŠ94, BKB96].

4.5   Notes and References

Convergence of asynchronous iterations: history
Sufficient conditions for the convergence of an asynchronous iteration have been given by several authors and can be broadly classified as follows: (a) the iteration operator is required to be a contraction in a vectorial norm [Bau78, Mie74, Rob76]; (b) the iteration operator is required to be a contraction in a weighted infinity norm ([ET82, BT89] and references therein); and (c) a Liapunov function of the weighted infinity norm type is used to prove asymptotic stability of the iteration ([KBŠ90]). Using a result [Mie74], it was pointed out in [ET82] that conditions of type (b) generalize (i.e., are weaker than) conditions of type (a), while the Liapunov approach (c) is essentially equivalent to approach (b). Brief surveys that classify many different results on asynchronism are [SBK98, SBKK98].

Synchronous computation is always a special case of chaotic asynchronous computation. In the linear case, the condition $(\rho(|H|) < 1)$ is shown to be necessary [CM69] for the convergence of asynchronous computations. This condition is stronger than (i.e., implies) the corresponding one for synchronous computations $(\rho(H) < 1)$, and, in fact, the necessity is interpreted as follows. If $\rho(|H|) \geq 1$, then it is possible to construct an *asynchronous* iteration sequence that does not converge. More details on this can be found in [BT89, SBKK98].

Models for asynchronous iterative methods
The model presented in section 4.2, being block and nonlinear, specializes to the linear point Chazan–Miranker model. It is similar to the linear state-space models in [LM86, Equation 4-9),p.140], and in [BEN88, Equation 2.19,p.184], but differs from both of the former in the way the delayed variables are handled. The class of nonlinear L_∞-contractions, considered in [TPM83], is a different class of nonlinear functions from the one considered in this chapter.

Asynchronous convergence results
A more general version of Theorem 3.5.4, stated in terms of weighted infinity norm contractions and proving r-absolute exponential stability (i.e., allowing possibly unbounded delays) may be found in [ET82, Theorem 3.4, Corollary 3.5]. The formulation in Section 3.5 is, however, simpler to use and adequate for the purposes of this book. The proof of the general result follows the analogous steps of the proofs in [CM69]. Note also that Miellou's result (cited in [ET82, Proposition 3.8]) that a so-called T-contraction (also referred to as a block-Lipschitz contraction) must also be a contraction in an appropriate weighted infinity norm shows, by the contraction mapping theorem, that the fixed point x^* exists and is unique. Abstract and very general versions of the Liapunov function approach applicable to both partial and total asynchronism are discussed in [BT89, KKKK83, KKKK84a, KKKK84b, AKKK92]. For generalizations of the Stein–Rosenberg theorem (that compares the convergence properties of the Jacobi and Gauss–Seidel methods), see [BM93].

Asynchronous team algorithms
An *A-team* (asynchronous team), described in [ST91], may be considered as a special case of a fully overlapped team algorithm with variable weights (where $w_i(k) \in \{0, 1\}$).

An overlapped team algorithm may dedicate a computer to the exclusive task of administrator (evaluation of a weighted sum of vectors). In most situations, the other computers will have more complex tasks of solving subproblems, so that such an allocation will lead to unbalancing of computer loads. One remedy is to include the administrator's task in the task list of one or more of the other computers.

These are versions with the so-called *implicit* administrators, in which the weighted sum of the different versions of a subvector is carried out as part of the algorithm to be applied and not as a special process (or another equation). When this is done, the dimension of the reference matrix H is reduced, eliminating a line whose elements sum to unity or more (this is conducive for the satisfaction of the convergence condition $\rho(H) < 1$). More details on different versions of asynchronous team algorithms can be found in [BKB96].

Preliminary but encouraging results on the use of team algorithms that combine genetic algorithms with LMI techniques and piecewise quadratic Liapunov functions in robust analysis and design problems can be found in [ABFK98].

Convergence of block-iterative Jacobi methods

For the class of linear, synchronous, block-iterative methods, the result [Rob69, Theorem 5] can be obtained by using Theorem 4.2.1 and Lemma 2.7.25. Also, for the particular case of constant blocks and no delays, the result of Theorem 4.2.1 can be compared with the results obtained in [Oku78, Theorem 5] which states that $\rho(H) < 1$ given that (i) all $h_{ii} \geq 0$, (ii) all diagonal blocks of $(I - H)$ are M-matrices and (iii)

$$\sum_{\substack{j \neq i}}^{n} \left\| (H_{ii} - I)^{-1} \right\|_\infty \| H_{ij} \|_\infty \frac{w_j}{w_i} < 1, \qquad (4.90)$$

for $i = 1, 2, \ldots, n$. Observe that, in the time-invariant delayless case, (4.12) is equivalent to

$$\sum_{\substack{j \neq i}}^{n} (1 - \| H_{ii} \|_\infty)^{-1} \| H_{ij} \|_\infty \frac{w_j}{w_i} < 1, \qquad (4.91)$$

for $i = 1, 2, \ldots, n$ and this condition assures that $\rho(H) < 1$ (using Lemma 2.7.25) without requiring the $(I - H_{ii})$, $i = 1, \ldots, n$, to be M-matrices. Note, however, that whenever $\| H_{ii} \|_\infty < 1$, the Banach lemma states that

$$(1 + \| H_{ii} \|_\infty)^{-1} \leq \| (I - H_{ii})^{-1} \|_\infty \leq (1 - \| H_{ii} \|_\infty)^{-1} \qquad (4.92)$$

so that, under this condition, (4.91) implies (4.90). However, the result in [Oku78] only holds under the additional conditions (i) and (ii) cited above.

5

Neural Networks, Circuits, and Systems

This chapter shows and discusses the occurrence of the diagonal structure, introduced in Chapter 1, in several classes of dynamical systems that include neural networks, digital filters, passive RLC circuits, and ecosystems. Some of the examples of Chapter 1, as well as new examples, are treated in greater detail here.

5.1 Continuous-Time Hopfield–Tank Neural Networks

In a series of influential papers, Hopfield and Tank [Hop82, Hop84, TH86] introduced a dynamical system model that now bears their name. This dynamical system has a simple physical interpretation as an electrical circuit containing nonlinear amplifiers, interconnected through a network of linear gains (= real numbers). This circuit was referred to as a neural circuit or network (often, and more accurately, as an *artificial* neural network) in view of the fact that each nonlinear amplifier could be seen as a model of a neuron activation function, and the *interconnection network* of linear gains as synaptic weights. Another way to regard such a dynamical system is to consider it as an analog computer to solve certain types of problems. In this chapter, the term neural network, now widely accepted to mean artificial neural network, will be used and throughout, the Hopfield–Tank neural network is the one being considered. The importance of these neural networks for the solution of many optimization problems in real time was discussed in [TH86, CU93] and also in [AFMM91] for the specific problem of analog-to-digital converter design. Within this context it is required that the neural network have a unique equilibrium point, stable and globally attractive, in order to avoid so-called spurious solutions.

This section points out that the mathematical model that describes Hopfield–Tank networks is actually a variant of the classical Persidskii nonlinear system, and as a consequence, the corresponding energy type Liapunov function is easily obtained from Theorem 3.2.3. Consequently, diagonal stability of the interconnection matrix leads to a simple proof of the existence, uniqueness and global asymptotic stability of the equilibrium in Hopfield–Tank neural networks, without making some common restrictive assumptions used in earlier results, such as symmetry of the interconnection matrix. In general, as will be seen below, once a diagonal-type Liapunov function is used, the whole question of absolute global stability of Hopfield–Tank neural networks is reduced to a matrix stability problem. In fact, the matrix stability classes that one is led to define are modifications of the well studied classes of diagonally and D-stable matrices, and one can benefit from the large body of results in the literature on these classes. It is also worth pointing out that the diagonal stability condition also guarantees structural stability, which ensures the desirable property of persistence of global asymptotic stability under a general class of perturbations.

The standard assumptions made for the continuous-time mathematical model of the Hopfield–Tank neural network are stated below. Details that were skipped in Chapter 1 are given here.

The model studied is given by the dynamical equations:

$$C_i \frac{du_i}{dt} = \sum_{j=1}^{n} t_{ij} v_j - \frac{u_i}{R_i} + I_i \tag{5.1}$$

$$v_i = g_i(\lambda u_i), \tag{5.2}$$

for $i = 1, 2, \cdots, n$, and where, $C_i > 0$ are the neuron amplifier input capacitances; $R_i > 0$ are the resistances; $T = (t_{ij})$ is the $n \times n$ real constant network interconnection matrix; I_i are the (constant) external current inputs to the ith neuron; $u_i = u_i(t) \in \mathbb{R}$ are the neural voltages; $\lambda > 0$ is a constant scaling factor, and $v_i = v_i(t) \in \mathbb{R}$ are the neuron output voltages.

Assumption 5.1.1 *The neuron activation functions $g_i(\cdot)$ have the following characteristics:*

(i) $g_i : \mathbb{R} \to (-1, +1)$ are Lipschitz-continuous maps.

(ii) $g_i(\cdot) \in \mathcal{S}_c$.

(iii) $g_i(\cdot)$ are strictly monotonically increasing; i.e., for all $x, y \in \mathbb{R}$, if $x > y$, then $g_i(x) > g_i(y)$.

In addition, the following notation, definitions and lemmas that are used in the subsections of this section, are collected here for convenience.

$C := \text{diag}(C_1, \cdots, C_n)$ and $R := \text{diag}(R_1, \cdots, R_n)$ are constant positive diagonal $n \times n$ matrices; $b := (I_1, \cdots, I_n)^T$ is a constant input vector; $g : \mathbb{R}^n \to H \subset \mathbb{R}^n$; $H = \{(h_1, \ldots, h_n)^T : h_i \in (-1, 1)\}$ is the diagonal nonlinear activation function. Note that some of the usual conditions adopted in the literature such as $t_{ii} = 0$ (no self-interactions); $t_{ij} = t_{ji}, \forall i, j$ (symmetry of the interaction matrix), co-operation ($t_{ij} > 0$), competition ($t_{ij} < 0$), sign-symmetry (sign $t_{ij} = $ sign $t_{ji}, \forall i, j$), and the sigmoidal shape of the $g_i(\cdot)$ functions are *not* assumed.

In order to pursue the stated objective of obtaining the largest class of interconnection matrices for which necessary and sufficient conditions for stability can be obtained, a new class of matrices, closely related to diagonally stable matrices, is defined.

Definition 5.1.2 *A real $n \times n$ matrix $A = (a_{ij})$ is said to belong to the class $\mathcal{I}_o$ if $(A - P_1)P_2$ is Hurwitz stable for all positive diagonal matrices P_1 and P_2.*

5.1.1 A Global Stability Result

This section occupies itself with the derivation of the global stability result announced in Section 1.5, for the Hopfield–Tank model (5.1), (5.2) subject to the assumptions above. The following preliminaries are needed.

The set of all $u \in \mathbb{R}^n$ that satisfy the equation

$$T_H g(u) - R^{-1} u + b = 0, \tag{5.3}$$

is called the set of equilibrium points. Rewriting this equation in a more convenient form;

$$g(u) + \Theta u = \bar{b}, \tag{5.4}$$

where

$$\Theta = (\theta_{ij}) = -T_H^{-1} R^{-1},$$

$$\bar{b} = (\bar{b}_1, \bar{b}_2, \ldots, \bar{b}_n)^T = -T_H^{-1} b.$$

Note that if T_H is assumed to be in $\mathcal{D}_c$, then it is Hurwitz stable and T_H^{-1} exists. Furthermore, by Lemma 2.1.5, the matrix T_H^{-1} is in $\mathcal{D}_c$ as well. Since the matrix R^{-1} is positive diagonal, Lemma 2.1.4 ensures that $T_H^{-1} R^{-1} = -\Theta$ also belongs to class $\mathcal{D}_c$ and therefore to class $\mathcal{D}_{co}$, whence from Lemma 2.1.16, it follows that $\Theta \in \mathcal{P}_0$ and $\det \Theta \neq 0$.

Equations of the type (5.4) arise in the study of equilibria of nonlinear transistor networks and the following theorem is relevant here.

Theorem 5.1.3 [SW69, Theorem 4, p. 1297] *For equation (5.4) under the assumption 5.1.1 on functions $g_i(\cdot)$, and for each $\overline{b} \in \mathbb{R}^n$, there is a unique equilibrium point $u^* \in \mathbb{R}^n$ if and only if $\Theta \in \mathcal{P}_0$ and $\det\Theta \neq 0$.* ☐

With these preliminaries, the global stability result is as follows.

Theorem 5.1.4 *Given the Hopfield–Tank model (5.1), (5.2) subject to the assumptions above, if the interconnection matrix $T_H = (t_{Hij})$ is Hurwitz diagonally stable, then there exists an unique (isolated) equilibrium point which is globally asymptotically stable.*

Proof. From Theorem 5.1.3 it is immediate that $T_H \in \mathcal{D}_c$ implies the existence and uniqueness of an equilibrium $u^* = (u_1^*, u_2^*, \ldots, u_n^*)^T$ for the neural circuit (5.1), (5.2). It remains to show global asymptotic stability.

By a shift of the origin to the unique equilibrium point u^*, equations (5.1), (5.2) are written in the following equivalent form:

$$\frac{dx_i}{dt} = \frac{1}{C_i} \left[\sum_{j=1}^{n} t_{Hij}\phi_j(x_j) - \frac{1}{R_i}x_i \right], \tag{5.5}$$

for $i = 1, 2, \ldots, n$, and where $x_i := u_i - u_i^*$; $\phi_i(x_i) = g_i(x_i + u_i^*) - g_i(u_i^*)$, and the functions $\phi_i(x_i)$ continue to satisfy the assumption 5.1.1. Let $\Phi = (\phi_1(x_1), \cdots, \phi_n(x_n))^T$, then:

$$\Phi : \mathbb{R}^n \to \overline{H} \subset \mathbb{R}^n : \overline{H} = \{(\overline{h}_1, \ldots, \overline{h}_n)^T : \overline{h}_i \in (-2, 2)\}.$$

Equation (5.5) can be written compactly as follows:

$$\frac{dx}{dt} = C^{-1}T_H\Phi(x) - (RC)^{-1}x, \tag{5.6}$$

which is model (3.17) from Chapter 3, for which the corresponding candidate Liapunov function has the following form:

$$V(x) = 2\sum_{i=1}^{n} p_i \int_0^{x_i} \phi_i(\tau)d\tau, \tag{5.7}$$

where the p_i are positive constants that are identified below. Along the trajectories of (5.6), the time-derivative of $V(x)$ is given by

$$\frac{dV(x)}{dt} = \Phi^T(x)\left[P(C^{-1}T_H) + (C^{-1}T_H)^T P\right]\Phi(x) - 2\Phi^T(x)P(CR)^{-1}x. \tag{5.8}$$

Since $T_H \in \mathcal{D}_c$, and C^{-1} is a positive diagonal matrix, by Lemma 2.1.4, $(C^{-1}T_H) \in \mathcal{D}_c$. Consequently, there exists a positive diagonal matrix $P = \text{diag}(p_1, \ldots, p_n)$ such that $P(C^{-1}T_H) + (C^{-1}T_H)^T P$ is negative definite. Since the matrix $P(CR)^{-1}$ is diagonal and the components of the function Φ satisfy assumption 5.1.1, the second term on the right-hand side of (5.8) is also negative for all $x \neq 0$. This means that dV/dt is negative definite.

Under assumption 5.1.1 on $g_i(\cdot)$, and therefore on $\phi_i(\cdot)$, one has that:

$$\lim_{|x_i| \to \infty} \int_0^{x_i} \phi_i(\tau)d\tau = \infty.$$

The latter condition ensures radial unboundedness of the Liapunov function $V(x)$ and consequently guarantees that $x = 0$ (or equivalently u^*) is globally asymptotically stable.   ∎

Several papers provide global stability results based on special classes of the interconnection matrix: for example, negative semidefinite [FMM92], M-matrix [Ros88], and lower triangular [AFMM91]. Some of these results [Ros88, Mat92] can be interpreted as particular cases of the diagonal stability result, since they require the interconnection matrix to belong to certain subsets of the set of diagonally stable matrices, which contains several types of matrices such as symmetric negative definite, asymptotically stable triangular, sign-stable pattern, quasidominant, M-, Z-, etc.: (see Chapter 2 for further details on these classes of matrices). Similarly, the result in [AFMM91] is also related, since it considers the class of stable (but not asymptotically stable) triangular interconnection matrices. For the case when the interconnection matrix is an M-matrix, vector Liapunov functions are used in [MFP89] to determine stability and stability regions of more general models that include perturbations.

5.1.2   Refinements of the Global Stability Result

The main objective in Section 5.1.1 was to derive a simple diagonal stability condition for global absolute stability using a diagonal-type Liapunov function that also provides a proof of structural stability, as discussed in the next section.

Thus Theorem 5.1.4 in Section 5.1.1 does not take into account the strict negative definiteness of the second term on the right-hand side of equation (5.8) for $(\dot{V})$, since $T_H \in \mathcal{D}_c$ guarantees the strict negative definiteness of the first term. It is the interplay between these two terms, while maintaining negative definiteness of $\dot{V}$, that is further discussed below.

In order to obtain a more general result than that of the previous section, and also to arrive at a conjecture about the largest class of interconnection matrices for which necessary and sufficient conditions for stability can be obtained, this section focuses on neural circuits of the type (5.1), (5.2), rewritten here as the following system of differential equations.

$$\dot{x} = -Dx + Tg(x) + I, \tag{5.9}$$

where $D = \text{diag}\,(d_1, \cdots, d_n)$, with $d_i > 0$, $i = 1, 2, \cdots, n$; $T = (t_{ij})$ is a $n \times n$ constant real matrix; $I = (I_1, I_2, \cdots, I_n)^T \in \mathbb{R}^n$ is a constant vector.

Assumption 5.1.5 *The activation functions g satisfy the following conditions.*

(i) $g(x) = (g_1(x_1), \cdots, g_n(x_n))^T$ *is a diagonal function* $g : \mathbb{R}^n \to \mathbb{R}^n$ *satisfying* $g_i : \mathbb{R} \to H_i \subset \mathbb{R}$ *where* $H_i = \{h_i \in \mathbb{R} : a_i < h_i < b_i\}$.

(ii) $g_i(\cdot)$ *are* C^1 *functions with* $dg_i(x_i)/dx_i > 0$, $\forall x_i \in \mathbb{R}$;

The function $g_i(\cdot)$ under the conditions above, belongs to class $\mathcal{S}_c$ for all i (see Section 3.1).

It is easy to see that equations (5.1), (5.2) can be written in the form (5.9) above, by choosing $T = C^{-1}T_H$ and $D = R^{-1}$.

A general definition of stability that allows perturbations in the activation functions (g_i), the resistance values (D), as well as the input vector (I) is now given. The stability result derived below allows the conclusion that small perturbations in the interconnection matrix T are also permissible.

Definition 5.1.6 *System (5.9) is said to be absolutely globally asymptotically stable with respect to* $\mathcal{S}_c$ *and parameters* I, D *if it possesses a globally asymptotically stable equilibrium point for every function* $g \in \mathcal{S}_c^n$, *every input vector* I, *and every diagonal matrix* $D > 0$.

To proceed with the analysis, once again the equilibrium point x^* of (5.9) is shifted to the origin by defining $z = x - x^*$. Thus, in the z coordinates, system (5.9) becomes:

$$\dot{z} = -Dz + Tf(z), \tag{5.10}$$

where $f(z) = (f_1(z_1), \cdots, f_n(z_n))^T, f_i(z_i) = g_i(x_i + x_i^*) - g_i(x_i^*)$ and $f(0) = 0$. Since the functions $g_i(x_i)$ are strictly increasing, $z_i f_i(z_i) > 0$, for all i and for all $z_i \neq 0$. Note that the functions $f_i(z_i)$ also satisfy assumption 5.1.5. On the other hand, since, for each i, the function $g_i(x_i)$ is bounded, by assumption 5.1.5, there exists a positive constant β_i such that

$$|f_i(z_i)| \leq \beta_i |z_i|, \forall i. \tag{5.11}$$

The main result of this section is now stated as follows.

Theorem 5.1.7 *Under assumption 5.1.5, if* T *is additively diagonally stable (i.e.,* $T \in \mathcal{A}_o$), *then the origin of system (5.10) is the unique equilibrium point, and the zero solution is absolutely globally asymptotically stable.*

Proof. In [FMM94, Theorem 1], it was shown that $-T \in \mathcal{P}_o$ is a necessary and sufficient condition for the existence and uniqueness of the equilibrium solution of system (5.10), for arbitrary D and I. Hence, by Lemma 2.1.16(iii), $T \in \mathcal{A}_o$ ensures the existence and uniqueness of the equilibrium point for system (5.10) for arbitrary D and I.

It remains to prove the global asymptotic stability of the zero solution of (5.10), for all D and I, under the condition $T \in \mathcal{A}_o$. Using the Liapunov function of Theorem 5.1.4; i.e.,

$$V(z) = 2 \sum_{i=1}^{n} p_i \int_0^{z_i} f(\tau) d\tau,$$

where the p_i are positive constants to be determined, the time-derivative of $V(z)$ along the trajectories of the system (5.10) is:

$$\dot{V}(z) \;=\; -2\sum_{i=1}^{n} d_i p_i f(z_i) z_i + 2\sum_{i=1}^{n}\sum_{j=1}^{n} p_i T_{ij} f(z_i) f(z_j)$$

$$=\; -2f^T(z)PDz + 2f^T(z)PTf(z).$$

In (5.11), let the β_i define the positive diagonal matrix $B := \operatorname{diag}(\beta_1, \cdots, \beta_n)$. Then the inequality (5.11) can be written as $z \succeq B^{-1}f(z)$. Using this majorization in the first term of the expression for $\dot{V}$ above, one gets

$$f^T(z)PDz \geq f^T(z)PDB^{-1}f(z)$$

and therefore, defining $\overline{D} := DB^{-1}$:

$$\dot{V} \;\leq\; -2f^T(z)P\overline{D}f(z) + 2f^T(z)PTf(z)$$

$$=\; f^T(z)[(T-\overline{D})^T P^T + P(T-\overline{D})]f(z)$$

$$<\; 0 \qquad \text{if } T \in \mathcal{A}_1.$$

Recalling that $\mathcal{A}_1 = \mathcal{A}_o$ (Fact 2.1.21), it follows that $T \in \mathcal{A}_o$ ensures that, for every D, I and $f(z)$, there exists a positive diagonal matrix P such that $\dot{V}(z)$ is negative definite. Also $\dot{V}(z) = 0$ only at $z = 0$. Moreover, $V(z)$ is radially unbounded since $V(z) \to \infty$ as $\|z\| \to \infty$. Now from standard Liapunov theory it follows that the zero solution of (5.10) or, equivalently, the equilibrium solution of (5.9) is globally asymptotically stable, for every D and I, so the zero solution of (5.9) is absolutely globally asymptotically stable. ∎

This theorem was first proved in [AT98], assuming $\beta_i = \beta$, for all i in (5.11), which led to the equivalent condition $T \in \mathcal{A}_o$.

A global stability conjecture

Given a differential equation

$$\dot{x} = f(x) \tag{5.12}$$

with $x \in \mathbb{R}^n$ and f a continuously differentiable map from $\mathbb{R}^n$ to $\mathbb{R}^n$, the *Markus–Yamabe* conjecture states that an equilibrium of (5.12) is globally asymptotically stable provided that the Jacobian matrix $J_f(x)(:= \partial f/\partial x)$ is Hurwitz stable for all x. This conjecture, also known as the global asymptotic stability Jacobian conjecture, was first made in [MY60]. Special cases of this conjecture, such as the Kalman conjecture [Vid93] also played a prominent part in the development of Liapunov stability theory and, in fact, counterexamples to the Kalman conjecture, for $n \geq 4$, provide counterexamples to the Markus–Yamabe conjecture for $n \geq 4$ (see [Bar88b, PS93, SW98] and references therein).

The Markus–Yamabe conjecture is now examined in the context of the system (5.10). Analyzing the linearized version of system (5.10), described around the equilibrium point one has:

$$\dot{y} = (Tf'(y) - D)y, \tag{5.13}$$

where $f'(y) = (df_1(y_1)/dy_1, \ df_2(y_2)/dy_2, \cdots, df_n(y_n)/dy_n)^T$ evaluated at the equilibrium $y = 0$, and where $y = x - x^o$. For convenience (5.13) is written in the form below

$$\dot{y} = (T - D_1)D_2 y, \tag{5.14}$$

where $D_2 := f'(y)$ is a positive diagonal matrix for all $y \in \mathbb{R}^n$, and $D_1 := D(f'(y))^{-1}$ is also a positive diagonal matrix. Note that $(f'(y))^{-1}$ is defined (and positive) for all $y \in \mathbb{R}$.

Clearly asymptotic stability of the zero solution of (5.14) for all positive diagonal matrices D_1, D_2 provides necessary conditions for absolute global asymptotic stability of (5.9) in the sense of definition 5.1.6, since local stability is necessary for global stability. Local stability means that matrix $(T - D_1)D_2$ has to have roots with negative real parts for arbitrary positive diagonal matrices D_1, D_2. In other words, T in $\mathcal{I}_o$ (Definition 5.1.2) is a necessary condition for absolute global asymptotic stability.

The following version of the Markus–Yamabe conjecture arises from the discussion above.

Conjecture 5.1.8 [KB95] $T \in \mathcal{I}_o$ *is a necessary and sufficient condition for absolute global asymptotic stability of* (5.9).

In other words, even though it is known that the general Markus–Yamabe conjecture is valid only for $n \leq 2$ ([Gut95, CvdEG+97]), it is being conjectured to be true for all n for the special class of nonlinearities describing neural circuit models of the type (5.9).

Evidence in support of the conjecture is as follows.

As with the Markus–Yamabe conjecture, Conjecture 5.1.8 is true for $n = 2$. The following lemma is needed.

Lemma 5.1.9 [LW99] *A matrix* $T = (t_{ij}) \in \mathbb{R}^{2 \times 2}$ *is in class* $\mathcal{I}_o$ *if and only if* $t_{ii} \leq 0, i = 1, 2$ *and* $\det T \geq 0$.

Proof. (*if*) Clearly, if $T_1 = (T - D_1)D_2$, where D_1 and D_2 are positive diagonal matrices, then, under the conditions on t_{ii} and $\det T$, it follows that the trace of T_1 is negative and $\det T_1 > 0$, so that T_1 is Hurwitz stable, implying that T is in class $\mathcal{I}_o$.

(*only if*) If $T \in \mathcal{I}_o$, then $-T \in \mathcal{P}_o$, by Lemma 5.1.17 below, implying that the principal minors of T must have the stated signs. ∎

Lemma 5.1.10 *For a real* 2×2 *matrix,* $A \in \mathcal{I}_o$ *if and only if* $A \in \mathcal{A}_o$.

Proof. The fact that $\mathcal{I}_o \subset \mathcal{A}_o$ follows directly from Lemma 5.1.9 and the definition of class $\mathcal{A}_o$. The reverse inclusion, true for matrices of any size, is proved in Corollary 5.1.11 below. ∎

From Lemma 2.1.17, the conjecture is true within the class $\mathcal{Z}$. In fact, several classes of matrices for which global asymptotic stability has been shown in the literature are subclasses of $\mathcal{I}_o$.

Fact 2.1.21 implies that $\mathcal{A}_o$ is a subset of $\mathcal{I}_o$, i.e., one has the following corollary.

Corollary 5.1.11 $\mathcal{A}_o \subset \mathcal{I}_o$.   □

Proof. Let $A \in \mathcal{A}_o$. Then, by Fact 2.1.21, $(A - D_1) \in \mathcal{D}_c, \forall D_1 > 0$. But matrices in $\mathcal{D}_c$ are D-stable (Lemma 2.1.4), thus $(A - D_1)D_2$ is stable, for all $D_2 > 0$.   ∎

The following facts hold [AT98].

Fact 5.1.12 *A matrix A belonging to any of classes $\mathcal{M}_o, \mathcal{C}_o$, or $\mathcal{D}_{co}$ implies that $A \in \mathcal{A}_o$ [BP79].*

Fact 5.1.13 *A matrix A of class $\mathcal{C}_o$ does not necessarily belong to class $\mathcal{D}_o$. For instance, the matrix $A = \begin{bmatrix} 0 & 1 \\ 0 & 0 \end{bmatrix}$ is a matrix that is in $\mathcal{C}_o$ but not in $\mathcal{D}_o$.*

Fact 5.1.14 *A matrix A of class $\mathcal{D}_o$ does not necessarily belong to $\mathcal{C}_o$. A matrix A with $a_{ii} = 0$ and $a_{ij} = -a_{ji}$ is an example: In particular, $A = \begin{bmatrix} 0 & -1 \\ 1 & 0 \end{bmatrix}$ belongs to $\mathcal{D}_o$ but not to $\mathcal{C}_o$.*

Fact 5.1.15 *A matrix A of class $\mathcal{A}_o$ does not necessarily belong to any of the classes $\mathcal{M}_o, \mathcal{C}_o, \mathcal{D}_o$ or to the class of positive semidefinite matrices.*

This is shown by the following example.

Example 5.1.16 *Consider the matrix:*

$$A = \begin{bmatrix} 0 & 1 & -1 \\ 0 & 1 & 0 \\ 1 & 0 & 0 \end{bmatrix},$$

where all principal minors of A are nonnegative and the eigenvalues of A are $\lambda_1 = 1, \lambda_2 = j, \lambda_3 = -j$. The symmetric part of A ($= 0.5(A + A^T)$) is not positive semidefinite. Since A has some positive off-diagonal entries, therefore $-A \notin \mathcal{M}_o$.

Since the comparison matrix of A is 3×3 and has positive determinant, it has a real negative eigenvalue, hence $-A \notin \mathcal{C}_o$.

For a positive diagonal matrix, $P = \text{diag}(p_1, p_2, p_3)$, the second leading principal minor of the matrix $A^T P + PA$ is $-p_1^2 < 0$, implying that $-A \notin \mathcal{D}_o$.

To see that $\alpha I + A$ is positive diagonally stable for all $\alpha > 0$, define $P_\alpha = \text{diag}(1, p(\alpha), 1)$, where $p(\alpha) > [4\alpha(1 + \alpha)]^{-1}$. Then, by calculation, the matrix $P_\alpha(\alpha I + A) + (\alpha I + A)^T P_\alpha$ has one eigenvalue equal to 2α while the other two have both their sum and product positive for all $\alpha > 0$. Thus *this matrix has all eigenvalues positive for all $\alpha > 0$, so that $-A \in \mathcal{A}_o$, as claimed.*

Finally, the following relation holds.

Lemma 5.1.17 [KB95] *If $T \in \mathcal{I}_o$, then $-T \in \mathcal{P}_o$.*

Proof. From the definition of the class $\mathcal{I}_o$, for any positive diagonal matrix D, the matrix $T - D$ has all eigenvalues with negative real parts, implying that $\det(K - T) > 0$. From Lemma 2.1.12, it follows that $-T \in \mathcal{P}_o$. ∎

Putting together all the comments and results by various authors [FMM94, KB94, KB95, AT98, LW98b, LW98a, LW99, KB99a], the inclusions that have been proved are the following:

$$\mathcal{M}_o \subset \mathcal{C}_o \subset \mathcal{A}_o \subset \mathcal{I}_o \subset \overline{\mathcal{P}}_o$$

$$\{\mathcal{M}_o \cap \{\text{irreducible}\}\} \subset \mathcal{D}_{co} \subset \mathcal{A}_o \subset \mathcal{I}_o \subset \overline{\mathcal{P}}_o,$$

where $\overline{\mathcal{P}}_o := -\mathcal{P}_o$. All inclusions are strict and though $\mathcal{C}_o$ and $\mathcal{D}_{co}$ have nonempty intersection, neither one contains the other. An example of a 3×3 matrix that is in $\mathcal{A}_o$, but not in any of the subsets $\mathcal{M}_o, \mathcal{C}_o, \mathcal{D}_o$, is given in Example 5.1.16.

The inclusion $\mathcal{A}_o \subset \mathcal{I}_o$ is strict, as is shown by the matrix

$$A = \begin{bmatrix} -1 & 0 & 50 \\ -1 & -1 & 0 \\ -1 & -1 & -1 \end{bmatrix},$$

which is D-stable, but not diagonally stable (Example 2.1.6). By Lemma 2.1.23 A is in $\mathcal{I}_o$. However, the characterization of diagonal stability for 3×3 matrices (Fact 2.8.2) shows that $A - \alpha I$ is not diagonally stable for $\alpha = 0.1$, so that $A \notin \mathcal{A}_o$.

Finally, the matrix

$$A = \begin{bmatrix} 0 & 1 & 0 \\ 0 & 0 & -1 \\ -1 & 0 & 0 \end{bmatrix},$$

which is clearly in $\mathcal{P}_o$, is not positive stable, and neither is $A + \alpha I$ for $\alpha \leq 0.5$, so that $-A$ is not in $\mathcal{I}_o$. This type of matrix was used in [FMM94] to illustrate the point that being in class $\mathcal{P}_o$ is not sufficient for absolute global asymptotic stability.

The complete picture, together with examples, is given in the Venn diagram in Figure 5.1.

It should be noted that obtaining a characterization of class $\mathcal{I}_o$ (algebraic or otherwise), is likely to be a hard problem since, so far, not even a characterization of class $\mathbb{D}_c$ is available for the general case. In terms of complexity, the characterization problem may well be NP- or co-NP-complete, since the problem of detecting whether a given matrix is in $\mathcal{P}$

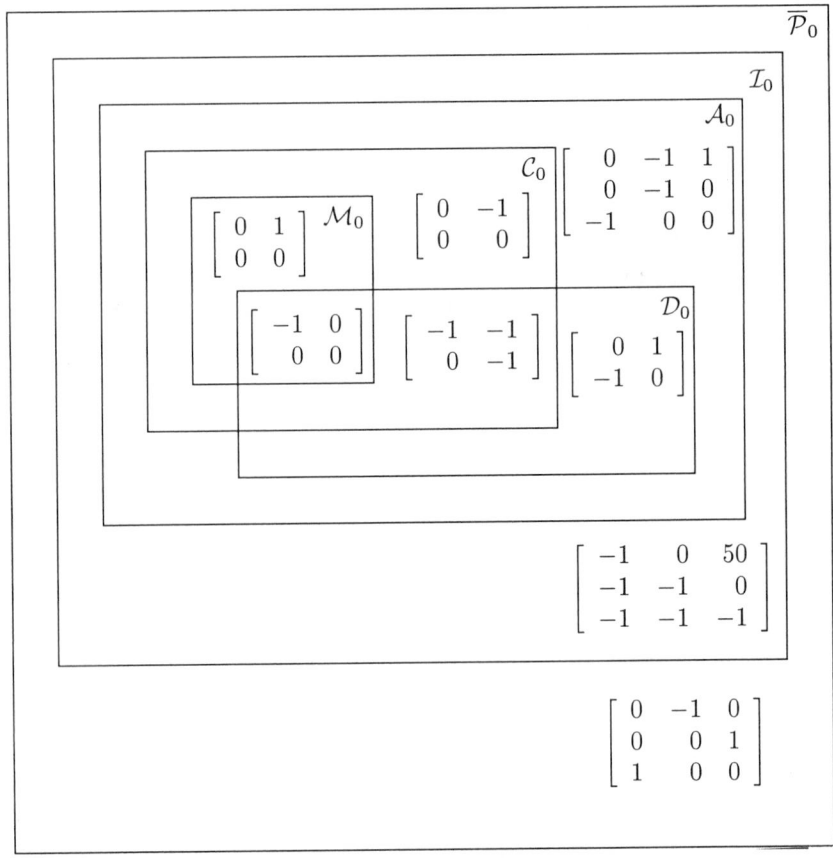

Figure 5.1. Venn diagram showing the relationship between different matrix stability concepts. The set identified by a letter is always to be taken as the largest rectangle that contains the letter in its upper right corner. The set $\overline{\mathcal{P}}_o$ is defined as $-\mathcal{P}_o$.

has been shown to be co-NP-complete [Cox94]. On the other hand, testing whether a matrix is in $\mathcal{D}_c$ or not has polynomial complexity [BGFB94].

5.1.3 Persistence of Global Asymptotic Stability

Looking at equation (5.6), it is natural to ask what can be said about the equilibrium solution if perturbations in the nonlinearities ϕ_i and in the parameters (matrices R, C, and T) occur. In the context of perturbations, it is useful to use the term *nominal* to refer to the unperturbed nonlinearities and parameters. Clearly, under the conditions of Theorem 5.1.4, global asymptotic stability is maintained under perturbations of $\phi_i(u_i)$s that keep the latter confined to the first and third quadrants as well as strictly monotonically increasing and Lipschitz-continuous. Actually, perturbations in both the nonlinearities and the parameters are tolerable under fairly mild conditions, namely, that the nominal interconnection matrix be diagonally stable and the nonlinear perturbations maintain the first-quadrant–third-quadrant, monotonic, Lipschitz-continuous character of the nonlinearities.

Under the additional hypothesis that the activation functions (and hence the ϕ_is) are C^1 (continuously differentiable), a more general answer to the question of persistence of stability under perturbations may be obtained by appealing to the concept of structural stability. Roughly speaking, given a C^1 vector field f, if for any sufficiently small perturbation of f (in the C^1 norm) that results in the map g, a homeomorphism exists that carries each trajectory of the original flow (of the differential equation $\dot{x} = f(x)$) onto a trajectory of the perturbed flow (of the differential equation $\dot{x} = g(x)$), then the vector field f is said to be *structurally stable* (see [HS74] for definitions and further details). Note that parameter and function perturbations are included as special cases of C^1 perturbations: if $g(x)$ is the result of a small C^1 perturbation of $h(x) := f(x, p)$ (where x denotes the state vector and p the parameter vector), all that is required is that $\|h(x) - g(x)\|$ and $\|Dh(x) - Dg(x)\|$ be small for all x in some open subset of $\mathbb{R}^n$.

The theorem below is needed to conclude structural stability of (5.6). In the sequel, let $D^n := \{x \in \mathbb{R}^n : \|x\| \leq 1\}$ and $\partial D^n := \{x \in \mathbb{R}^n : \|x\| = 1\}$.

Theorem 5.1.18 [HS74, p.314] *Let $f : W \to \mathbb{R}^n$ be a C^1 vector field on an open set $W \supset D^n$ with the following properties:*

(i) *f has exactly one equilibrium $0 \in D^n$ and 0 is a sink (i.e., all eigenvalues of $Df(0)$ have negative real parts);*

(ii) *f points inwards along the boundary ∂D^n of D^n, i.e., $\langle f(x), x \rangle < 0$, if $x \in \partial D^n$;*

(iii) *$\lim_{t \to \infty} \phi_t(x) = 0$, $\forall x \in D^n$, where ϕ_t is the flow of f.*

Then f is structurally stable on D^n.   □

Using the Liapunov function $W(u) = \frac{1}{2}u^T RCu$, which has a negative definite time derivative *outside* a ball B, item (ii) of Theorem 5.1.18 was proved in [SWC91, Theorem 3, p. 198], for (5.1), (5.2). However, since W is not guaranteed to be negative definite inside B, property (iii) cannot be concluded, hence neither can global asymptotic stability be concluded. Using the global Liapunov function V in (5.7) implies properties (ii) and (iii) for D^n, while (i) follows from the assumption that T is diagonally stable (see the proof of Theorem 5.1.4 and Lemma 2.1.4), so that Theorem 5.1.18 allows one to conclude that the Hopfield–Tank vector field, described by the right-hand side of equation (5.1) and equation (5.2), is structurally stable on D^n. Similar structural stability results can be obtained for the corresponding stability results in Section 5.1.2.

In summary, this section imposed the conditions of diagonal and additive diagonal stability on the interconnection matrix leading to a simple proof of the existence, uniqueness and global asymptotic stability of the equilibrium of the Hopfield–Tank neural circuit, generalizing some earlier results. More importantly, these conditions also lead to a proof of structural stability of the equilibrium solution, a concept that is arguably more appropriate than global asymptotic stability in the context of imperfect physical realizations of a neural circuit model since it ensures persistence of global asymptotic stability in the face of parameter perturbations, unequal scaling factors and (nonidentically) perturbed neuron activation functions.

In order to estimate a region of parametric stability around a given equilibrium value of the parameter vector, the following problem should be solved: Given $A \in \mathcal{D}_c$, determine the largest r such that for all ΔA such that $\|\Delta A\| \leq r$, in a given norm $\| \cdot \|$, $(A + \Delta A) \in \mathcal{D}_c$. Aspects of this problem were discussed in Section 2.4.

5.2   Discrete-Time Neural Networks

Takeda and Goodman [TG86] introduced two different continuous-valued discrete-time versions of the Hopfield–Tank neural network [Hop84]. They also introduced the concept of asynchronous transition modes in which "one particular neuron i need not wait for the last neuron n (in order) to decide its new state: It makes use of (available) information about states of other neurons that have already renewed their states." The motivation cited by Takeda and Goodman for the introduction of asynchronism was that their digital computer simulations showed that "asynchronous transition modes greatly reduced oscillatory or wandering behavior." From the point of view of modeling, since no evidence has been found for the existence of a central synchronizing clock in biological neural nets [Sej86, p.383], it is of interest to consider asynchronism in artificial neural networks. Additional motivation for the study of the Takeda-Goodman model is given in [Kun93, p.64 ff.] which mentions the recent advent of the switched capacitor device technology, which would facilitate implementation and the potential application to optimization problems.

The synchronous and asynchronous transition mode models of Takeda and Goodman are analyzed in this section with regard to existence, uniqueness, and stability of equilibria.

The synchronous model is written as follows :

$$x(k + 1) = Tf(x(k)) + (I - B)x(k) + u, \qquad (5.15)$$

where it is assumed that (i) the interconnection matrix T belongs to a class of Schur D-stable matrices, which includes, but is not restricted to, the class of symmetric stable matrices; (ii)f, in addition to being a diagonal and sigmoidal function, is continuously differentiable, with slope-limited component functions; and (iii) $B = I$, so that only the so-called direct transition mode is considered.

Under these assumptions, it is shown that equation (5.15) admits a unique locally asymptotically stable equilibrium. When the hypothesis on the interconnection matrix is strengthened to diagonal stability, a diagonal quadratic Liapunov function shows the global asymptotic stability of the unique equilibrium, using the results of Chapter 3.

This stability result is also extended to the asynchronous or desynchronized case and, as in Chapter 3, a diagonal-type Liapunov function is used to derive a condition that ensures global asymptotic stability under asynchronism, without the common assumption of symmetry of the interconnection matrix.

The synchronous case
Consider the synchronous transition mode model written as:

$$x(k+1) = Tf(x(k)) + u. \qquad (5.16)$$

Assumption 5.2.1 *The interconnection matrix $T = (t_{ij}) \in \mathbb{R}^{n \times n}$, the vector input $u \in \mathbb{R}^n$ and the nonlinearities f satisfy the following conditions:*

(i) the interconnection matrix T is Schur D-stable.

(ii) the input vector u is constant.

(iii) f is a slope-limited, real-valued diagonal map of class C^1, i.e., $f : (x_1, \ldots, x_n)^T \mapsto (f_1(x_1), \ldots, f_n(x_n))^T$; where: (a) for all i, the function $f_i(x_i)$ is continuously differentiable in x_i and tends to finite limits as x_i tends to $\pm\infty$; and, (b) for all i, for all $x_i \in \mathbb{R}$, $0 < (df_i(x_i)/dx_i) \le 1$.

Note that in item (iii) above the requirement on the nonlinearity f_i is a little stronger than merely requiring $f_i \in \mathcal{S}_d$.

The following lemma is needed in the sequel.

Lemma 5.2.2 [OR70, Result 5.3.9, p.137] *Suppose that $f : \mathbb{R}^n \to \mathbb{R}^n$ is continuously differentiable on all of $\mathbb{R}^n$, and that $f'(x)$ is nonsingular for all x in $\mathbb{R}^n$. Then f is a homeomorphism from $\mathbb{R}^n$ onto $\mathbb{R}^n$ if and only if $\lim_{\|x\| \to \infty} \|f(x)\| = \infty$.* $\square$

Using the above result, local stability is immediate.

Theorem 5.2.3 *Under the assumption 5.2.1: (a) the neural network represented by (5.16) admits an equilibrium $(x(k) = x^e$, for all $k)$ that is uniquely determined by the constant $u \in \mathbb{R}^n$ and depends continuously on u; (b) furthermore, this unique equilibrium is locally asymptotically stable.*

Proof. Existence and uniqueness follows by use of Lemma 5.2.2, and local asymptotic stability by linearization. ∎

A strengthening of assumption 5.2.1(i) on the interconnection matrix T, enables the derivation of the following global stability result:

Theorem 5.2.4 *Let assumptions 5.2.1 (i) and (ii) hold, and let the interconnection matrix T be Schur diagonally stable. Under these assumptions, the neural network represented by (5.16) admits an equilibrium ($x(k) = x^e$, for all k) that is uniquely determined by the constant $u \in \mathbb{R}^n$, depends continuously on u and is globally asymptotically stable.*

Proof.   Since diagonal stability of the matrix T implies its D-stability, Theorem 5.2.3 ensures existence and uniqueness of the equilibrium. To prove global asymptotic stability, observe that the dynamical equation (5.16) can be written in the form (3.32), after a shift to the unique equilibrium, and that hypotheses of Theorem 3.4.2 are satisfied.   ∎

For applications in which a unique equilibrium is of interest, such as analog-to-digital (A–D) conversion, the above theorem provides a design guideline as to the choice of an adequate interconnection matrix.

An interesting feature of the above theorem is that the global asymptotic stability persists under perturbations in the interconnection matrix (since the set of diagonally stable matrices is known to be an open set) as well as to perturbations in the activation functions (since, in order to apply Theorem 3.4.2, all that is required of the latter is that they satisfy the conditions $\phi_i(0) = 0$ and $|\phi_i(x_i(k))| \le |x_i(k)|$, i.e., belong to class $\mathcal{S}_d(k)$). In other words, the asymptotic stability is absolute in the sense defined in Chapter 1. Of course, it should be noted that, in order to maintain the uniqueness of the equilibrium, the perturbed activation functions must satisfy assumption 5.2.1 (iii).

The asynchronous case
Using the notation of Section 4.3, the asynchronous version of equation (5.16) is as follows:

$$x_i(k+1) = \sum_{j=1}^{n} t_{ij} f_j(x_j(d_{ij}(k))) + u_i, \qquad (5.17)$$

for $i = 1, \cdots, n$, and where the delays $d_{ij}(k)$ are subject to assumption 3.5.2. Using Theorem 3.5.4, global stability conditions under partial asynchronism can be deduced immediately.

Assumption 5.2.5 *The neural net (5.17) has the following properties.*

(i) *Each activation function $f_i(x_i)$ are Lipschitz-continuous, with constant ℓ_{f_i}, i.e.,*

$$|f_i(x) - f_i(y)| < \ell_{f_i}|x - y|;$$

(ii) *there exists a unique equilibrium.*

Note that if assumption 5.2.1 (iii) holds, then assumption 5.2.5 (i) holds with all the ℓ_{f_i} equal to 1, but the converse is not true, so that 5.2.5 (i) is the weaker assumption.

Theorem 5.2.6 *Consider the following cases:*

(a) assumption 5.2.1 (iii) holds and $|T| \in \mathcal{D}_d$.

(b) assumption 5.2.5 holds and $C \in \mathcal{D}_d$, where $c_{ij} := |t_{ij}|\ell_{f_j}$.

In both cases ((a) and (b)), the unique equilibrium of (5.17) is globally asymptotically stable for all admissible asynchronisms.

Proof. Follows immediately from Theorem 3.5.4 by calculation of the constants ℓ_{f_i}. ∎

This theorem also provides a guideline on the choice of an interconnection matrix for a neural network operating in the asynchronous mode and for which it is desired to have a unique globally asymptotically stable equilibrium for all threshold inputs. The assumption 5.2.1 (iii) has been relaxed to 5.2.5 (i). Note that an asynchronous convergence result equivalent to case (a) was proved in [TBT90].

Application and robustness issues
For certain applications, such as the design of analog-to-digital (A–D) converters [AFMM91, TH86, Vid92] and optimization [CU93], neural networks that admit a unique equilibrium are of interest. The results in this chapter provide a guideline for the design of such networks: Their interconnection matrices should belong to classes of diagonally stable matrices, depending on the mode of operation and the type of stability desired.

The design of a continuous-valued continuous-time analog to digital converter that is based on a Hopfield–Tank network with a strictly lower triangular interconnection matrix is described in [AFMM91]. Such matrices are both diagonally and D-stable in the discrete-time case as was shown in Chapter 2. Furthermore, the interesting point about their design is that it is not based on an energy function approach. It is observed that a certain structure (in this case triangular): (i) leads to a unique globally stable equilibrium, which implies correct operation of the A–D converter, since there are no so-called "spurious states"; (ii) is conducive to a particular design (A–D conversion). An inspection of the design equations for the A–D converter in [AFMM91] shows that they can be used for the Takeda–Goodman direct transition model (5.16) as well.

The discussion in the preceding paragraph and the results of this section lead to the following conclusions: (a) triangular matrices are just one example of the larger class of diagonally stable matrices that ensure a unique globally stable equilibrium; (b) designs which use diagonally stable matrices enjoy strong robustness properties. For example, a continuous-valued discrete-time implementation of the A–D converter circuit proposed in [AFMM91] could operate correctly under a large class of asynchronous modes and is not affected by small errors in implementation of the interconnection matrix and/or activation functions; (c) since, for both the synchronous and asynchronous case, a Liapunov function approach is used, it is possible to derive estimates for the rate of convergence of the circuit to its equilibrium (see Section 4.2.1). In other words, for the A–D converter example, one can derive estimates for the speed of conversion, which is useful for applications to time-varying signals.

Robustness of the conditions of Theorems 5.2.4 and 5.2.6 may be seen from Lemma 2.7.25 and from the fact that the set of diagonally stable matrices is open, so that small perturbations do not affect the property of diagonal stability. Now, since these theorems require diagonal stability of the interconnection matrix (or one derived from it), it is clear the theorems hold under small perturbations in the elements of the interconnection matrix T. This is analogous to the result in section 5.1.3 where the condition for structural stability was shown to be robust in a similar manner.

5.3 Passive RLC Circuits

In this section it is shown that there is a direct relationship between a class of diagonally stable and, in particular, sign-stable matrices and passive RLC circuits. Ladder circuits that are commonly used as passive filters provide examples as well as a physical interpretation of the robust stability property of sign-stable and more generally of diagonally stable matrices, for it is well known that passive RLC circuits are asymptotically stable for arbitrary values of the passive elements that constitute these circuits. In this respect, the use of an appropriate state-space description of the mentioned circuits, is enough to display the properties described above.

Using this state space description and Kalman's state-space formulation of passivity [Kal63] it is easy to certify that these circuits are passive in the sense of Kalman. One way to do this is to use the fact that sign-stable matrices belong to class $\mathcal{D}_c$ (see Lemma 2.2.14). This fact raises a natural and more general question: Are all linear dynamical systems with sign-stable system matrices, or, more generally, with diagonally stable system matrices passive? If true, this would provide a simple explanation for the strong robust stability properties enjoyed by these systems.

The discussion below will show that, in general, this is not true and thus the robust stability property associated with systems possessing a structure that can be associated with a matrix in class $\mathcal{D}_c$ is not merely a consequence of passivity.

Consider a ladder network in the classical representation [Hol69]. The series elements are given by their impedances (Z_i) and the parallel elements by their admittances (Y_i); from the Kirchhoff voltage and current laws, written alternately, the following network equations result:

$$
\begin{aligned}
Y_1 V_1 &= i_o & &+i_1 \\
Z_1 i_1 &= -V_1 & 0 \quad &+V_2 \\
Y_2 V_2 &= & -i_1 \quad 0 \quad &+i_2 \\
Z_2 i_2 &= & -V_2 \quad 0 \quad &+V_3 \\
&\vdots & &\ddots \\
Z_{q-1} i_{q-1} &= & -V_{q-1} \quad 0 \quad &+V_q \\
Y_q V_q &= & -i_{q-1} \quad & 0.
\end{aligned}
\tag{5.18}
$$

For most RLC networks, when the dynamical equations of system (5.18) are written down, this basic structure is preserved, i.e., the system retains the tridiagonal form. In the case of RLC networks, the system matrix is sign-stable while for RC networks, it is only diagonally stable.

One of the classical patterns of the ladder LC networks corresponding to "all pole" transfer functions is the so-called low pass filter circuit [Hol69]. By simple substitution of the corresponding elements in equations (5.18) one gets the state space representation below:

$$
\begin{aligned}
\dot{x} &= Ax + bu \\
y &= cx,
\end{aligned}
\tag{5.19}
$$

where $u = V_1$; $q = (n+1)/2$, for odd n, $x^T = (V_1, i_1, V_2, i_2, \ldots, i_{q-1}, V_q)$, $b^T = [(R_1 C_1)^{-1}, 0, 0, \ldots, 0, 0]$, $c = [0, 0, 0, \ldots, 0, 1]$, and where

$$
A = \begin{bmatrix}
-(R_1 C_1)^{-1} & +C_1^{-1} & & & & \\
-L_1^{-1} & 0 & +L_1^{-1} & & & \\
& -C_2^{-1} & 0 & +C_2^{-1} & & \quad O \\
& & & \ddots & & \\
& O & & -L_{q-1}^{-1} & 0 & +L_{q-1}^{-1} \\
& & & & -C_q^{-1} & 0
\end{bmatrix}.
$$

Since all the parameters R_i, L_i, C_i are positive, the sign pattern of matrix A is such that A is sign-stable and in addition is in Schwarz form (see Sections 2.2 and 2.3).

Another example of an RLC circuit corresponding to a sign-stable structure is related to the representation of port immittances $H(s)$ in ladder form. In these cases, given that the rational function $H(s) = n(s)/d(s)$ is positive real, the realization problem is that of obtaining an RLC circuit in a ladder form, with the corresponding state-space representation. The latter can be obtained using the algorithms discussed in Section 3.3 and the results are as follows:

$x^T = (V_1, i_1, \ldots, i_{q-1}, V_q)$, $b^T = [C_1^{-1}, 0, \ldots, 0]$, $c = [1, 0, 0, \ldots, 0]$,

$q = (n + 1)/2$, ; $n \geq 1$, odd, and

$$A = \begin{bmatrix} -(R_1C_1)^{-1} & +C_1^{-1} & & & & \\ -L_1^{-1} & -R_2L_1^{-1} & +L_1^{-1} & & & O \\ & -C_2^{-1} & -R_3C_2^{-1} & +C_2^{-1} & & \\ & & & \ddots & & \\ & & & -L_{q-1}^{-1} & & +L_{q-1}^{-1} \\ O & & & & -C_q^{-1} & -R_nC_q^{-1} \end{bmatrix}$$

and again matrix A is easily identified to be sign-stable (by Lemma 2.2.14).

Another case, similar to the one above, occurs when one sets $Z_i = R_i$ and $Y_i = C_i$ $(i = 1, 2, \ldots, n)$ in equation (5.18). The resulting system of equations describes an RC network:

$$\begin{aligned} \frac{d}{dt} C_1 V_1 &= & -i & +i_1 & \\ R_1 i_1 &= & -V_1 & 0 & +V_2 \\ \frac{d}{dt} C_2 V_2 &= & & -i_1 & 0 & +i_2 \\ &\vdots & & & \ddots \\ \frac{d}{dt} C_n V_n &= & & & -i_{n-1} & 0 \end{aligned} \qquad (5.20)$$

and using voltages (V_i) as state variables, the corresponding state-space representation is given by:

$$\begin{aligned} \dot{x} &= D^{-1}Ax + D^{-1}bu \\ y &= cx, \end{aligned} \qquad (5.21)$$

where $u = i$, $D = \text{diag}(C_1, C_2, C_3, \ldots, C_n)$, $x^T = (V_1, V_2, \ldots, V_n)$, $b^T = [1, 0, 0, \ldots, 0]$, $c = [1, 0, 0, \ldots, 0]$, and

$$A = \begin{bmatrix} -R_1^{-1} & +R_1^{-1} & & & \\ +R_1^{-1} & -K_2 & +R_2^{-1} & & O \\ & +R_1^{-1} & -K_3 & & \\ & & & \ddots & R_1^{-1} \\ O & & & +R_{n-1}^{-1} & -K_n \end{bmatrix}.$$

Also,

$$K_i = (R_{i-1} + R_i)/(R_{i-1}R_i); \ 1 < i \leq n.$$

Clearly matrix A is not sign-stable; however it is easily verified that this matrix satisfies the conditions $(-1)^{i+1}d_i(A) > 0$, for $i = 1, 2, \ldots, n$ where $d_i(A)$ are the leading principal minors of A and consequently (by Theorem 2.2.7) A is in class $\mathcal{D}_c$ (actually it is an M-matrix) for arbitrary values of the RC elements of the network.

An interesting feature of the examples discussed above is related to the existence of a quadratic diagonal Liapunov function obtained from the relationship (2.10) that, in turn, is directly associated with the energy stored in the inductors and capacitors of the circuit [Kal63], that is:

$$E = (1/2) \sum_{i=1}^{n_\ell} L_i x_i^2 + (1/2) \sum_{j=1}^{n_c} C_j x_j^2,$$

where x_i represents the current through ith inductor, and x_j is the voltage across jth capacitor. Other examples also show that the closer one gets to a true "energy" function of a given system, which is essentially diagonal, the closer one gets to a nonconservative stability condition.

The relationship between electrical circuits and sign-stable, or class $\mathcal{D}_c$, matrices is not limited to ladder structures, and is also useful in RC active filter design [KH80, RHK82]. The examples discussed above serve to illustrate the connection between a system known to be robustly stable from physical principles and the corresponding state space representation that is mathematically known to be diagonally, sign, and hence robustly, stable.

From the above discussion, the following question arises naturally: What is the relationship between diagonally stable linear systems and passive linear systems? The answer is provided by the state-space characterization of passive linear systems known as the Kalman–Popov–Yakubovich lemma, a version of which is given below.

Lemma 5.3.1 [Kal63] *Consider the dynamical system:*

$$\begin{aligned} \dot{x} &= Ax - Bu \\ y &= Cx \\ u &= \Phi(y). \end{aligned} \tag{5.22}$$

Let the matrices $A \in \mathbb{R}^{n \times n}$, $B \in \mathbb{R}^{n \times n_i}$, $C \in \mathbb{R}^{n_o \times n}$ be such that the pair (A, B) is controllable and the pair (A, C) is observable. Then, the following statements are equivalent.

(i) *There exists a symmetric positive definite matrix P and a symmetric nonnegative definite matrix Q such that:*

$$\begin{aligned} PA + A^T P &= -2Q \tag{5.23} \\ B^T P &= C. \tag{5.24} \end{aligned}$$

(ii) *The quadratic form $V = x^T P x$ is a Liapunov function for the time-invariant dynamical system (5.22) such that $\dot{V}(x) \leq 0$, where Φ is any continuous function such that $\Phi(0) = 0$ and, for all y, $y^T \Phi(y) \geq 0$.* □

If either of the equivalent statements above holds, then the linear system (A, B, C) is said to be passive.

Note that the matrix relationship (5.24) is fundamental in this discussion. Whenever it is assumed that matrix P in (5.23) is diagonal, then, in order to meet passivity conditions, matrices B^T and C must have the same structure since they differ only by a scaling positive diagonal matrix P. Therefore, in general, diagonally stable systems are not passive, since this property is strongly dependent on the input–output properties of the system, which, in turn, depend strongly on the system matrices B and C.

In the context of this discussion, the Persidskii system (3.5) may be viewed from the perspective of generalized Lur'e type systems of the type (5.22). In fact, taking $C = I$, $A = 0$ in (5.22), the matrix $-B$ represents the system matrix associated to the Persidskii form, i.e., $\dot{x} = -B\Phi(x)$, where $\Phi(x) \in \mathcal{S}_c^n$ and consequently $x^T\Phi(x) > 0$.

In this case, once again, the state-space system (A, B, C) belongs to the class of passive systems, provided that the crucial relationship (5.24) is satisfied. Clearly, in the case of an arbitrary Persidskii system, this relationship is not satisfied. Thus, although one can express a Persidskii-type system in the Lur'e form (5.22), Theorem 3.2.3 is not a consequence of passivity.

In particular, if the system matrix $-B$ is symmetric and Hurwitz, the corresponding Persidskii system happens to be passive with $V(x) = x^T B^{-1} x$ being the associated nondiagonal quadratic Liapunov function, and consequently

$$\dot{V}(x) = x^T B^{-1} \dot{x} = -x^T B^{-1} B\Phi(x) = -x^T \Phi(x)$$

is negative definite. The relationship (5.23) is satisfied for $A = 0$, and the relationship (5.24), in this particular case, is satisfied with $P = B^{-1}$, i.e., $B^T P = BB^{-1} = C = I$.

Another connection between passivity and Liapunov functions was derived in [BY89], where it was shown that the single-input, single-output, strictly proper system $\dot{x} = Ax + bu$, $y = c^T x$ is passive if and only if there exists a simultaneous quadratic Liapunov function for the set of matrices $\{A, -bc^T\}$. In addition, several other interesting results on these connections are also derived in [BY89].

5.4   Digital Filters in State-Space Description

In most applications of digital filtering, the original data or signal to be filtered are analog in nature. Thus an analog-to-digital converter is usually an integral part of a digital filtering system. Analog-to-digital conversion consists of the processes of sampling a continuous-time signal and converting the samples (real numbers) into a sequence of finite-length binary numbers. The error that occurs in analog-to-digital conversion is caused by the finiteness of the binary representation, since an exact conversion would require an infinite length, and is referred to as *quantization error*. This error depends on the type of binary representation of real numbers and several formats known by names such as two's-complement, one's-complement, and signed magnitude are used. Another phenomenon closely related to finite length representation is known as *overflow* and occurs when the number to be represented is larger (or smaller) than the largest (respectively, smallest) number representable in the chosen format. Overflows generally create large errors and one method commonly used to handle them is to reset to the largest (or smallest) number representable. This is also known as a *saturation characteristic* and is clearly of the sector type (belongs to $\mathcal{S}_d$). This is also true for the nonlinearities representing several types of quantization. The introduction of the overflow nonlinearity has an important consequence. For instance, with a two's-complement overflow characteristic, a disastrous effect can occur after an internal overflow: The output of the filter can (depending on the "poles" of the filter) become independent of the input sequence. This condition is called an *overflow oscillation* and, in more technical terms, is a *limit cycle oscillation* in the nonlinear dynamical system that results from the introduction of the overflow (saturation) nonlinearity into a linear digital filter. An excellent discussion of the points briefly mentioned above is in [RM87].

The problem of obtaining conditions that ensure the absence of undesirable limit cycles in zero-input digital filters with signal quantization has been studied for several years ([MM85, EM85, RM87, Mer94, Jac96] and references therein). There are recurring discoveries of asymptotic stability criteria for digital filters subject to nonlinearities which belong to a class of sector-bounded nonlinearities. In this section it is shown that the class of nonlinear time-varying discrete-time models discussed in Chapter 3 can, in particular, describe zero-input digital filters subject to most combinations of the commonly occurring nonlinearities that represent different classes of signal quantizations. The corresponding stability results of Chapter 3, based on Schur diagonal stability, are then immediately applicable.

The nonlinearities associated with quantization (truncation and rounding, but not two's-complement) and overflow operations are all sector functions belonging to the class $\mathcal{S}_d(k)$ (see [EM85] for a detailed discussion of these nonlinearities and the sectors to which they belong) and mathematical models associated with these types of digital filters are described by the following nonlinear time-varying discrete-time models analyzed in Chapter 3.

$$x_i(k+1) \;=\; Q_i\left(\sum_{j=1}^{n} a_{ij}x_j(k), k\right) \tag{5.25}$$

$$x_i(k+1) \;=\; \sum_{j=1}^{n} Q_{ij}(a_{ij}x_j(k), k) \tag{5.26}$$

for $i = 1, \ldots, n$, and where all nonlinearities Q belong to the class $\mathcal{S}_d(k)$, and are the scalar functions associated to the particular type of quantization. The above models represent, respectively, the quantization operations performed *after* and *before addition* of the state variables in the corresponding digital filter.

Thus, in order to study the problem of robust stability (seeking conditions for the absence of self-sustained oscillations), Theorems 3.4.4 and 3.4.3 can be utilized directly, and the conditions $|A| \in \mathcal{D}_d$ and $A \in \mathcal{D}_d$ guarantee (respectively) this stability.

Theorem 5.4.1 *If $A \in \mathcal{D}_d$ then the zero solution of (5.25) is globally asymptotically stable. In other words, if there is a positive diagonal matrix P for which $P - A^T P A$ is positive definite, then the only periodic solution for the system (5.25) is the identically zero solution and all other solutions tend to zero asymptotically.* □

Since $|A|$ is a nonnegative matrix, $\rho(|A|) < 1$ if and only if $|A| \in \mathcal{D}_d$, by Lemma 2.7.27. For the class of models (5.25) that describe quantization after addition, Theorem 5.4.1 is a generalization of the result in [MMR78] to the time-varying case.

Theorem 5.4.1 was stated and proved as Theorem 3.4.3 in the previous chapter. The condition $\rho(|A|) < 1$ implies the sufficient condition ($A \in \mathcal{D}_d$) for stability of the model (5.25) but is not implied by it and is therefore more restrictive. If only two's-complement overflow in direct form filters is considered, then the result is not conservative. This condition arises naturally as a sufficient condition for the stability of the zero solution of (5.26), using, once again, a diagonal Liapunov function. Theorem 5.4.2 was stated and proved as Theorem 3.4.4 and is restated below for ease of reference.

Theorem 5.4.2 *If the spectral radius of $|A|$ is less than one, or, equivalently, $|A| \in \mathcal{D}_d$, then the zero solution of (5.26) is globally asymptotically stable.* □

As far as state-space models of digital filters are concerned, neither (5.25) nor (5.26) need to be time-varying models. In addition, it is usual to assume that for all $i, Q_i = Q$ in (5.25) and that for all $i, j, Q_{ij} = Q$ in (5.26)—in other words, all quantizers, saturation nonlinearities, etc., possess exactly the same functional representation. Neither this assumption nor the assumption of time-invariance is used here, so that all results in this section are generalizations of results in [MMR78, VL87, Bro92].

If it is assumed that the zero solution of (5.26) is globally asymptotically stable *for all $Q_{ij} \in \mathcal{S}_d(k)$*, then the converse of Theorem 5.4.2 holds (see Theorem 3.4.26). In other words, if stability is to be maintained for the whole class of sector-bounded nonlinearities (and not exclusively for the subset of saturation, overflow, truncation, and rounding nonlinearities, cf. [Rit89]), then the sufficient condition also becomes necessary.

Stability criteria for direct and normal form digital filters
In order to fix ideas, asymptotic stability conditions are given below for the

most commonly occurring digital filters, namely, direct and normal form filters with two by two coefficient matrices.

In the direct or canonical form and in the normal form, the filter coefficient matrices are, respectively:

$$A_d = \begin{bmatrix} 0 & 1 \\ b & a \end{bmatrix}, \qquad A_n = \begin{bmatrix} \sigma & \omega \\ -\omega & \sigma \end{bmatrix}.$$

Using Condition 2.34 of Proposition 2.7.20, the following results are obtained:

Matrix	in $\mathcal{D}_d$ if and only if	
A_d	$|b| < 1,$	$|a| < \min\{1 - b, 1 + b\}$
$|A_d|$	$|b| < 1,$	$|a| < |1 - |b||$
A_n	$|\sigma^2 - \omega^2| - 1 < 0,$	$|\sigma| < (1 + \sigma^2 - \omega^2)/2$
$|A_n|$	$|\sigma^2 - \omega^2| - 1 < 0,$	$|\sigma| < |1 + \sigma^2 - \omega^2|/2$

A word of caution about the results in this section is in order here. Since the results here apply to a general class of systems, it is natural that they are more conservative than results that are derived for the specific nonlinearities that occur in those filters. The papers [BL91, DBR98], which use computer-aided exhaustive search, the paper [EM85], which uses computer-generated Liapunov functions tailored to fit the various combinations of nonlinearities that occur in digital filtering, and the book [LM94] contain good examples of such results. On the other hand, as pointed out in [Reg92], the use of direct form and normal form digital filters is so widespread that easy-to-use analytical stability information is also very useful for design procedures. Another remark pertinent here is that stability conditions of the type $|A| \in \mathcal{D}_d$ actually guarantee stability of "asynchronous" digital filters, i.e., filters that may have time-varying delays in the interconnections, as in the model (3.68).

5.5 Two-Dimensional (2D) Dynamical Systems

Two-dimensional (2D) dynamical systems are discrete-time systems in which the evolution of a vector, that can be thought of as the analog of the state vector of ordinary one-dimensional dynamical systems, is governed by an equation in which two integer indices occur. For instance, many vector variables that occur in digital image processing do not belong to the time domain, but rather to a two-dimensional spatial domain. Several methods are used to represent the operations that need to be performed on two-dimensional variables (often called signals in the jargon of signal and image processing). In this section, the focus is on a class of *partial difference equations* that represent the evolution of a two-dimensional dynamical system (which is often referred to as a *2D digital filter*). Once again, as in the case of one-dimensional digital filters studied in Section 5.4, a basic problem is to choose the filter parameters so that the zero solution is globally asymptotically stable, which implies the nonexistence of undesirable limit cycles, when the filter is implemented with the usual overflow and fixed-point nonlinearities. In this section the stability analysis of two-dimensional dynamical systems described by the so called Roesser model is carried out in terms of the diagonal stability of a matrix associated with this two-dimensional system.

The state-space Roesser model for a 2D dynamical system with zero input was first introduced in [GR72], further developed in [Roe75, KLMK77], and described by the equations below.

$$\begin{pmatrix} x^h(i+1,j) \\ x^v(i,j+1) \end{pmatrix} = \begin{bmatrix} A_{11} & A_{12} \\ A_{21} & A_{22} \end{bmatrix} \begin{pmatrix} x^h(i,j) \\ x^v(i,j) \end{pmatrix}, \qquad (5.27)$$

for $i \geq 0$, $j \geq 0$, and where $x^h \in \mathbb{R}^m$, $x^v \in \mathbb{R}^n$, $A_{11} \in \mathbb{R}^{m \times m}$, $A_{12} \in \mathbb{R}^{m \times n}$, $A_{21} \in \mathbb{R}^{n \times m}$, $A_{22} \in \mathbb{R}^{n \times n}$. Let

$$A := \begin{bmatrix} A_{11} & A_{12} \\ A_{21} & A_{22} \end{bmatrix} \in \mathbb{R}^{(m+n) \times (m+n)}, \quad x(i,j) := \begin{pmatrix} x^h(i,j) \\ x^v(i,j) \end{pmatrix} \in \mathbb{R}^{n+m},$$

and $D(\ell) = \{(i,j) : i + j = \ell, i \geq 0, j \geq 0\}$.

In the context of 2D image processing applications, the superscripts h and v are suggested by the terms "horizontal" and "vertical," respectively, while $D(\ell)$ refers to indices along a diagonal line in the (i,j) plane.

In the implementation of linear digital filters, signals are usually both represented and processed in a so-called finite wordlength format, which gives rise to nonlinear effects such as overflow and quantization. Applying a diagonal nonlinearity $f \in \mathcal{S}_d^{m+n}$ (this includes the overflow nonlinearities known as saturation and truncation), implies that the model (5.27) becomes:

$$\begin{pmatrix} x^h(i+1,j) \\ x^v(i,j+1) \end{pmatrix} = f\left(\begin{bmatrix} A_{11} & A_{12} \\ A_{21} & A_{22} \end{bmatrix} \begin{pmatrix} x^h(i,j) \\ x^v(i,j) \end{pmatrix} \right), \qquad (5.28)$$

for $i \geq 0$, $j \geq 0$. This model, as in the 1D filter case, represents the overflow or quantization operation performed *after* addition of the variables x^h and x^v.

A *finite set of boundary conditions* is also assumed: i.e., there exist two positive integers I and J such that

$$\begin{cases} x^h(i,0) = 0 & \text{for } i \geq I \\ x^v(i,0) = 0 & \text{for } i \geq I \\ x^h(0,j) = 0 & \text{for } j \geq J \\ x^v(0,j) = 0 & \text{for } j \geq J. \end{cases}$$

The main stability result for the Roesser model is as follows.

Theorem 5.5.1 [EAF79] *The zero solution of (5.28) is asymptotically stable if A is in $\mathcal{D}_d$.*

Proof. (outline only) Let $A \in \mathcal{D}_d(P)$ where it is assumed that $P = \text{diag}\,(P_1, P_2)$, where P_1 and P_2 are positive diagonal matrices in $\mathbb{R}^{m \times m}$ and $\mathbb{R}^{n \times n}$ respectively. Define the diagonal Liapunov function:

$$V(i,j) = V^h(i,j) + V^v(i,j) = (x^h(i,j))^T P_1 x^h(i,j) + (x^v(i,j))^T P_2 x^v(i,j).$$

Clearly $V(i,j) > 0$, when $x(i,j) \neq 0$ and $V(i,j) = 0$ only when $x(i,j) = 0$. Denoting

$$x_{11} := Ax(i,j); \qquad\qquad \overline{x}_{11} := f(Ax(i,j)),$$

$$V_{11} := V^h(i+1,j) + V^v(i,j+1) = \overline{x}_{11}^T(i,j) P \overline{x}_{11}(i,j)$$

$$\Delta V(i,j) := V_{11}(i,j) - V(i,j) = \overline{x}_{11}^T(i,j) P \overline{x}_{11}(i,j) - x^T(i,j) P x(i,j).$$

To show asymptotic stability, it first needs to be shown that $\Delta V(i,j)$ is nonpositive, and that it is zero only when $x(i,j) = 0$. Denoting the kth component of a vector $x(\cdot,\cdot)$ as $x_k(\cdot,\cdot)$, and writing out the expression for ΔV in full (and adding and subtracting the term $x_{11}^T P x_{11}$) gives:

$$\begin{aligned}
\Delta V(i,j) &= x_{11}^T(i,j) P x_{11}(i,j) - x^T(i,j) P x(i,j) \\
&\quad - \sum_{k=1}^{m} p_k^h [x_k^2(i+1,j) - f_k(x_k(i+1,j))^2] \\
&\quad - \sum_{k=1}^{n} p_k^v [x_k^2(i,j+1) - f_k(x_k(i,j+1))^2] \\
&= -x^T(i,j)(P - A^T P A)x(i,j) \\
&\quad - \sum_{k=1}^{m} p_k^h [x_k^2(i+1,j) - f_k(x_k(i+1,j))^2] \\
&\quad - \sum_{k=1}^{n} p_k^v [x_k^2(i,j+1) - f_k(x_k(i,j+1))^2],
\end{aligned}$$

where p_k^h and p_k^v are the kth diagonal elements of P_1 and P_2 respectively. Now since $A \in \mathcal{D}_d(P)$ and $f_k \in \mathcal{S}_d$, it follows that $\Delta V(i,j)$ is nonpositive, and that it is zero only when $x(i,j) = 0$. The proof is then completed by showing that V decreases along consecutive diagonal lines $D(\ell)$ and $D(\ell+1)$:

$$\sum_{(i,j) \in D(\ell)} \sum V^h(i,j) + V^v(i,j) \leq \sum_{(i,j) \in D(\ell+1)} \sum V^h(i,j) + V^v(i,j)$$

and equality holds only when $V(i,j) = 0, (i,j) \in (D(\ell) \cup D(\ell+1))$. From this it clearly follows that $V(i,j)$ tends to zero asymptotically as $m \to \infty$, which, in turn, means that $\lim_{(i \text{ or } j) \to \infty} x(i,j) = 0$. ∎

The Fornasini–Marchesini (FM) model and variants

The zero input linear Fornasini–Marchesini [FM78] state-space model below is another popular 2D system model:

$$x(i+1, j+1) = Ax(i+1, j) + Bx(i, j+1) + Cx(i, j). \tag{5.29}$$

It is also used to model 2D state-space digital filters [BJ90], discretization of some partial differential equations with initial and boundary conditions [Kac85], etc.

Several methods have been proposed to study the stability of the model (5.29): for example, using the so-called 2D eigenvalues, [Kac85, Chapter 4], [FM76, FM78], etc.; using Liapunov functions, [Hin93, Lu94], etc., and using infinity or generalized infinity norms, or the usual 1D eigenvalues.

Actually, a diagonal stability condition also ensures the asymptotic stability of the zero solution of this class of systems.

Theorem 5.5.2 [BT92, SB98] *Consider the 2D linear system*

$$x(i+1, j+1) = Ax(i+1, j) + Bx(i, j+1) + Cx(i, j),$$

where $x \in \mathbb{R}^{n \times 1}$, $A, B, C \in \mathbb{R}^{n \times n}$ *and* i, j *are positive integers. The initial conditions satisfy* $\lim_{i \to \infty} x(i, 0) = 0$ *and* $\lim_{j \to \infty} x(0, j) = 0$. *If*

$$(|A| + |B| + |C|) \in \mathcal{D}_d, \tag{5.30}$$

then $x(i, j) \to 0$ *as* $i, j \to \infty$.

Proof. See the proof of the more general Theorem 5.5.3.   □

Consider a more general 2D FM model with shifts and quantization type nonlinearities. Suppose $C = 0$, for simplicity, so that the model considered is the following:

$$x(i+1, j+1) = Q(Ax(i+1, j - d_2(i, j)) + Bx(i - d_1(i, j), j+1)), \tag{5.31}$$

where, for all i, j, $d_1(i, j)$, and $d_2(i, j)$ are positive integers, referred to as variable shifts, since both depend on the indices i, j; and the function $Q(x) = (q_1(x_1), \cdots, q_n(x_n))^T \in \mathcal{S}_d^n$ represents one of the nonlinearities occurring in digital filtering, as discussed in Section 5.4. The 2D system (5.31) is assumed to have boundary conditions satisfying:

$$\lim_{i \to \infty} x(i, 0) = \lim_{j \to \infty} x(0, j) = 0. \tag{5.32}$$

Examples of other boundary conditions found in the literature are as follows. Finite memory boundary conditions

$$x(i, 0) = x(0, j) = 0 \quad \text{for } i, j \text{ large enough} \tag{5.33}$$

are used in most papers related to 2D models and the finite sum boundary conditions

$$\sum_i \|x(i, 0)\| < +\infty, \text{ and } \sum_j \|x(0, j)\| < +\infty, \tag{5.34}$$

are also used to prove asymptotic stability using a Liapunov function along the diagonal lines of the (i, j) lattice. Obviously, the finite memory boundary conditions imply the finite sum boundary conditions which, in turn, imply the asymptotic stability boundary conditions assumed in this section.

The stability definition used here is that of asymptotic stability of the zero solution of the dynamical system (5.31) [LM94].

The motivation for introducing shifts into 2D models is that they arise, for example, in asynchronous iterative methods for the solution of partial differential equations [AAI$^+$93]. The models with shifts and the corresponding stability results are natural generalizations of known results for 1D systems (the shifts are usually referred to as delays in the 1D case: cf. Sections 3.5 and 3.6).

Condition (5.30) actually ensures asymptotic stability of the zero solution of the FM model (5.31) with variable shifts and sector nonlinearities. The variable shifts are assumed to be bounded, i.e., there exists a positive integer M such that

$$0 \le d_1(i, j) \le M - 1; \qquad 0 \le d_2(i, j) \le M - 1. \qquad (5.35)$$

Some notation is needed to proceed further. The partial order relations $\preceq$ and $\prec$ defined in Section 2.1 are extended in the obvious way by interpreting objects with double indices (i, j) as one by two matrices and applying the same partial order relations to them.

Recall that, for a positive vector $v \succ 0$, a weighted infinity norm is defined as:

$$\|x\|_v = \max_i \frac{|x_i|}{v_i},$$

where x_i is the ith component of the vector x, and furthermore that

$$\|x\|_v \le \beta \text{ if and only if } |x| \preceq \beta v.$$

Theorem 5.5.3 *Consider the 2D system (5.31) with variable shifts, subject to boundary conditions satisfying (5.32). Under the condition*

$$(|A| + |B|) \in \mathcal{D}_d, \qquad (5.36)$$

the zero solution of (5.31) is asymptotically stable.

Proof. Since $(|A| + |B|) \in \mathcal{D}_d$, from Lemma 2.7.25(vi), there exists a positive vector $v \succ 0$ and a scalar α: $0 \le \alpha < 1$, such that

$$(|A| + |B|)v \preceq \alpha v.$$

For the positive integer M, define a sequence $S(k)$ of sets of indices for $k = 1, 2, \ldots$ as follows:

$$S(k) = \{(i, j) \mid 1 \le i \le kM, \ (k-1)M + 1 \le j \le kM \\ \text{or } 1 \le j \le kM, \ (k-1)M + 1 \le i \le kM\}$$

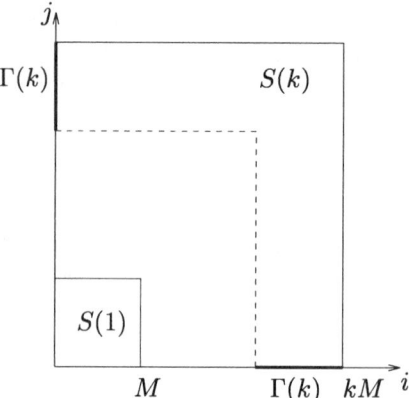

Figure 5.2. The sets $S(k)$ and $\Gamma(k)$.

and a sequence of sets of boundary indices for $k = 1, 2, \ldots$ as follows (see Figure 5.2):

$$\Gamma(k) = \{(i, 0), (0, j) \text{ such that } (k-1)M + 1 \le i, j \le kM\}.$$

For $k = 1, 2, \ldots$, define the diagonal-type Liapunov functions:

$$\beta(k) := \max_{(i,j) \in S(k)} \|x(i, j)\|_v,$$

$$\gamma(k) := \max_{(i,j) \in \Gamma(k)} \|x(i, j)\|_v,$$

and let

$$c(k) := \max\{\beta(k), \gamma(k), \gamma(k+1)\}.$$

Let $(i_0 + 1, j_0 + 1)$ be an index in $S(k+1)$. From (5.35), it is clear that $(i_0 + 1, j_0 - d_2(i_0, j_0))$ and $(i_0 - d_1(i_0, j_0), j_0 + 1)$ are indices in $S(k+1) \cup S(k) \cup \Gamma(k+1) \cup \Gamma(k)$. Suppose that for all (i, j) such that $(i, j) \in S(k+1)$ and $(i, j) \prec (i_0 + 1, j_0 + 1)$, the inequality $\|x(i, j)\|_v \le c(k)\alpha$ holds. Then

$$
\begin{aligned}
|x(i_0 + 1, &j_0 + 1)| \\
&= |Q\left((Ax(i_0 + 1, j_0 - d_2(i_0, j_0)) + Bx(i_0 - d_1(i_0, j_0), j_0 + 1))\right)| \\
&\preceq |Ax(i_0 + 1, j_0 - d_2(i_0, j_0)) + Bx(i_0 - d_1(i_0, j_0), j_0 + 1)| \\
&\preceq |A| |x(i_0 + 1, j_0 - d_2(i_0, j_0))| + |B| |x(i_0 - d_1(i_0, j_0), j_0 + 1)| \\
&\preceq c(k)|A|v + c(k)|B|v \\
&\preceq c(k)\alpha v,
\end{aligned}
$$

which is equivalent to

$$\|x(i_0 + 1, j_0 + 1)\|_v \le \alpha c(k).$$

So, by induction on $S(k+1)$ in both indices i and j, it follows that for all $(i,j) \in S(k+1)$, $\|x(i,j)\|_v \leq \alpha c(k)$, i.e.,

$$\beta(k+1) \leq \alpha c(k). \tag{5.37}$$

Now, without loss of generality, suppose that $\gamma(k+1) \leq \gamma(k)$. Otherwise, use

$$\tilde{\gamma}(k) = \max\{\gamma(k), \gamma(k+1), \ldots\}$$

to replace $\gamma(k)$. Thus (5.37) can be written as

$$\beta(k+1) \leq \alpha \max\{\beta(k), \gamma(k)\}. \tag{5.38}$$

Expanding (5.38) for $k = 1, 2, \cdots$ gives

$$\begin{aligned}
\beta(1) &= \beta(1); \\
\beta(2) &\leq \max\{\alpha\beta(1), \gamma(1)\}; \\
\beta(3) &\leq \max\{\alpha^2\beta(1), \alpha\gamma(1), \gamma(2)\}; \\
&\vdots \\
\beta(k+1) &\leq \max\{\alpha^k\beta(1), \alpha^{k-1}\gamma(1), \ldots, \alpha\gamma(k-1), \gamma(k)\}; \\
&\vdots
\end{aligned}$$

For any arbitrary fixed $\epsilon > 0$, there exists some K (which depends on ϵ) such that:

$$\gamma(k) < \epsilon, \quad \alpha^k < \epsilon \quad \text{for all } k \geq K,$$

and therefore, for all $k \geq 2K - 1$,

$$\begin{aligned}
\beta(k+1) &\leq \max\{\alpha^k\beta(1), \alpha^{k-1}\gamma(1), \ldots, \alpha^K\gamma(k-K), \\
&\qquad \alpha^{K-1}\gamma(k-K+1), \ldots, \gamma(k)\} \\
&\leq \max\{\epsilon\beta(1), \epsilon\gamma(1), \epsilon\} \\
&= \epsilon \max\{\beta(1), \gamma(1), 1\}.
\end{aligned}$$

Thus $\lim_{k\to\infty} \beta(k) = 0$. ∎

Condition (5.36) is a generalization to the 2D case of the Chazan–Miranker condition for asymptotic stability of the zero solution of asynchronous or 'chaotic' 1D systems discussed in Chapter 4. In other words, variable shift 2D systems can also be thought of as asynchronous 2D systems. It is probable that many characteristics of 1D asynchronous systems are also true of 2D systems as well, although this is an area where little research has been done so far. Condition (5.36) is also necessary in the following sense: if $B = 0$, and $\rho(|A|) \geq 1$, then there exists a 2D system (with a specially chosen sequence of shifts and boundary conditions) for which convergence to the zero solution does not occur. Note that, since the stability condition derived here (condition (5.36)) actually coincides with the stability condition derived above for the unshifted system, it can be concluded that the latter is a very conservative stability condition (for the unshifted systems).

The proof of Theorem 5.5.3 is also valid for the stability of the more general shifted 2D system:

$$\begin{aligned}
x(i+1, j+1) \quad = \quad &Q(Ax(i - d_1(i,j), j - d_2(i,j)) \\
&+ Bx(i - d_3(i,j), j - d_4(i,j)) \\
&+ Cx(i - d_5(i,j), j - d_6(i,j))),
\end{aligned}$$

under the conditions that all shifts are bounded and the condition $(|A| + |B| + |C|) \in \mathcal{D}_d$, thus justifying the claim that it includes Theorem 5.5.2 as a special case.

5.6 Trophic Chains and Communities with Vertical Structure

This section discusses the stability problem of a variant of the Lotka–Volterra model introduced in Section 1.3. For this variant, however, the nonlinear change of variables used in Section 1.3 does not lead to a Persidskii-type system. Matrix diagonal stability does nevertheless have a useful role to play in obtaining a local stability result, given the fact that the linearized basic system matrix is tridiagonal and, for this class of systems, diagonal solutions to the Liapunov equation are known, as discussed in Section 2.8. First, a brief explanation of the biology leading to the model is given, closely following [SL83].

A community structure that involves the transfer of energy, trapped in food, from one species to another through predator–prey type relationships, is known in ecology as a *trophic chain*.

Trophic chains are not isolated from each other and, taking interconnections into account, they form a trophic graph, also called a web. An example of a trophic web is the ecosystem of a small stream. It is an open ecosystem with a part of the basic resource (energy) coming from fallen leaves (1) and other organic residues (2), carried by the flow of the stream. The system has three trophic levels: the species green algae (3) and diatoms (4) make up the producers level; the *Protonemura* species (5), blackflies and nonbiting midges (6), caddisfly (*Philopotamus*) (7), and mayfly (*Ecdyonurus*) (8) constitute the level of primary consumers, while stonefly (*Perla*) and caddisfly (*Dinocras*) (9) make up the secondary consumers level. The caddisfly species (9), which build a catching net, and the caddisfly (*Rhyacophila*) occupy an interstitial level. Evidently, the sequences of species $3 \rightarrow 6 \rightarrow 12$ or $2 \rightarrow 7 \rightarrow 11$ form trophic chains (see Figure 5.3).

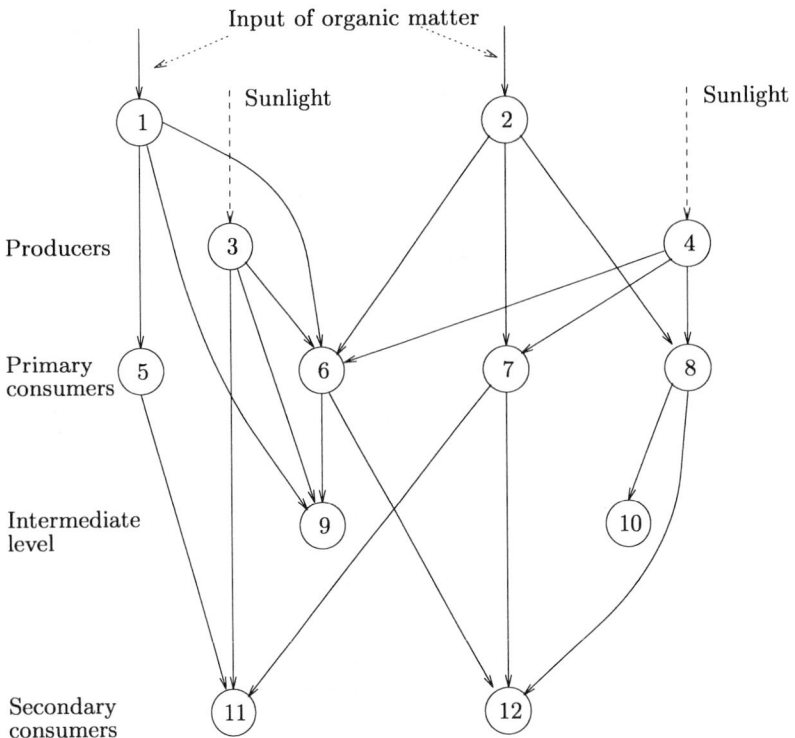

Figure 5.3. A trophic chain with three trophic levels. Producers level, (1): fallen leaves, (2): organic residues, (3): green algae, (4): diatoms; Primary consumers level, (5): *Protonemura* species, (6): blackflies and nonbiting midges, (7): caddisfly (*Philopotamus*), and (8): mayfly (*Ecdyonurus*); secondary consumers level, (9): stonefly (*Perla*) and caddisfly (*Dinocras*). The caddisfly species (9), which build a catching net, and the caddisfly (*Rhyacophila*) occupy an interstitial level. The sequences of species 3 → 6 → 12 or 2 → 7 → 11 form trophic chains. From [Jon49].

A trophic chain is thus a community in which every two adjacent species make up a prey-predator pair and is often also called a *vertically structured community*.

The notion of resource in real ecosystems is taken to be either energy or a vitally important substance of some kind such as carbon, nitrogen, or phosphorus. In respect of energy, any ecosystem is open. For example, sunlight flows through the system and disperses at the same time as heat (respiration, vital activity, etc.). As far as biological substances are concerned, ecosystems are closed to some extent, since a certain closure is performed by the decomposers (microorganisms, fungi, worms), which decompose dead organic material into mineral components, thus providing nutrients to the primary trophic levels. The same is true of trophic chains. There are two types: open (flow-through) and closed (cyclic).

To model these systems, let r be a resource, consumed first by a species of biomass x_1. The specific rate of uptake $f_0(r)$ is the amount of resource consumed by a unit biomass of the first species (one species) in unit time. Only the k_1th fraction of the total consumed resource $f_0(r)$ goes into the reproduction of the first species biomass, the rest being needed to maintain vital activity. In addition, the first species biomass dies at a constant rate m_1, serves as a resource for a second species, and is consumed at a specific rate $f_1(x_1)$; a third species consumes the biomass of the second and so on. The chain comes to an end with the nth species, whose biomass is not consumed by any species. The functions f_i are also known as *trophic functions* and model the predator response to the prey population density. Figure 5.4 shows block diagrams (schematic diagrams) of open and closed chains.

Suppose that the whole ecosystem approaches some state of equilibrium in which the steady-state population of only the first j species is nonzero. In this state of stable equilibrium, the ecosystem is termed a *trophic chain of j members*.

Let the supply rate of the external resource to the ecosystem be constant and equal to q. The first question of interest is as follows: What should this rate be (for given trophic functions $f_0(r)$, $f_i(x_i)$, and parameters k_i, m_i and a_i) for this community to have a stable nontrivial equilibrium with nonzero populations for the first j species? In other words, what are the conditions for a trophic chain of j members to exist?

The mass–balance equations for species in an open chain are the following:

$$\begin{aligned}
\dot{r} &= q - f_0(r)x_1 \\
\dot{x}_1 &= -m_1 x_1 + k_1 f_0(r)x_1 - f_1(x_1)x_2 \\
\dot{x}_i &= -m_i x_i + k_i f_{i-1}(x_{i-1})x_i - f_i(x_i)x_{i+1} \\
\dot{x}_n &= -m_n x_n + k_n f_{n-1}(x_{n-1})x_n,
\end{aligned} \tag{5.39}$$

for $i = 2, 3, \cdots, n-1$.

For a closed chain, the equations are all the same except for the equation for $\dot{r}$, which has the additional input $\sum_{i=1}^{n} a_i m_i x_i$ on the right-hand side. From biological considerations, clearly the parameters a_i, k_i lie in the interval $(0, 1]$. If it is assumed that the predator is always hungry, this means that the trophic functions can be assumed to be linear, which is reasonable for small prey populations. Thus, setting

$$f_0(r) = \alpha_0 r, \quad f_i(x_i) = \alpha_i x_i$$

for $i = 1, \cdots, n$, the mass–balance equations become the Lotka–Volterra equations (except for the first equation, which contains the term q), and they are also called *almost Volterra* equations. Now using the notational convention $r = x_0$ and $x_{n+1} = 0$, one gets the following dynamical equations for open and closed trophic chains:

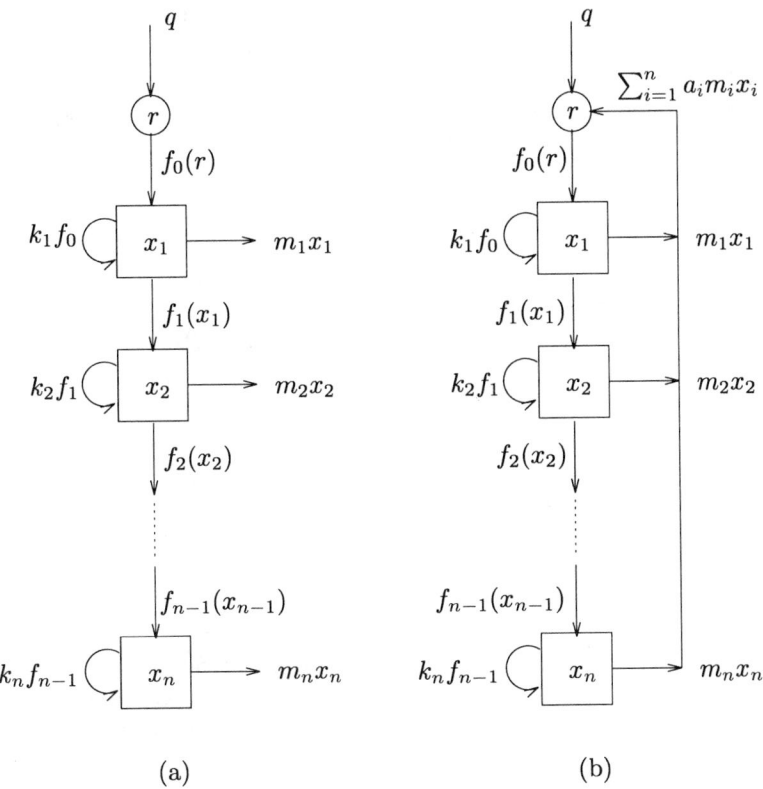

(a)   (b)

Figure 5.4. (a) An open or "flow-through" trophic chain with n members. (b) A closed trophic chain (or cycle) with n members. The coefficients a_i $(0 \leq a_i \leq 1)$ are the fraction of the resource reproduced by the destructors present in the dead biomass of the ith species. The coefficients specify the degree of closure in the trophic chain.

Open chain

$$\dot{x}_0 = q - \alpha_0 x_0 x_1$$
$$\dot{x}_i = x_i(-m_i + k_i \alpha_{i-1} x_{i-1} - \alpha_i x_{i+1}) \tag{5.40}$$

for $i = 1, \cdots, n$.

Closed chain

$$\dot{x}_0 = q - \alpha_0 x_0 x_1 + \sum_{i=1}^n a_i m_i x_i$$
$$\dot{x}_i = x_i(-m_i + k_i \alpha_{i-1} x_{i-1} - \alpha_i x_{i+1}), \tag{5.41}$$

for $i = 1, \cdots, n$.

For both open and closed chains, assuming that $q > 0$, it is possible to calculate conditions for the existence of equilibria of the form

$$(x_0^*, \ x_1^*, \ \cdots \ , x_j^*, \ 0, \ \cdots \ 0), \tag{5.42}$$

and such formulas are worked out in detail for odd and even values of j in [SL83]. Assuming that the values of x_i^* have been evaluated, the key question is to determine when such an equilibrium is asymptotically stable. In this section, this analysis will be pursued for linearized versions of these chains, thus leading to local stability results. For open chains the linearized system matrix has the following form

$$F^o = \begin{bmatrix} A_j & 0 \\ 0 & D_{n-j} \end{bmatrix}, \tag{5.43}$$

where A_j is a tridiagonal matrix

$$A_j := \begin{bmatrix} -b_0 & -d_0 & & & & \\ b_1 & -h_1 & -d_1 & & O & \\ & \cdot & \cdot & \cdot & & \\ & & \cdot & \cdot & \cdot & \\ O & & \cdot & \cdot & \cdot & \\ & & b_{j-1} & -h_{j-1} & -d_{j-1} \\ & & & b_j & -h_j \end{bmatrix} \tag{5.44}$$

with entries defined as: $b_0 = \alpha_0 x_1^*$, $b_i = k_i \alpha_{i-1} x_i^*$, $d_0 = \alpha_0 x_0^*$, $d_i = \alpha_i x_i^*$, $h_i = 0$, for $i = 1, \cdots, j$ and the negative diagonal matrix $D_{n-j} :=$ diag $(-m_{j+1} + k_{j+1}\alpha_j x_j^*, -m_{j+2}, \cdots, -m_n)$. The fact that the first entry of the matrix D_{n-j} is negative follows from the reasonable assumption that there is an upper bound on the size of the jth equilibrium population: $x_j^* < m_{j+1}/(\alpha_j k_{j+1})$. In fact, from this it is not hard to show that the necessary and sufficient condition for an open trophic chain with j members to exist is that the supply rate q of the external resource be bounded from above and below [SL83].

For closed chains the linearized system matrix is modified as follows:

$$F^c = \begin{bmatrix} A_j^1 & C \\ 0 & D_{n-j} \end{bmatrix}, \tag{5.45}$$

where A_j^1 is no longer tridiagonal, but defined as

$$A_j^1 := \begin{bmatrix} -b_0 & c_1 - d_0 & c_2 & \cdots & c_j \\ b_1 & 0 & -d_1 & \cdots & \\ & & \ddots & & \ddots & O \\ O & & b_{j-1} & 0 & d_{j-1} \\ & & & b_j & 0 \end{bmatrix}, \quad C := \begin{bmatrix} c_{j+1} & c_{j+2} & \cdots & c_n \\ & O & \end{bmatrix}, \tag{5.46}$$

where $c_i := a_i m_i > 0$.

At this point it is possible to apply the conventional analysis and calculate the eigenvalues of the matrices F^o, F^c and find the conditions under which all have real negative part. Alternatively, the results of Section 2.2 can be used, taking advantage of the structure of the linearized system.

It is easily seen that matrix A_j is structurally similar to the tridiagonal Schwarz form. Thus, a direct application of the arguments of Section 2.3 to this system shows that F^o is diagonally semistable ($F^o \in \mathcal{D}_{c2} \cap \mathcal{D}_{co}$) and asymptotic stability can be concluded using LaSalle's theorem.

In the case of closed chains, although the linearized overall system matrix F^c is no longer tridiagonal, structural information can still be taken advantage of by rewriting the linearized system in incremental form:

$$\dot{y} = \tilde{A}_j y, \qquad y_i := x_i - x_i^*; \quad i = 0, \cdots, j. \tag{5.47}$$

The following notation is needed: $z, b, c^T \in \mathbb{R}^j$ and $B \in \mathbb{R}^{j \times j}$ are defined as follows. $z := (y_1 \ y_2 \ \cdots \ y_j)^T$, $c := (c_1 - d_0 \ c_2 \ \cdots \ c_j)^T$, $b = (b_1 \ 0 \ \cdots \ 0)^T$ and

$$B := \begin{bmatrix} -h_1 & -d_1 & & & & \\ b_2 & -h_2 & -d_2 & & \mathcal{O} & \\ & \cdot & \cdot & \cdot & & \\ & & \cdot & \cdot & \cdot & \\ & & & \cdot & \cdot & \cdot \\ \mathcal{O} & & b_{j-1} & -h_{j-1} & -d_{j-1} \\ & & & b_j & -h_j \end{bmatrix}.$$

Consequently the system $\dot{y} = \tilde{A}_j y$ can be written as:

$$\begin{aligned} \dot{z} &= Bz + y_0 b \\ \dot{y}_0 &= cz - b_0 y_0. \end{aligned} \tag{5.48}$$

The above decomposition of the closed chain system is motivated by the observation that the subsystem matrix B is tridiagonal, with the property that the sub- and superdiagonal entries satisfy the inequality $d_i b_{i+1} < 0$. If it can be assumed that all the h_is are positive, then Corollary 2.2.11 shows that $B \in \mathcal{D}_{c2}$, with $B^T P + PB = -Q$, where $P = \text{diag}(p_1, \cdots, p_j)$ and Q are positive diagonal matrices. Taking advantage of this, the diagonal Liapunov function, based on the tridiagonal part of system (5.48), is

$$V(z, y_0) := z^T P z + (1/2) y_0^2 \tag{5.49}$$

and shows the asymptotic stability of the zero solution of (5.48), and hence the local asymptotic stability of the equilibrium solution of (5.41). In [SL83], the explicit calculations showing the truth of these claims are carried out.

For most closed chains, however, the h_is are all zero, and in these cases, the same Liapunov function as above works, with an additional assumption: $c_1 \neq 0$, $c_2 = \cdots = c_j = 0$. This ensures that the matrix $\tilde{A}_j$ is still acyclic-3 and in class $\mathcal{D}_{co}$. Notice that under the condition that D_{n-j} is a negative diagonal matrix, the stability of $\tilde{A}_j$ implies that of F^c. Calculating the time derivative of the function (5.49), under these conditions, one gets:

$$\dot{V} = (p_1 b_1 - 0.5(d_0 - c_1)) y_0 y_1 - b_0 y_0^2.$$

The choice $p_1 = (d_0 - c_1)/(2b_1) > 0$ (since $d_0 - c_1$ is known to be positive) makes $\dot{V}$ negative semidefinite, and LaSalle's theorem can be used to show the asymptotic stability of the zero solution $(z(t), y_0(t)) \equiv 0$ of system (5.48).

The aim in this section was to show that trophic chains with vertical structure have a natural basic tridiagonal structure that can be made use of in the context of matrix diagonal stability. Using a diagonal-type Liapunov function which is directly related to this structure has the following advantages: (i) The conclusion reached is qualitative and does not depend on the exact values of the entries of B, but rather just on their signs; (ii) estimates of domains of attraction of the equilibrium may be obtained from the level sets of this Liapunov function [Kha92].

5.7   Notes and References

Conditions for no limit cycles in digital filters

For the model (5.25), the condition $A \in \mathcal{D}_d$ was derived earlier in [MMR78] in the time-invariant case, and, for the model (5.26) ($|A| \in \mathcal{D}_d$), in [KHH84]), using diagonal Liapunov functions. A discussion of recurrent discoveries of stability conditions derivable from the diagonal stability condition is in [KB92], on which Section 5.4 is based.

Global stability results for continuous-time neural networks

In the context of robustness of the global stability result derived in this chapter, observe that the condition of diagonal stability of the interconnection network also guarantees structural stability which, in turn, ensures parametric asymptotic stability. The latter concept was introduced in [IOŠ90] to deal with the joint problem of existence and stability of equilibrium states in nonlinear systems with uncertain parameters, the case of artificial neural networks being mentioned as an example of such a system.

The result in [For94, Theorem 4 (i)] is essentially the same as the absolute stability part of Theorem 5.1.7. Related earlier results on global asymptotic stability of neural Hopfield-Tank circuits are as follows. The result in [Ros88] ensures global asymptotic stability under the condition that the interconnection matrix T is off-diagonally isotone (i.e., $t_{ij} \geq 0, i \neq j$, $t_{ii} < 0$ and $-T$ is row-sum dominant). These conditions are known to imply that T is an M-matrix [FP62, Theorem 3,4, p.387] and such matrices constitute a well-known subclass of class $\mathcal{D}_c$ [Moy77, Theorem 3,p.57]. Thus Theorem 5.1.7 generalizes the result in [Ros88]. The result in [Mat92] relies on a particular (unweighted) Persidskii-type Liapunov function.

The scaling factor $\lambda > 0$ does not alter the analysis in Section 5.1.1; increasing λ only makes the functions $g_i(\cdot)$ and consequently $\phi_i(\cdot)$ steeper, which does not affect the stability results. In the context of global stability of neural circuits, the idea of looking at the class of additively diagonally stable matrices $\mathcal{A}_1 = \mathcal{A}_o$ was first noted in [KB94, Remark 4, p. 173], and then subsequently by other authors [FT95, FK98].

The results of [Mat92, FK98, LW98b, LW98a] are easily seen to be special cases of Theorem 5.1.7, which, in turn, is contemplated in Conjecture 5.1.8, first made in [KB95].

Discrete-time neural networks

The model (5.15) introduced in [TG86] has been studied in the synchronous parallel updating mode in [MW89, Hor95] and in [TBT90, HM93, Koi94, BKK96] in the asynchronous updating mode. A result similar to Theorem 5.2.6 was derived in [TBT90], using different methods and slightly different hypotheses.

Additive Schur diagonal stability is also useful in getting less conservative global stability results and has been used (though not with this name) in [BK95, JGN95].

The stability of continuous-time, continuous-state Hopfield–Tank neural networks with delays in the interconnections has also been studied using diagonal-type Liapunov functions. In [HQ98], a quasidominance condition on the interconnection matrix ensures the global asymptotic stability of the model

$$\dot{u}_i = -\frac{u_i}{\tau_i} + \sum_{j=1}^n t_{ij} f_j(u_j(t - \tau_{ij}(t))) + I_i.$$

The same model is studied using various diagonal-type Liapunov functions in [DZ98], which contains many additional references as well as an application of the theory of monotone dynamical systems [Smi95] to this problem.

Trophic chains

The material in Section 5.6 is from [SL83], where the reader can find many additional references as well as examples of the applications of diagonal and D-stability in ecological models. Also see [Jef74] for a discussion of trophic chains from the perspective of diagonal and sign-stability.

Compartmental systems also belong to the class of biological models that have a diagonally stable M-matrix structure, and these systems are also shown to have robust stability properties [Lad76].

6

Interconnected Systems: Stability and Stabilization

This chapter discusses various applications of dynamical systems in which diagonally stable structures can be used advantageously. These examples are in the area of stability and stabilization of interval systems, introduced in Chapter 3, and of stability of interconnected systems, also known as large scale systems. An application to the stability and stabilization of electrical energy systems is also discussed.

6.1 Diagonal Stability in the Large Scale Systems Approach

In the so-called large scale systems approach, a system that is large or complex in some sense is decomposed into a number of smaller or simpler systems, which are first analyzed separately. The overall system is then regarded as the interconnection of the subsystems, and the aim of the large scale systems approach is to use information obtained from the analysis of the subsystems as well the interconnection structure in order to draw conclusions on the (qualitative) behavior of the overall interconnected system.

There is an extensive literature on the large scale system approach, generated during the period starting from the early sixties, with a peak in the late seventies/early eighties [MM77, Šil78], and continuing into the nineties [Šil91]. It should be pointed out that the phrase "large scale system approach" is used in the following sense: "the phrase 'large scale' has nothing to do with the intrinsic size of the problem, but rather with the approach taken to solving it...The emphasis...is on treating a given system not as a whole, but as an interconnection of smaller (and simpler) parts"(the quote is from [Vid86, p. 1086]).

Stability is one of the major structural properties that has been studied and many approaches and techniques have been used. The Liapunov approach has been, and continues to be, one of the most powerful tools for this analysis. In particular, the partitioning and aggregation approach combined with Liapunov and comparison principle techniques, has been very popular with practitioners of large scale system theory. In this approach, the "large" system (S) is partitioned into a set of subsystems (S_i); individual Liapunov functions (V_i) are associated to each of these subsystems and constitute, in a manner to be made precise below, the Liapunov function for the overall system.

The exact way in which this overall system Liapunov function is constructed varies from one method to other (see the appendix to this chapter). Stability of the overall system is based on the study of this Liapunov function along the trajectories of a reference system [Šil78, MM77].

In the process of constructing this Liapunov function, one is led to use a diagonal-type Liapunov function, which essentially results from "diagonal assembly" (a type of weighted combination, where the weights are the elements of the diagonal solution of the Liapunov matrix equation) of the individual Liapunov functions.

The natural questions are: Why is this so? Do these methods really need to use diagonal-type Liapunov functions, or are they used merely on account of their simplicity and convenience?

It is shown that if the underlying technique is based on a comparison principle then a diagonal assembly of an overall Liapunov function from the individual Liapunov functions (i.e., those associated to the subsystems) is induced by the necessity of diagonal stability of a certain test matrix, thus leading to a diagonal-type Liapunov function. Similarly, in the so-called scalar Liapunov function approach, the use of a diagonal-type Liapunov function for the overall system leads to the necessity of diagonal stability of another type of test matrix.

Finally, it should be pointed out that many of the results presented below may be viewed independently of the context of large scale interconnected systems. However, since it is a context strongly associated with these kinds of results, the discussion that follows is set in the context of large scale systems and their connection with matrix diagonal stability.

In brief, the so-called vector Liapunov approach used in the analysis of large scale systems is based on the following ingredients (for further details see the appendix to this chapter):

1. a vector function $V = (V_1, V_2, \ldots, V_\ell)^T$; where $V_i > 0$ are positive definite functions and $V_i : \mathbb{R}^{n_i} \to \mathbb{R}_+$; $V : \mathbb{R}^n \to \mathbb{R}_+^\ell$; $n = \sum_i^\ell n_i$.

2. a vector differential inequality, i.e.,

$$\dot{V} \preceq h(V) ; \quad V(0) = V_0 \tag{6.1}$$

$$h : \mathbb{R}^\ell \rightarrow \mathbb{R}^\ell \tag{6.2}$$

3. a differential equation associated to the inequality (6.1), which is called the *comparison system*, i.e.,

$$\dot{r} = h(r) ; \quad r(0) = r_0. \tag{6.3}$$

Uniqueness of the solutions of (6.3) is assumed for convenience in the discussion that follows, although, in general, this is not necessary [RHL77, LL69]. It is also assumed that $r^* = 0$ is the equilibrium solution of (6.3) that is of interest in the stability analysis in this approach.

In order to illustrate the use of the ingredients above, the following interconnected system (6.4) will be used. It is described by the following set of state-space differential equations.

$$\dot{x}_i(t) = f_i(x_i(t), t) + g_i(x(t), t), \tag{6.4}$$

for $i = 1, \ldots, \ell$ and where $x_i(t) \in \mathbb{R}^{n_i}$ is the state vector of the ith subsystem, $x(t) = (x_1(t)^T \cdots x_\ell(t)^T)^T \in \mathbb{R}^n$, and $f_i : \mathbb{R}^{n_i} \times \mathbb{R}_+ \rightarrow \mathbb{R}^{n_i}$; $g_i : \mathbb{R}^n \times \mathbb{R}_+ \rightarrow \mathbb{R}^{n_i}$ are continuous. As is customary in Liapunov theory, it is assumed that this set of differential equations has a unique solution for each initial condition, which depends continuously on the initial condition. The system

$$\dot{x}_i(t) = f_i(x_i(t), t) \tag{6.5}$$

is called the ith *(isolated) subsystem*, and the term $g_i(x(t), t)$ is called the ith *interconnection* or *interaction term*.

Given the decomposition of the overall system (6.4), individual Liapunov functions (V_i positive definite) are obtained for the subsystems (6.5) and consequently a vector $V = (V_1, V_2, \ldots, V_\ell)^T$ is determined. From the interconnection characteristics of the overall system, using norm majorizations, a differential inequality of the type (6.1) is constructed. Replacing the inequality in (6.1) by equality, a differential equation (6.3) results and is used as a comparison system.

In this context, in order to apply the comparison principle, the function $h(\cdot)$ has to have the additional specific property of quasimonotonicity. The class of quasimonotone functions is defined below.

Definition 6.1.1 *A function $f(x) : \mathbb{R}^n \rightarrow \mathbb{R}^n$ is said to be* quasimonotone *increasing (often abbreviated to quasimonotone) if, for all $x, y \in \mathbb{R}^n$ such that $x_j \leq y_j$, $x_i = y_i$, $(i = 1, 2, \ldots, n; \ i \neq j)$, the following inequality holds:*

$$f_i(x) \leq f_i(y), \quad i = 1, 2, \ldots, n.$$

The following property is needed to state the comparison principle.

Property 6.1.2 *(a) the solutions of the comparison system (6.3) remain in the nonnegative orthant $\mathbb{R}^\ell_+$ for all $t > 0$, for all nonnegative initial conditions (this is called "nonnegative orthant invariance");*
(b) the solutions of the (6.3), for all nonnegative initial conditions, tend asymptotically to the equilibrium (this is global asymptotic stability in the nonnegative orthant).

The comparison principle: *If Property 6.1.2 holds, then all the solutions of (6.1) satisfy $V(t) \preceq r(t), \forall t > 0$, provided $V(0) = r(0)$.*

This principle ensures the asymptotic stability of the overall system (6.4), since the vector Liapunov function V tends asymptotically to zero (see [LL69, RHL77] for more details).

Thus quasimonotonicity of $h(\cdot)$, invariance of $\mathbb{R}^\ell_+$ with respect to the flow of the mapping $h(\cdot)$; and asymptotic stability of $r^* = 0$, are the essential ingredients that permit the application of the comparison principle in order to ensure asymptotic stability of a given system (6.4) using the vector Liapunov approach.

Many different types of comparison or reference systems are used: linear, nonlinear, time-varying, etc. The most commonly used, taken as sample cases, are the following (see [Šil78] and references therein):

$$\dot{r} \;=\; Wr; \qquad r(0) = r_0 \tag{6.6}$$

$$\dot{r} \;=\; Wq(r); \qquad r(0) = r_0, \tag{6.7}$$

where $W = (w_{ij})$ is a real square matrix called the *comparison matrix* associated to the system (6.4) and $q(\cdot)$ is a diagonal nonlinear function such that $q(\cdot) : \mathbb{R}^\ell \to \mathbb{R}^\ell$; $q(r) = (q_1(r_1), q_2(r_2), \ldots, q_\ell(r_\ell))^T$. Furthermore, each function q_i arises in such a manner that actually $q_i : \mathbb{R}_+ \to \mathbb{R}_+$ and is monotonically increasing with $q_i(0) = 0$ (see appendix to this chapter). Consequently, it can be concluded that the functions q_i belong to the positive infinite sector, i.e., $q \in \mathcal{S}^\ell_c$.

In view of this, the reader's attention is called to the fact that the system (6.7) is actually a Persidskii-type system. For this system to work as a comparison system, it must possess nonnegative solutions for nonnegative initial conditions (nonnegative invariance property) and asymptotic stability of the equilibrium for all functions $q_i(\cdot) \in \mathcal{S}_c$ (absolute stability); this is discussed below.

First consider the linear comparison system (6.6). The nonnegative invariance property is characterized in the result below:

Lemma 6.1.3 [BNS89] *The nonnegative orthant $\mathbb{R}^\ell_+$ is positively invariant with respect to the linear mapping $W : \mathbb{R}^\ell \to \mathbb{R}^\ell$ as well as with respect to its flow if and only if $W \in \mathcal{Z}$.* □

In other words, in the case of the linear system (6.6), positive invariance is ensured if and only if matrix W belongs to class $\mathcal{Z}$. In addition, in order to apply the comparison principle using the system (6.6) as a comparison system, the quasimonotonicity property is also necessary, and this is characterized as follows.

Lemma 6.1.4 [Šil78] *For the linear mapping* $W : \mathbb{R}^\ell \to \mathbb{R}^\ell$ *to be quasimonotone, it is necessary and sufficient that* $W \in \mathcal{Z}$. $\qquad\square$

Notice that this condition coincides with the condition for positive invariance.

In order to possess property 6.1.2 (b), the system (6.6) has to be Hurwitz stable ($W \in \mathcal{H}$). Since, by Lemma 2.1.17, for a matrix $W \in \mathcal{Z}$, stability is equivalent to diagonal stability, thus the results above show that $W \in (\mathcal{Z} \cap \mathcal{D}_c)$ is a necessary and sufficient condition for the use of the comparison principle based on system (6.6) as a reference system. This is stated formally as follows:

Proposition 6.1.5 *The comparison principle based on system (6.6) as a reference system may be applied if and only if W is a diagonally stable matrix with nonnegative off-diagonal elements.* $\qquad\square$

This proposition explains why diagonal stability of the comparison matrix W is necessary and consequently induces the use of a diagonal-type Liapunov function.

An analogous result is now given for the nonlinear comparison system (6.7). Clearly the necessary and sufficient conditions on the linear system (6.6) provide necessary conditions for the nonlinear system (6.7) to possess properties 6.1.2 (a) and (b); therefore W has necessarily to belong to $\mathcal{Z}$ in order to ensure nonnegative invariance of $\mathbb{R}_+^\ell$. Furthermore, to ensure (local) stability of the equilibrium $r^* = 0$ of (6.7), W has to be in $\mathcal{H}$ as well, i.e., $W \in \mathcal{Z} \cap \mathcal{H}$, which, by Lemma 2.1.17, is equal to $\mathcal{Z} \cap \mathcal{D}_c$.

To ensure Property 6.1.2 (a) for the nonlinear case (6.7); namely, nonnegative invariance of $\mathbb{R}_+^\ell$ under the flow of the nonlinear mapping $Wq(\cdot) :$ $\mathbb{R}^\ell \to \mathbb{R}^\ell$, a result on monotone flows, interesting in its own right, is stated after the following definition.

Definition 6.1.6 *The flow $\phi_t(\cdot)$ of $\dot{x} = f(x)$, where $f \in C^1(U)$, for an open convex $U \subset \mathbb{R}^\ell$, preserves the partial ordering of vectors in the nonnegative orthant $\mathbb{R}_+^\ell$ if, given any two vectors $x \preceq y$ in $\mathbb{R}^\ell$, $\phi_t(x) \preceq \phi_t(y)$, for all $t > 0$. Such a flow is called* order preserving.

Lemma 6.1.7 [Smi88] *The flow of $\dot{x} = f(x)$, where $f \in C^1(U)$, for an open convex $U \subset \mathbb{R}^\ell$, preserves the partial ordering of vectors in the nonnegative orthant $\mathbb{R}_+^\ell$ if and only if the Jacobian $Df(x)$, in U, is in class $\mathcal{Z}$.* $\qquad\square$

A more general statement, valid for arbitrary orthants in $\mathbb{R}^\ell$, may be found in [Smi88, Lemma 2.1]. An immediate consequence of this result is the following.

Corollary 6.1.8 *The positive orthant $\mathbb{R}_+^\ell$ is positively invariant under the flow of the mapping $Wq(\cdot) : \mathbb{R}^\ell \to \mathbb{R}^\ell$ if $W \in \mathcal{Z}$.*

Proof. Use the order preserving property of the flow with respect to zero, noting that: (i) the Jacobian of $Wq(\cdot)$ is WD, where D is a positive diagonal matrix, since $q \in \mathcal{S}_c^\ell$, and, in addition, each $q_i : \mathbb{R}_+ \to \mathbb{R}_+$ is monotonically increasing with $q_i(0) = 0$; and (ii) the class $\mathcal{Z}$ is closed under multiplication by a positive diagonal matrix.   ∎

Thus the condition $W \in \mathcal{Z}$ guarantees Property 6.1.2 (a) for the nonlinear comparison system. In order to ensure that Property 6.1.2 (b) holds, Persidskii's theorem 3.2.3 is used. For ease of reference, it is restated here, in the notation of this section.

Theorem 3.2.3 [Per69]   *The zero solution, $r^* = 0$, of the nonlinear system (6.7) is globally asymptotically stable for all $q(\cdot) \in \mathcal{S}_c^\ell$ if $W \in \mathcal{D}_c(P)$. Equivalently, for each member of the class of nonlinear systems $\{\dot{r} = Wq(r), \quad r(0) = r_0 : q \in \mathcal{S}_c^\ell\}$, the positive definite function*

$$V(r) = 1/2 \sum_{i=1}^{\ell} p_i \int_0^{r_i} q_i(\tau)d\tau \tag{6.8}$$

is a diagonal-type Liapunov function establishing global asymptotic stability of the zero solution of the corresponding system.   □

Therefore, $W \in \mathcal{D}_c$ is a sufficient condition for the absolute stability of the equilibrium of (6.7). Putting together the results of Corollary 6.1.8 (which guarantees property 6.1.2 (a)) and Theorem 3.2.3 (which guarantees property 6.1.2 (b)) gives the following result.

Theorem 6.1.9 *The system (6.7) functions as a comparison system for the corresponding vector differential inequality (6.1) if and only if $W \in (\mathcal{Z} \cap \mathcal{D}_c)$ (i.e., W is a diagonally stable matrix with nonnegative off-diagonal elements).*   □

From equation (3.7), it is clear that $W \in \mathcal{D}_c$ is also a necessary condition for system (6.7) to admit a diagonal-type Liapunov function of the form (6.8).

Notice that it is a scalar diagonal-type Liapunov function that actually proves the stability of the system (6.4) in the so-called vector Liapunov approach. To the authors' knowledge, the Persidskii diagonal-type Liapunov function (6.8) was first exploited in the analysis of comparison systems in [KB99b].

Other diagonal-type Liapunov functions are used in the large scale systems approach [MM77, Šil78, Šil91] although they are not referred to as being of diagonal-type. One of the points being made here is that they are all associated to a diagonal matrix solution of a Liapunov matrix inequality. More specifically, the positive numbers d_i in equations (3.2) through (3.4) are the diagonal elements of this solution.

It should also be emphasized that comparison matrices that satisfy the conditions of Theorem 6.1.9 are diagonally stable and must, in addition, have the prescribed sign pattern of nonnegative off-diagonal elements. The class of M-matrices possesses these properties (see Theorem 2.2.1) and this is the reason for their widespread use in large scale systems analysis.

An alternative approach to the stability analysis of large scale systems is the so-called scalar Liapunov function approach (see the appendix to this chapter for further details) and its essential features are discussed below. In this approach, the overall Liapunov function is usually of the form

$$V(x,t) = \sum_{i=1}^{\ell} d_i V_i(x_i, t), \tag{6.9}$$

which is a diagonal-type Liapunov function. This, in turn, entails the necessity of diagonal stability of a certain matrix which is obtained from the interconnection structure of the overall system (6.4).

One commonly used test matrix derived in [MM77] and taken here as a sample, is denoted by $T = (t_{ij})$, where

$$t_{ij} = \begin{cases} d_i(\sigma_i + a_{ii}) & i = j \\ (d_i a_{ij} + d_j a_{ji})/2 & i \neq j. \end{cases}$$

A sufficient condition for the stability of the interconnected system (6.4) is based on the negative definiteness of this test matrix T. Notice that the elements a_{ij} and σ_i are derived from the subsystems and the interconnection structure of the system (6.4) and the elements $d_i > 0$ are related to the Liapunov function (6.9). It is easy to verify that the matrix T is negative definite if and only if the matrix B is diagonally stable, where the entries of matrix $B = (b_{ij})$ are: $b_{ii} = (\sigma_i + a_{ii})$ and $b_{ij} = a_{ij}$, $i \neq j$.

In other words, the function (6.9) is a Liapunov function for the system (6.4) if and only if B is a diagonally stable matrix. It is clear that the necessity of diagonal stability of B results from the use of the diagonal-type Liapunov function (6.9) for the overall system (6.4).

An additional point is that, in the scalar Liapunov approach, in principle, nonnegativity of the off-diagonal entries is not required, as opposed to the vector approach where this is required for nonnegative invariance and quasimonotonicity. This, in principle, allows a larger class of test matrices that ensure stability of the interconnected system. However, both the scalar and vector approaches require diagonal stability of a certain matrix and a necessary condition for this is that the diagonal entries of this matrix be negative. The interpretation of this is that the isolated subsystems must necessarily be stable for the analysis to proceed. The approach in [Djo83] relaxes this requirement using nondiagonal scalar Liapunov functions. More details on this topic may be found in [MM77, BT82, Djo83, Šil91].

The perspective above, first presented in [KB99b], aimed to give a concise and unified view of the role of diagonal stability and consequently of diagonal-type Liapunov functions in the so-called large scale systems approach. More specifically, in the framework of the comparison principle, it was pointed out above that, since diagonal stability is an essential property, this, in turn, induces the use of diagonal-type Liapunov functions, where the diagonal elements of the solution to the Liapunov matrix inequality are the weights defining the diagonal-type Liapunov function.

The use of the Persidskii-type Liapunov function clearly exemplifies this inducing effect. In addition, diagonal-type Liapunov functions are natural candidates for the role of simultaneous Liapunov functions in the stability analysis for a large class of perturbations (for example, those related to connective stability [Šil78]) which is another attractive feature of diagonal stability, as discussed in Chapter 3.

6.2 Absolute Stability of Interconnected Systems

This section presents various examples in which structural properties of classes of nonlinear interconnected systems are exploited in the analysis of the absolute stability problem. It is verified that a priori identification of diagonally stable structures leads to less conservative results than those obtained by the large scale systems approach using vector or scalar Liapunov functions (discussed in the previous section). Furthermore, one may obtain simpler ways of determining stability criteria; the examples illustrate that, even when the system does not have the desired structure explicitly, one may find an equivalent mathematical representation of the system (by a change of variables) that displays a diagonally stable structure.

Given the general character of the large scale systems methods discussed in the previous section, it is natural that their application to particular classes of nonlinear systems leads to overly sufficient conditions. In addition if the vector Liapunov function approach is employed, it is clear that the stability criteria are associated with the diagonal stability of a nonnegative matrix. In many cases, a careful analysis of such systems, that takes into account some inherent structural properties, leads to less conservative results based directly on the diagonal stability of a system matrix that is not necessarily nonnegative.

It is shown how the results of Chapters 2 and 3 are used in order to identify classes of matrices associated with the analysis of a system from the viewpoint of diagonal stability and the related diagonal-type Liapunov functions. Generally speaking, this happens in two cases: (i) when the system already presents a diagonally stable structure; (ii) when the system does not have a diagonally stable structure, but can be mathematically represented (via change of variables) as one that does, as shown in the examples in Sections 1.3 and 1.7.

In the sequel these aspects are illustrated, using classical examples from the literature and, more specifically, the class of Lur'e systems viewed as interconnected systems.

Let the standard Lur'e system be given by the equations

$$\begin{aligned} \dot{y} &= By + cf(\sigma) \\ \sigma &= d^T y, \end{aligned} \tag{6.10}$$

where $y^T = (y_1, y_2, \ldots, y_{n+1})$; B, c, and d are constant matrices of appropriate dimensions; B is Hurwitz stable; σ is a scalar; and $f(\cdot)$ is a nonlinear function. If B has a real eigenvalue $-\rho_n < 0$ and (B, d) is an observable pair, then there is a nonsingular transformation $(y_1, \ldots, y_{n+1}) \to (x_1, \ldots, x_n, \sigma)$ which reduces (6.30) to the form:

$$\begin{aligned} \dot{x} &= Ax + f(\sigma) \\ \dot{\sigma} &= -\rho_n \sigma - rf(\sigma) + a^T x, \end{aligned} \tag{6.11}$$

where $x \in \mathbb{R}^n$ and A is Hurwitz stable. In [MP72, PR67], the system (6.11) is regarded as an interconnection of the subsystems $\dot{x} = Ax$ and $\dot{\sigma} = -\rho_n \sigma - rf(\sigma)$, with the additional convenient property that the argument of the nonlinear function is a single state variable, which allows advantage to be taken of the structure related to the class of systems (3.5).

Example 6.2.1 *Consider the following system:*

$$\begin{aligned} \dot{x}_i &= -\rho_i x_i + \sigma \\ \dot{\sigma} &= \sum_{i=1}^4 \beta_i x_i + rp_2 \sigma - f(\sigma) \end{aligned} \tag{6.12}$$

for $i = 1, 2, 3, 4$ and where $\rho_i, r, -p_2 > 0$; $f(0) = 0$, $\sigma f(\sigma) > 0$. This system was considered in [PR67, MM77] and represents the longitudinal motion of an aircraft. In both the cited references (6.12) is treated as a linear interconnection of the subsystems below:

$$\begin{aligned} \dot{x}_i &= -\rho_i x_i \\ \dot{\sigma} &= rp_2 \sigma - f(\sigma) \end{aligned}$$

for $i = 1, 2, 3, 4$, with the interconnection matrices $C_{12} = [1, 1, 1, 1]^T$, $C_{21} = [\beta_1, \beta_2, \beta_3, \beta_4]$. A sufficient stability condition obtained in [PR67], via the vector Liapunov function method is:

$$\sum_{i=1}^4 \xi_i^2 < 1/4 \; ; \; \xi_i = (2\beta_i)/\rho_1 p_2 r, \quad \text{with } \rho_1 \le \rho_2 \le \rho_3 \le \rho_4.$$

A less conservative condition obtained in [MM77] via the scalar Liapunov function method is the following:

$$\sum_{i=1}^4 \xi_i^2 < 1. \tag{6.13}$$

In fact, (6.12) can be rewritten in Lur'e form:

$$\dot{z} = Az + bf(\sigma),$$

where

$$A = \begin{bmatrix} -\rho_1 & & & & 1 \\ & -\rho_2 & O & & 1 \\ & & -\rho_3 & & 1 \\ & O & & -\rho_4 & 1 \\ \beta_1 & \beta_2 & \beta_3 & \beta_4 & rp_2 \end{bmatrix} ; \ z = \begin{pmatrix} x_1 \\ x_2 \\ x_3 \\ x_4 \\ \sigma \end{pmatrix} ; \ b = - \begin{bmatrix} 0 \\ 0 \\ 0 \\ 0 \\ 1 \end{bmatrix}.$$

$$(6.14)$$

On examining the overall system, it turns out to be of the class (3.10) with an acyclic-3 structure (see Section 2.3). Thus Theorems 2.2.7 and 3.2.4 ensure absolute stability in the positive infinite sector, under the condition $(-1)^i \det_i(A^+) > 0$. Since $\rho_i, i = 1, 2, 3, 4$ are positive, this is equivalent to $\det(A^+) < 0$.

For comparison, consider the case in which all the interconnection elements (β_i) are positive, then the sufficient condition for absolute stability is $\det(A^+) = \det(A) < 0$, which, in turn, is equivalent to the inequality below

$$rp_2 + \sum_{i=1}^{4} \beta_i/\rho_i < 0. \tag{6.15}$$

The strict inequality above can be relaxed to nonstrict inequality:

$$rp_2 + \sum_{i=1}^{4} \beta_i/\rho_i \leq 0. \tag{6.16}$$

Analyzing inequality (6.16) one verifies that, in the case considered above, this condition is not only sufficient but also necessary for the absolute stability in the infinite sector of system (6.12). This is so because inequality (6.16) is equivalent to the condition $\det A \leq 0$, which is also necessary for the asymptotic stability of this system with $f(\sigma) = \epsilon\sigma$ for $\epsilon \to 0_+$. Therefore condition (6.16) is also necessary for absolute stability and consequently less restrictive than the ones obtained by the so-called large scale systems approach.

Another case in which necessary and sufficient conditions for absolute stability are obtained for this system is when all coefficients β_i are negative; and consequently matrix A is sign-stable. In this situation the condition $\det(A^+) < 0$ is trivially satisfied and, once again, necessity and sufficiency are represented by condition (6.16). The Liapunov function for this system comes directly from Theorems 2.2.7 and 3.2.4 and is as follows:

$$V(x, \sigma) = \sum_{i=1}^{4} |\beta_i| x_i^2 + \sigma^2.$$

Example 6.2.2 *Suppose that all eigenvalues of matrix B in (6.10) are real negative and distinct. As in [PR67], this system can be brought to the form below:*

$$\begin{pmatrix} \dot{x} \\ \dot{\sigma} \end{pmatrix} = \begin{bmatrix} D & b \\ a^T & 0 \end{bmatrix} \begin{pmatrix} x \\ f(\sigma) \end{pmatrix} - \begin{bmatrix} 0 \\ 1 \end{bmatrix} \hat{f}(\sigma), \qquad (6.17)$$

where $D = \text{diag}\,(-\rho_1, -\rho_2, \ldots, -\rho_n)$; $\hat{f}(\sigma) = \rho_{n+1}\sigma + rf(\sigma)$; $r \geq 0$; $b^T = (1, 1, \ldots, 1)$; $\rho_i > 0$, $i = 1, \ldots, n+1$; $a^T = (a_1, a_2, \ldots, a_n) = (\gamma_1(\rho_{n+1} - \rho_1), \gamma_2(\rho_{n+1} - \rho_2), \ldots, \gamma_n(\rho_{n+1} - \rho_n)); \gamma_1, \gamma_2, \ldots, \gamma_n$ *constants. Using a vector Liapunov approach the following inequality that ensures absolute stability was obtained in [PR67]:*

$$\sum_{i=1}^{n} \gamma_i^2 \,(\rho_{n+1} - \rho_i)^2 < \frac{\rho_{n+1}^2 \rho_1^2}{n\,k^2} \qquad (6.18)$$

where it was assumed that $\rho_1 = \min_i \{\rho_i\}$ $(1 \leq i \leq n)$.

Since, as in the previous example, system (6.17) has acyclic-3 structure, Theorem 2.2.7 is applied in conjunction with Theorem 3.2.4. The following sufficient condition for absolute stability in the sector $[0, k], k > 0$, is obtained:

$$\sum_{i=1}^{n} \frac{\gamma_i(\rho_{n+1} - \rho_i)}{\rho_i} - \frac{\rho_{n+1}}{k} < r. \qquad (6.19)$$

Again, as in the previous example, under the condition that the elements a_i are either all positive or all negative, inequality (6.19) turns out to be equivalent to the necessary condition for asymptotic stability in case the f's are linear "gains" in the interval $[0, k]$, i.e.,

$$(-1)^{n+1}\,\det \begin{bmatrix} D & kb \\ a^T & -(\rho_{n+1} + rk) \end{bmatrix} > 0. \qquad (6.20)$$

Thus, in this case, inequality (6.19) is a necessary and sufficient condition for absolute stability for nonlinearities in the sector $[0, k]$ and is clearly less restrictive than (6.18).

The diagonal-type Liapunov function for this example is given by:

$$V(x, \sigma) = \sum_{i=1}^{n} |\gamma_i(\rho_{n+1} - \rho_i)|x_i^2 + 2\int_0^{\sigma} f(\xi)\,d\xi.$$

The examples given above show how diagonal stability and the related results can be exploited in the stability analysis of some classes of nonlinear systems previously studied in the literature by the large scale systems approach. These examples illustrate that if the system under study has a diagonally stable structure, then significant improvements in the stability results are generally obtained when compared with those derived by large scale systems methods. In addition, it also becomes apparent that the results are usually based on relatively simple algebraic matrix stability conditions.

6.3 Linearly Interconnected Lur'e Systems

As shown in the previous section, for special classes of nonlinear systems, a skillful utilization of the systems interconnection structure can be taken advantage of to derive nonconservative stability results. In particular, these stability results are easily derived when the diagonally stable structure of the particular system can be exploited.

In this spirit, this section gives results for the global stability problem of linearly interconnected Lur'e systems. The results are based on the assumption that each (nonlinear) Lur'e subsystem may have a convenient state-space realization, as well as special properties of the matrices that represent the interconnection structure between these subsystems, mainly properties of the class $\mathcal{D}_c$. If each subsystem can be represented in a special state-space structure, then it is sufficient to check the properties of the interconnection matrix. These properties can be tested by algebraic conditions, so that this method can be used as a synthesis tool as exemplified in the power systems control application described in Section 6.5.

Examples taken from the literature show that the results are less conservative than others that treat the same class of problems by resorting, for example, to Popov's method.

The class of systems considered in this section is represented by the linear interconnection of N Lur'e type subsystems, in which the nonlinearities $\psi_i(\cdot)$ are continuous real-valued functions satisfying, for $i = 1, 2, \ldots, N$

$$\psi_i(y_i)y_i > \alpha_i y_i^2; \quad \psi_i(0) = 0, \text{and } \alpha_i \geq 0 \tag{6.21}$$

and where the transfer function corresponding to the linear part of each of the N isolated Lur'e subsystem is represented by:

$$g_i(s) = \theta_i n_i(s)/d_i(s) \; ; \; \theta_i > 0. \tag{6.22}$$

Considering a minimal state-space realization of each of these transfer functions $g_i(s)$, given by the triplet (A_i, b_i, c_i), the interconnected system being considered is represented by:

$$\begin{aligned} \dot{x}_i &= A_i x_i + b_i v_i \\ y_i &= c_i x_i; \quad v_i = u_i - \psi_i(y_i) \\ u_i &= -\sum_{j=1}^{N} k_{ij} y_j, \end{aligned} \tag{6.23}$$

where $i = 1, 2, \ldots, N$; $A_i \in \mathbb{R}^{n_i \times n_i}$; $b_i \in \mathbb{R}^{n_i \times 1}$; $c_i \in \mathbb{R}^{1 \times n_i}$ are constant matrices and $x_i \in \mathbb{R}^{n_i}$; $x_i^T = (x_{1i}, x_{2i}, \ldots, x_{n_i i})$.

Conditions for existence and uniqueness of solutions of this system are assumed to hold. Calling x the state vector of the interconnected system, where x is the direct sum of the vectors $x_1, x_2, \ldots, x_N$, the interconnected system is said to be *absolutely stable* if the zero solution $x = 0$ of (6.23) is stable for all nonlinearities $\psi_i(\cdot)$ that satisfy (6.21).

By setting $u_i = 0$ $i = 1, 2, \ldots, N$ in (6.23) one has the state realization of the i isolated subsystems, and it is also assumed that each one is absolutely stable. It follows that a necessary condition for each subsystem to be absolutely stable for nonlinearities in the infinite sector is that the relative degree, $\delta g_i(s) := $ degree $d_i(s) - $ degree $n_i(s) \leq 2$.

The derivation of the stability results is based on a scalar diagonal-type Liapunov function of the form (6.9):

$$V(x) = \sum_{i=1}^{N} \gamma_i V_i(x_i); \ \gamma_i > 0, \quad i = 1, 2, \cdots, N, \tag{6.24}$$

where $V_i(x_i)$ is the Liapunov function corresponding to the ith isolated subsystem. Assuming that each subsystem can be represented in a suitable state-space realization, the time derivative of $V(x)$ along the solution of the interconnected system can be written in the form:

$$\dot{V}(x) = \sum_{i=1}^{N} \gamma_i \dot{V}_i(x_i) + \Delta \dot{V}(y), \tag{6.25}$$

where $y^T = (y_1, y_2, \ldots, y_N)$; $\dot{V}_i(x_i)$ is the time-derivative of $V_i(x_i)$ corresponding to the ith isolated subsystem and $\Delta \dot{V}(y)$ depends on the output coupling structure between subsystems. Since each isolated subsystem is absolutely stable, to prove absolute stability of the interconnected system it is sufficient to ensure that $\Delta \dot{V}(y) \leq 0$ (is negative semidefinite). Alternately, if $\Delta \dot{V}$ is negative definite, then it suffices to guarantee that $\sum_{i=1}^{N} \gamma_i \dot{V}_i(x_i)$ is negative semidefinite.

In cases where $\Delta \dot{V}(y)$ happens to be a quadratic form, the stability results are derived on the basis of simple matrix properties. Actually, keeping this basic idea in mind, one also adds some of the so-called excess negative definite terms (see Definition 6.3.5) of $\sum_{i=1}^{N} \gamma_i \dot{V}_i(x_i)$ to the term $\Delta \dot{V}(y)$. These terms, as shown below, permit the stability criteria to be also based on a sort of stability margin of each subsystem, and not strictly on the interconnection structure. Some definitions given in Section 3.2 for isolated systems are now repeated here, for convenience, for the interconnected systems.

Definition 6.3.1 *The ith isolated subsystem of (6.23) belongs to class $\mathcal{R}_1$ if: (i) it has a minimal state-space realization (A_i, b_i, c_i) such that $A_i = (a_{jk})_i$ is an $n_i \times n_i$ tridiagonal matrix; $b_i^T = [\theta_i, 0, 0, \ldots, 0]$ and $c_i = [1, 0, 0, \ldots, 0]$; and (ii) $A_i \in \mathcal{D}_c$.*

If the ith subsystem belongs to class $\mathcal{R}_1$, the corresponding transfer function $g_i(s)$ is also said to belong to *class $\mathcal{R}_1$*. It should be noticed that if

$g_i(s) \in \mathcal{R}_1$ then necessarily $\delta g_i(s) = 1$. The procedure to check whether a given transfer function $g_i(s)$ belongs to the class $\mathcal{R}_1$ has two steps: (i) the given transfer function is subjected to the tridiagonal realization procedure described in Section 3.3; (ii) the resulting tridiagonal system matrix is tested for diagonal stability by Theorem 2.2.7.

Definition 6.3.2 *An isolated subsystem of (6.23) belongs to class $\mathcal{R}_2$ if: (i) for this subsystem exists a minimal state-space realization (A_i, b_i, c_i) such that $A_i = (a_{jk})_i$ is an $n_i \times n_i$ tridiagonal matrix; $b_i^T = [0, 1, 0, \ldots, 0]$; $c_i = [1, 0, \ldots, 0]$; and (ii) $A_i \in \mathcal{D}_c$.*

As in Definition 6.3.1, whenever the ith subsystem is in class $\mathcal{R}_2$, then its transfer function $g_i(s)$ is also said to belong to *class $\mathcal{R}_2$*. If $g_i(s) \in \mathcal{R}_2$, then necessarily the relative degree $\delta g_i(s) = 2$ and $\theta_i = (a_{12})_i$.

With the above definitions, the following results of Section 3.2 can be restated.

Theorem 6.3.3 [KR83]   *If the ith subsystem is in class $\mathcal{R}_1$, then its zero solution is absolutely stable for all nonlinearities in the positive infinite sector, and the corresponding Liapunov function that shows this is given by*

$$V_i(x_i) = x_i^T P_i x_i, \tag{6.26}$$

where P_i for each $i = 1, 2, \ldots, N$ is a positive diagonal solution of

$$P_i A_i + A_i^T P_i = -Q_i, \quad Q_i > 0, \tag{6.27}$$

which is given by:

$$
\begin{aligned}
P_i &= \operatorname{diag}(p_{1i}, p_{2i}, \ldots, p_{n_i i}) \\
p_{1i} &= |(a_{21})_i/(a_{12})_i| \, ; \, p_{2i} = 1 \\
p_{ki} &= \prod_{j=3}^{k} |(a_{j-1,j})_i/(a_{j,j-1})_i|
\end{aligned}
\tag{6.28}
$$

and

$$\dot{V}_i = x_i^T (P_i A_i + A_i^T P_i) x_i - 2 p_{1i} \theta_i \psi_i(y_i) y_i < 0. \tag{6.29}$$

□

Theorem 6.3.4 *If the ith subsystem is in class $\mathcal{R}_2$, then its zero solution is absolutely stable for all nonlinearities in the positive infinite sector, and the corresponding diagonal-type Liapunov function that shows this is given by*

$$V_i(x_i) = x_i^T P_i x_i + (2/\theta_i) \int_o^{y_i} \psi_i(\tau) \, d\tau, \tag{6.30}$$

where P_i are also given by (6.28) and

$$\dot{V}_i(x_i) = x_i^T (P_i A_i + A_i^T P_i) x_i - 2(p_{2i} |(a_{11})_i| / \theta_i) \psi_i(y_i) y_i < 0. \tag{6.31}$$

□

Theorems 6.3.3 and 6.3.4 are also valid (with similar Liapunov functions) when classes $\mathcal{R}_1$ and $\mathcal{R}_2$ are defined for matrices $A_i \in \mathcal{D}_c$ which are not necessarily tridiagonal.

The definition of excess coefficients is necessary to state the results for the interconnected system (6.23).

Definition 6.3.5 *A positive number r_i for which it is true that $(A_i + \text{diag}\,(r_i, 0, 0, \ldots, 0)) \in \mathcal{D}_c$ holds is called an* excess coefficient *of the ith subsystem in $\mathcal{R}_1$ corresponding to Definition 6.3.1.*

Definition 6.3.6 *Positive numbers r_{i_1} and r_{i_2} for which it is true that $(A_i + \text{diag}\,(r_{i_1}, r_{i_2}, 0, \cdots, 0)) \in \mathcal{D}_c$ holds are called* excess coefficients *of the ith subsystem in $\mathcal{R}_2$ corresponding to Definition 6.3.2. The matrix R_i is defined as*

$$R_i := \text{diag}\,(r_{i_1}, r_{i_2}, 0, \cdots, 0).$$

Given these definitions, stability results for the interconnected system (6.23) are as follows.

Theorem 6.3.7 *The zero solution of the interconnected system (6.23) is absolutely stable if each isolated subsystem is in class $\mathcal{R}_1$ and $H = (h_{ij}) \in \mathcal{D}_{co}$, where*

$$\begin{aligned} h_{ii} &= -(k_{ii} + \alpha_i + r_i/\theta_i); & i = 1, 2, \ldots, N \\ h_{ij} &= -k_{ij}; \ i, j = 1, 2, \ldots, N; & i \neq j. \end{aligned} \tag{6.32}$$

Proof. Since $g_i(s) \in \mathcal{R}_1$ then, for the ith subsystem there exists a function $V_i(x_i)$ of the form (6.26) with time-derivative given by (6.29). For the interconnected system (6.23), a Liapunov function of the form (6.24) is used. Since the state-space realization of the ith subsystem puts it in class $\mathcal{R}_1$, given the interconnection structure, the time-derivative along the solution of the interconnected system can be written as the sum of two parts:

$$\dot{V}(x) = \dot{V}_a(x) + \dot{V}_b(x) \tag{6.33}$$

where

$$\dot{V}_a(x) = \sum_{i=1}^{N} \gamma_i [\dot{V}_i(x_i) + 2p_{1i}\theta_i(r_i/\theta_i + \alpha_i)x_{1i}^2],$$

and, following Theorem 6.3.3, one has $\dot{V}_a(x) < 0$. From the interconnection structure of the system (6.23) it follows that:

$$\dot{V}_b = y^T(DH + H^T D)y, \tag{6.34}$$

where $D = \text{diag}\,(d_1, \cdots, d_N)$; $d_i = \gamma_i \theta_i p_{1i}$; $i = 1, 2, \cdots, N$.

Since $H \in \mathcal{D}_{co}$, this implies that there exists a positive diagonal matrix $\overline{P} = \text{diag}\,(\overline{p_1}, \cdots, \overline{p_N})$ that makes $\dot{V}_b \leq 0$. Setting $d_i = \overline{p_i}$, $i = 1, 2, \ldots, N$, one obtains the coefficients $\gamma_i = \overline{p_i}/(p_{1i}\theta_i)$ for the Liapunov function (6.24) that ensure that $\dot{V}(x)$ is negative definite and the proof is completed. ∎

Theorem 6.3.8 *The zero solution of the interconnected system (6.23) is absolutely stable if each isolated subsystem belongs to class $\mathcal{R}_2$ and*

$$M = (M_{ij}) \in \mathbb{R}^{2N \times 2N} \text{ is in class } \mathcal{D}_{co}, \qquad (6.35)$$

where the corresponding diagonal matrix $D = \text{diag}(d_1, d_1, d_2, d_2, \ldots, d_N, d_N)$, and

$$M_{ii} = \text{diag}((m_{11})_i, (m_{22})_i)$$

$$(m_{11})_i = -[(k_{ii} + \alpha_i)|(a_{11})_i| + r_{1i}|(a_{21})_i|]/\theta_i; (m_{22})_i = -r_{2i}$$

$$M_{ij} = \begin{bmatrix} 0 & 0 \\ k_{ij} & 0 \end{bmatrix}; \quad i \neq j$$

Proof. Since $g_i(s) \in \mathcal{R}_2$, this implies that for each subsystem there exists a function $V_i(x_i)$ of the form (6.30) with its time-derivative given by (6.31). As in the previous theorem, a candidate diagonal-type Liapunov function of the form (6.24) is taken for the interconnected system. Due to the tridiagonal state-space realization of each subsystem which characterizes class $\mathcal{R}_2$, and due to the interconnection structure, the expression (6.31) can be written in the form (6.33), where in this case

$$\dot{V}_a(x) = \sum_{i=1}^{N} \gamma_i[\dot{V}_i(x_i) + 2x_i^T P_i R_i x_i + 2\alpha_i x_{1i}^2 |(a_{11})_i|/\theta_i],$$

where R_i is given by definition 6.3.6 and the matrix P_i corresponds to the Liapunov function of the ith subsystem, given by equation (6.28). By Theorem 6.3.4, $\dot{V}_a(x) < 0$. Using (6.35) one has that:

$$\dot{V}_b(x) = z^T(DM + M^T D)z, \qquad (6.36)$$

where $z^T = (y_1, x_{21}, y_2, x_{22}, \ldots, y_N, x_{2N})$ and

$$D = \text{diag}(\gamma_1, \gamma_1, \gamma_2, \gamma_2, \ldots, \gamma_N, \gamma_N).$$

Since, by condition (6.35), $M \in \mathcal{D}_{co}$, it implies that there exists a positive diagonal matrix T of the form $T = \text{diag}(t_1, t_1, t_2, t_2, \ldots, t_N, t_N)$ that makes $\dot{V}_b(x) \leq 0$, then, as in the previous result, $d_i = t_i; i = 1, 2, \ldots, N$, is a choice that makes $\dot{V}(x)$ negative definite, and the proof is complete. ∎

Note that condition (6.35) is actually more restrictive than $M \in \mathcal{D}_{co}$, since a special type of diagonal matrix is needed.

The results given above are based on two major conditions: Each of the subsystems must belong to class $\mathcal{R}_1$ (or $\mathcal{R}_2$) which is algebraically verifiable; and the interconnection matrix must have the diagonal stability property.

Methods to determine whether a given matrix belongs to class $\mathcal{D}_c$ are discussed in Section 2.8. Most of these methods also provide the corresponding diagonal matrix and therefore the diagonal Liapunov functions for the interconnected system.

Although the last two theorems given in this section could be condensed into a single one, considering the interconnection of subsystems of classes $\mathcal{R}_1$ and $\mathcal{R}_2$, they were, however, stated separately for ease of comprehension.

Simple examples are given below in order to illustrate the application of the results above and compare them with those obtained by Popov methods in [BR80, MI75] for the same systems. In particular, Example 6.3.11, shows how one utilizes these results when systems of both classes $\mathcal{R}_1$ and $\mathcal{R}_2$ are present in the interconnected system.

Example 6.3.9 *Consider the interconnected system composed of two Lur'e systems where* $g(s) = (s + a)^{-1}$; $g_2(s) = (s + b)^{-1}$; $a, b > 0$; $k_{ii} = 0$, $i = 1, 2$. *Since* $g_1(s)$ *and* $g_2(s) \in \mathcal{R}_1$ ($A_1 = -a$; $A_2 = -b$), *by Theorem 6.3.7 a condition for absolute stability is given by* $H \in \mathcal{D}_c$ *where* $h_{ii} = -\alpha_i - r_i$; $h_{ij} = -k_{ij}$, $i \neq j$; $i, j = 1, 2$. *In this case, one can choose* $r_1 = a$; $r_2 = b$.

If $k_{12} k_{21} < 0$, *then* $H \in \mathcal{D}_c$ *without any further conditions on the moduli of* k_{12} *and* k_{21}. *If* $k_{12} k_{21} > 0$, *then* $H \in \mathcal{D}_c$ *under the condition*

$$(\alpha_1 + a)(\alpha_2 + b) > k_{12} k_{21} \tag{6.37}$$

that is less restrictive than the stability condition given in [MI75] for this system, which requires $\alpha_1 \alpha_2 > k_{12} k_{21}$.

Note that the condition derived above is also necessary for the asymptotic stability of the linear system considering $\psi_i(y_i) = \alpha_i y_i$. Therefore it is a necessary and sufficient condition for absolute stability of this system.

Example 6.3.10 *Consider the interconnected system composed of three Lur'e systems where* $g_1(s) = (s + a)^{-1}$; $g_2(s) = (s + b)^{-1}$; $g_3 = (s + c)^{-1}$. *Taking the subsystems interconnected in a special structure as in [BR80] where* $k_{ij} = +1$, $i \neq j$ *and* $k_{ii} = 0$, $i = 1, 2, 3$, *consequently matrix* H *is given by*

$$H = \begin{bmatrix} -(\alpha_1 + a) & -1 & -1 \\ -1 & -(\alpha_2 + b) & -1 \\ -1 & -1 & -(\alpha_3 + c) \end{bmatrix}.$$

Since H *is a symmetric matrix, checking the condition* $H \in \mathcal{D}_c$, *is equivalent to checking whether it is negative definite, i.e.,* $H \in \mathcal{D}_c(I)$. *Thus, by Theorem 6.3.7, absolute stability holds under the conditions*

$$(\alpha_1 + a)(\alpha_2 + b) > 1 \tag{6.38}$$

$$(\alpha_1 + a)(\alpha_2 + b)(\alpha_3 + c) + 2 > \alpha_1 + \alpha_2 + \alpha_3 + a + b + c. \tag{6.39}$$

As in Example 6.3.9, it turns out that conditions (6.38) and (6.39) are also necessary for the asymptotic stability in the linear case, when $\psi_i(y_i) = \alpha_i y_i$ *and therefore not conservative. This example can also be analyzed using*

Theorem 3.2.4 since it can be written in the form of equation (3.8) and the same stability result is obtained.

Example 6.3.11 *Let* $g_1(s) = (s^2 + 2s + 3)^{-1}$; $g_2(s) = (s+1)^{-1}$; $k_{ii} = 0$; $i = 1, 2$. *In this case* $g_1(s) \in \mathcal{R}_2$ *and* $g_2(s) \in \mathcal{R}_1$ *and the corresponding tridiagonal sign-stable realizations chosen are:*

$$A_1 = \begin{bmatrix} -1 & 1 \\ -2 & -1 \end{bmatrix}; \ b_1 = \begin{bmatrix} 0 \\ 1 \end{bmatrix}; \ c_1^T = \begin{bmatrix} 1 \\ 0 \end{bmatrix} \text{ and } A_2 = -1 = -b_2 = -c_2.$$

Choosing $P_1 = \text{diag}(2, 1)$, $P_2 = 1$, $r_{11} = 1$, $r_{21} = 1$, $r_2 = 1$ *one gets:*

$$M = \begin{bmatrix} -(\alpha_1 + 2) & 0 & 0 \\ 0 & -1 & -k_{12} \\ -k_{21} & 0 & -(\alpha_2 + 1) \end{bmatrix}.$$

Let $D = \text{diag}(\gamma_1, \gamma_1, \gamma_2)$, $\gamma_i > 0$. *Then, according to Theorem 6.3.8 it is required that the matrix* $Q := DM + M^T D$ *be negative definite. From the problem data, it is easy to see that the first two leading principal minors are negative and positive, respectively. Thus the matrix* Q *is negative definite if and only if* $\det Q$ *is negative, and defining* $\gamma := (\gamma_2/\gamma_1) > 0$, *this holds if and only if*

$$k_{21}^2 \gamma^2 - 4(\alpha_1 + 2)(\alpha_2 + 1)\gamma + (\alpha_1 + 2)k_{12}^2 < 0. \tag{6.40}$$

The left-hand side of (6.40) is a quadratic function in γ *and the problem is to choose* γ *such that the inequality (6.40) is respected and, at the same time, the ranges for the interconnection parameters* k_{12} *and* k_{21} *maximized. Rewriting (6.40) as*

$$k_{21}^2 < \frac{(\alpha_1 + 2)[4(\alpha_2 + 1)\gamma - k_{12}^2]}{\gamma^2},$$

it is clear that if the right-hand side is maximized in terms of γ, *then this will lead to the largest range for* k_{21}^2. *It is not hard to see that, in this particular case, the maximizing* $\gamma := \gamma_0$ *also leads to the largest range for* k_{12}^2. *By calculus,* $\gamma_0 = k_{12}^2/2(\alpha_2 + 1)$ *and substituting this value in (6.40), one obtains the following condition for absolute stability:*

$$|k_{12}k_{21}| < 2(\alpha_2 + 1)\sqrt{\alpha_1 + 2}. \tag{6.41}$$

In this case, the necessary and sufficient condition for asymptotic stability for $\psi(y_i) = \alpha_i y_i$ *is*

$$-2\alpha_2^2 - 8\alpha_2 - 2\alpha_1 - 12k_{12}k_{21} < (\alpha_1 + 3)(\alpha_2 + 1). \tag{6.42}$$

By utilizing the method proposed in [MI75], one can obtain the following condition for the stability of this system:

$$0 < k_{12}k_{21} < \alpha_1\alpha_2. \tag{6.43}$$

Condition (6.41) is less restrictive than (6.43) except for large values of
α_1 and, in addition, condition (6.41) allows k_{12} and k_{21} to have opposite
signs, in contrast to (6.43).

It should be noted that the stability results obtained for these classes of
systems are dependent on the particular realizations chosen for each of the
subsystems. Thus it may happen that with the same approach but with
different subsystem realizations, the results obtained will be different (i.e.,
more conservative for some realizations and less so for others).

The basic ideas of analysis and synthesis of interconnections of Lur'e
type systems taking advantage of the interconnection structure (in class
$\mathcal{D}_c$) are exploited in Section 6.5 in the case of multiarea interconnected
power systems.

6.4   Stabilization by Linear Feedback

Stabilization of a linear control system with uncertain parameters by state
or output feedback, briefly referred to in Section 1.1, is studied in this
section in the context of *robust stabilization*, i.e., it is required to find a
feedback matrix that stabilizes a class of linear systems. The classes of
interval matrices and parameter-dependent system and feedback matrices
are considered here from the perspective of achieving diagonal stability of
a closed-loop system matrix in order to ensure robust stability.

Interval systems
The problem of modeling uncertain systems and guaranteeing their stability
has been an area of intense research activity, currently known as the area
of robust control. One popular model of uncertain systems is the class of
interval systems introduced in Section 3.4.1 and slightly extended here to
include an input.

The problem of stability and stabilization of discrete-time interval
systems has attracted the attention of many authors; see for example
[BP90, KYF89, Man88, SH91, YN89, ZD89, Reg92, Dus97]. Here it is
shown that the exponential stability results of Section 3.4.1 can be used
to get a stabilization result for discrete-time interval systems. Once again,
the key concepts are diagonal-type Liapunov functions and the diagonal
stability of a nonnegative majorant matrix associated to the interval closed-
loop matrix. The use of a simultaneous diagonal-type Liapunov function
leads to simpler proofs, when compared, for example, to those that use
Perron–Frobenius theory [SH91], or comparison systems [BP90], with the
additional benefit that it can be used in the nonlinear and time-varying
cases, ensuring exponential stability.

To simplify the presentation, assume that the system under consideration
is single-input and that there is no uncertainty in the input matrix b. In
other words, using the language of Section 3.4.1 a little loosely, for any

time-dependent selection $A(k) \in \mathcal{A}(\underline{A}, \overline{A})$ it is required to find a single matrix K that defines a linear state feedback:

$$u(k) = v(k) + Kx(k), \qquad (6.44)$$

that stabilizes

$$x(k+1) = A(k)x(k) + bu(k). \qquad (6.45)$$

To be more precise, the *closed-loop matrix* is defined as $\mathcal{A}^c(\underline{A}^c, \overline{A}^c)$, where

$$\underline{A}^c := \underline{A} + bK = (\underline{a}_{ij}^c) \; ; \; \overline{A}^c := \overline{A} + bK = (\overline{a}_{ij}^c). \qquad (6.46)$$

The problem is to determine a feedback matrix K such that the difference inclusion system associated to (6.46) becomes exponentially stable. Clearly such a matrix K may not always exist and a sufficient condition is given below together with some remarks on necessary conditions.

Theorem 6.4.1 *There exists a feedback matrix K such that the difference inclusion system associated to (6.46) becomes exponentially stable if the nonnegative matrix $U_K \in \mathcal{D}_d$, where*

$$U_K := (u_{ij}^K) := \max\left\{ |\underline{a}_{ij}^c|, \; |\overline{a}_{ij}^c| \right\}. \qquad (6.47)$$

Proof. Follows immediately from Theorem 3.4.17. $\qquad \square$

Notice that this sufficient condition for the existence of a matrix K is *implicit*, in the sense that the test matrix U_K is dependent on matrix K. It is this dependence that is being emphasized by the subscript K. Below it will be shown how to check this implicit condition by an explicit algorithm.

An important distinction between the above result and Theorem 4 in [SH91] is that constraints of nonnegativity and time-invariance have been removed. The advantage of relaxing nonnegativity constraints will be clarified in the examples which follow. It suffices to say here that the increased computational burden (caused by having piecewise smooth constraint functions) may be justified by the larger feasible sets as well as greater tolerance to perturbations or even the existence of a solution in cases where the method proposed in [SH91] fails (as in Example 6.4.4).

If the input vector b in (6.45) is replaced by an interval matrix $\mathcal{B}(\underline{B}, \overline{B})$, then Theorem 6.4.1 still holds, provided that $\underline{A}^c$ and $\overline{A}^c$ are redefined appropriately, since it will no longer be true that they are equal to $\underline{A} + \underline{B}K$ and $\overline{A} + \overline{B}K$ respectively.

In the time-invariant case, an obvious necessary condition is that for each element A of the interval matrix, the pair (A, B) be stabilizable. It is harder to give necessary conditions in the time-varying case [Eng90] and, from this point of view, Theorem 6.4.1 may be regarded as a somewhat strong sufficient condition for the exponential stabilizability of the class of difference inclusion (interval) systems. However, the following example adapted from one in [Eng90] shows that the condition in Theorem 6.4.1

can, in some cases, also be necessary even when the interval matrix does not belong to any of the classes listed in Lemma 3.4.18.

Example 6.4.2 *Consider the difference inclusion:*

$$x(k+1) \in \mathcal{A}x(k), \tag{6.48}$$

where

$$A = \begin{bmatrix} 0 & [\underline{a}_{12}, 1] \\ [0, 1/2] & 0 \end{bmatrix}, \quad \underline{a}_{12} \le -1.$$

The majorant matrix is given by:

$$U = \begin{bmatrix} 0 & -\underline{a}_{12} \\ 1/2 & 0 \end{bmatrix}.$$

From Theorem 6.4.1 a sufficient condition for the exponential stability of this interval system is that

$$I - U = \begin{bmatrix} 1 & \underline{a}_{12} \\ -1/2 & 1 \end{bmatrix}$$

should have all minors positive, which happens if and only if $\underline{a}_{12} > -2$. On the other hand, the interval system (6.48) contains the following time-varying system

$$x(k+1) = A(k)x(k),$$

where

$$A(k) = \begin{bmatrix} 0 & a_{12} \\ 0 & 0 \end{bmatrix}, \quad a_{12} \le 1, \quad \text{for } k \text{ even}$$

$$A(k) = \begin{bmatrix} 0 & 0 \\ 1/2 & 0 \end{bmatrix}, \quad \text{for } k \text{ odd}.$$

The initial condition $x(0) = (0, 1)^T$ generates the following sequence of states:

$$\begin{pmatrix} 0 \\ 1 \end{pmatrix}, \begin{pmatrix} a_{12} \\ 0 \end{pmatrix}, \begin{pmatrix} 0 \\ a_{12}/2 \end{pmatrix}, \begin{pmatrix} a_{12}^2/2 \\ 0 \end{pmatrix}, \begin{pmatrix} 0 \\ a_{12}^2/4 \end{pmatrix}, \begin{pmatrix} a_{12}^2/4 \\ 0 \end{pmatrix}, \begin{pmatrix} 0 \\ a_{12}^3/8 \end{pmatrix}, \cdots .$$

Since $a_{12} \le 1$, this sequence converges if and only if $a_{12} \in (-2, 1]$ so that the sufficient condition $\underline{a}_{12} > -2$, is also necessary in this example.

Computation of a feedback matrix for an interval system
In order to formulate the problem of computing a stabilizing feedback matrix K whenever it exists, denote the set of all feedback matrices that stabilize (6.45) by $\mathcal{F}$, where

$$
\begin{aligned}
\mathcal{F} \quad &:= \quad \{K : U_K := \max\{|\underline{a}_{ij}^c|, |\overline{a}_{ij}^c|\} \in \mathcal{D}_d\} & (6.49) \\
&= \quad \{K : \text{all leading principal minors of } I - U_K \text{ positive}\} & (6.50) \\
&= \quad \{K : \rho(U_K) < 1\}. & (6.51)
\end{aligned}
$$

It is the second characterization of $\mathcal{F}$ that is most useful from the viewpoint of design. This type of condition appears frequently in the large scale systems approach [Šil89] (see Section 6.1).

Indeed, suppose that an objective function, $\phi(K)$, is given and that it depends on the feedback matrix K. Two common examples are: $\phi(K)$ is some norm of K; or ϕ is the length of the interval to which a particular entry of A belongs. In the first case, it is usual to minimize ϕ (to avoid saturation at plant inputs and save energy) and in the second to maximize ϕ (to make the system robust to the largest possible structured uncertainty (in norm) when using a fixed controller). Of course, other specifications and multicriterion objective functions are also common, but, to simplify exposition, consider the following mathematical programming problem:

$$\min_{K \in \mathcal{F}} \{\phi(K)\} \tag{6.52}$$

Clearly, the optimal (norm minimizing or robustness maximizing, etc.) feedback, K_{opt} is given by

$$K_{\text{opt}} = \operatorname{argmin}_{K \in \mathcal{F}} \{\phi(K)\}$$

In what follows, the next result is important.

Lemma 6.4.3 *For the interval matrix* $\mathcal{A}(\underline{A}, \overline{A})$, *the matrix* U_K *can be rewritten as:*

$$(U_K)_{ij} = \frac{|\underline{a}_{ij} - \bar{a}_{ij}|}{2} + \left| \boldsymbol{b}_i^T \boldsymbol{k}_j + \frac{\underline{a}_{ij} + \bar{a}_{ij}}{2} \right|; \quad i, j \in \{1, \dots, n\}, \tag{6.53}$$

where $\boldsymbol{k}_j \in \mathbb{R}^m$ *is the* jth *column of the feedback matrix* K, *and* $\boldsymbol{b}_i \in \mathbb{R}^m$ *is the* ith *row of matrix* B.

Proof. The terms $\underline{a}_{ij}^c$ and $\bar{a}_{ij}^c$ (defined in (6.46)) are

$$\underline{a}_{ij}^c = \underline{a}_{ij} + \boldsymbol{b}_i \boldsymbol{k}_j, \quad \bar{a}_{ij}^c = \bar{a}_{ij} + \boldsymbol{b}_i \boldsymbol{k}_j,$$

where $\boldsymbol{b}_i, \boldsymbol{k}_j \in \mathbb{R}^m$. Using the identity

$$\max\{|x|, |y|\} = \frac{1}{2}|x - y| + \frac{1}{2}|x + y|$$

and identifying $x = \underline{a}_{ij}^c = \underline{a}_{ij} + \boldsymbol{b}_i \boldsymbol{k}_j$ and $y = \bar{a}_{ij}^c = \bar{a}_{ij} + \boldsymbol{b}_i \boldsymbol{k}_j$, one obtains (6.53). ∎

To fix ideas, consider the following difference inclusion, also known somewhat imprecisely as a *time-varying interval system*, first studied in the time-invariant case in [SH91]:

$$x(k+1) \in \begin{bmatrix} [\underline{a}_{11}, \bar{a}_{11}] & 0 & 0.5 \\ 1 & 0.5 & 1 \\ 0.5 & 0 & [\underline{a}_{33}, \bar{a}_{33}] \end{bmatrix} x(k) + \begin{bmatrix} 0 \\ 0 \\ 1 \end{bmatrix} u(k). \tag{6.54}$$

Set $\underline{a}_{11} = 0.5$, $\bar{a}_{11} = 0.6$, $\underline{a}_{33} = 0.55$, $\bar{a}_{33} = 0.65$, and $K = [\ k_1 \quad k_2 \quad k_3\ ]$.

Example 6.4.4 *Consider system (6.54) where the problem is to maximize the interval to which the parameter $a_{33}(k)$ may belong when the bound $\underline{a}_{33}$ is fixed (e.g., at the value 0.55), while preserving exponential stability, by the choice of an appropriate linear state feedback matrix K.*

$$x(k+1) \in \begin{bmatrix} [0.5, 0.6] & 0 & 0.5 \\ 1 & 0.5 & 1 \\ 0.5 & 0 & [0.55, \bar{a}_{33}] \end{bmatrix} x(k) + \begin{bmatrix} 0 \\ 0 \\ 1 \end{bmatrix} u(k),$$

and

$$\underline{A} = \begin{bmatrix} 0.5 & 0 & 0.50 \\ 1 & 0.5 & 1.00 \\ 0.5 & 0 & 0.55 \end{bmatrix}, \quad \bar{A} = \begin{bmatrix} 0.6 & 0 & 0.5 \\ 1 & 0.5 & 1 \\ 0.5 & 0 & \bar{a}_{33} \end{bmatrix}, \quad b = \begin{bmatrix} 0 \\ 0 \\ 1 \end{bmatrix}.$$

From (6.53):

$$U_K = \begin{bmatrix} 0.6 & 0 & 0.5 \\ 1 & 0.5 & 1 \\ |0.5 + k_1| & |k_2| & u_{33} \end{bmatrix},$$

where $u_{33} = \dfrac{|0.55 - \bar{a}_{33}|}{2} + \left| \dfrac{0.55 + \bar{a}_{33}}{2} + k_3 \right|.$

This problem leads to the following mathematical programming formulation:

$$\begin{cases} \text{maximize} & \bar{a}_{33} \\ \text{subject to} & -0.25|0.5 + k_1| - 0.9|k_2| - 0.1|0.55 - \bar{a}_{33}| \\ & -0.1|0.55 + \bar{a}_{33} + 2k_3| + 0.2 > 0. \end{cases}$$

The optimal solution in this case is $\bar{a}_{33} = 2.55 - \epsilon$; whereas the maximum found by the method proposed in [SH91] *is $\bar{a}_{33} = 1.55$.*

Example 6.4.5 *Now consider system (6.54) with the choices $\phi(K) = \|K\|_2$, $\underline{a}_{33} = -3$ and $\bar{a}_{33} = -2$. Consequently one has*

$$x(k+1) \in \begin{bmatrix} [0.5, 0.6] & 0 & 0.5 \\ 1 & 0.5 & 1 \\ 0.5 & 0 & [-3, -2] \end{bmatrix} x(k) + \begin{bmatrix} 0 \\ 0 \\ 1 \end{bmatrix} u(k).$$

It is easy to see that the feasible region obtained by the method proposed in [SH91] *will be empty (because the interval system will no longer have the property of nonnegativity that the method requires). On the other hand, from Lemma 6.4.3*

$$I - U_K = \begin{bmatrix} 0.4 & 0 & -0.5 \\ -1 & 0.5 & -1 \\ -|0.5 + k_1| & -|k_2| & 0.5 - |k_3 - 2.5| \end{bmatrix},$$

and the corresponding mathematical programming problem is:

$$\begin{cases} \text{minimize} & (k_1^2 + k_2^2 + k_3^2) \\ \text{subject to} & -0.25|0.5 + k_1| - 0.9|k_2| - 0.2|k_3 - 2.5| + 0.1 > 0. \end{cases}$$

In this particular example, it is easy to see that the optimal choice of k_2 is zero, so that the optimal solution is $(k_1^*, k_2^*, k_3^*) = (-0.32, 0, 2.26)$, as illustrated in Figure 6.1 below.

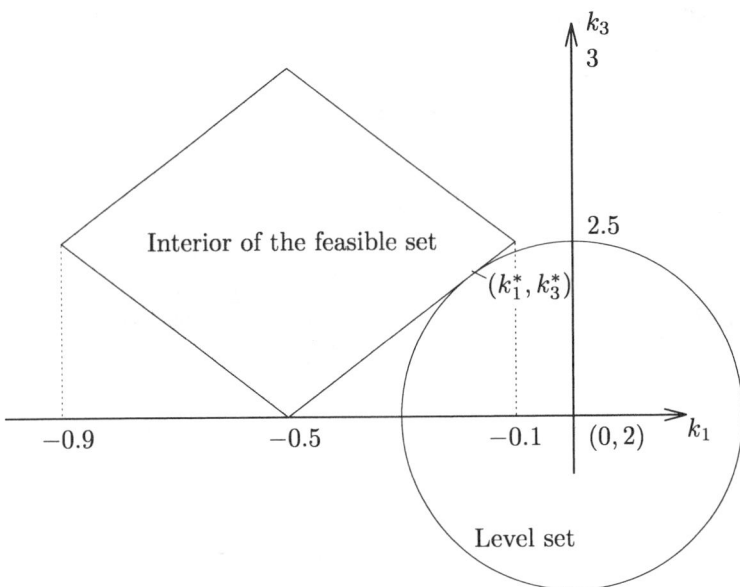

Figure 6.1. The constraint set (parallelogram) and the level set of the objective function (circle) in the k_1-k_3 space, showing the optimal solution (k_1^*, k_3^*).

In general, since the method discussed above imposes fewer constraints on the optimization problem than the one proposed in [SH91], the feasible region will be larger and will contain that of [SH91], so that the optimal values are at least as good, if not better (see [MKB92] for additional numerical results). The result above actually guarantees exponential stability of the feedback system even in the time-varying case—a fact that is not ensured when eigenvalue-based analyses are used. For general methods to solve nonlinear programming problems of the type above, see [GMW81, BSS93].

One disadvantage of the introduction of absolute values in the definition of the majorant matrix (U_K) is that this leads to constraint functions that are piecewise smooth (i.e., smooth almost everywhere). This is to be contrasted with the nonnegativity constraints in [SH91], which are smooth, but lead to smaller feasible regions.

Application of Theorem 3.4.17 to the robust quadratic stabilization of time-varying discrete interval systems gives a sufficient condition for stabilizability in the time-varying case and leads to a nonlinear programming problem to find the optimal stabilizing feedback whenever it exists. It is also shown, by an example, that the sufficient condition for stabilizability of an interval time-varying system can become a necessary condition. It should also be pointed out that, to date, computable necessary and sufficient conditions for stabilizability of an uncertain discrete-time (interval) system are not available [GPS95].

Feedback stabilization using diagonal Liapunov functions: continuous-time examples

As in the example given in Section 1.2, examples that follow illustrate how the identification, whenever possible, of a diagonally stable structure of a given system leads to robust stability results that are less conservative than those obtained by other methods. On the other hand, it should be kept in mind that results obtained with parameter-dependent Liapunov functions are valid only for uncertainties that do not vary with time, which is a reasonable assumption in many applications.

Example 6.4.6 *Consider the linear system with output feedback analyzed in* [ZK87, Šil89]:

$$\dot{x} = \begin{bmatrix} -1 & 0 & 0 \\ 0 & -2 & 0 \\ 0 & 0 & -3 \end{bmatrix} x + \begin{bmatrix} 1 & 0 \\ 0 & 1 \\ 1 & 1 \end{bmatrix} u \qquad (6.55)$$

$$y = \begin{bmatrix} 1 & 0 & 1 \\ 0 & 1 & 0 \end{bmatrix} x \qquad (6.56)$$

with the decentralized output feedback

$$u = -\begin{bmatrix} 1 - k_1 & 0 \\ 0 & 1 - k_2 \end{bmatrix} y. \qquad (6.57)$$

The corresponding closed-loop system is given by

$$\dot{x} = \begin{bmatrix} -2 + k_1 & 0 & -1 + k_1 \\ 0 & -3 + k_2 & 0 \\ -1 + k_1 & -1 + k_2 & -4 + k_1 \end{bmatrix} x, \qquad (6.58)$$

and for this system it is desired to identify the regions for the feedback gains k_1, k_2, k_3 *for which exponential stability of the closed-loop system is maintained.*

The parameter-dependent system (6.58) can be viewed as the interconnection of three first-order subsystems. In addition, by a permutation of coordinates $z^T = (z_1, z_2, z_3) := (x_2, x_1, x_3)$, this closed-loop system can be rewritten in the equivalent form:

$$\dot{z} = \begin{bmatrix} -3 + k_2 & 0 & 0 \\ 0 & -2 + k_1 & -1 + k_1 \\ -1 + k_2 & -1 + k_1 & -4 + k_1 \end{bmatrix} z. \qquad (6.59)$$

The system matrix in the form above clearly displays an acyclic-3 decomposable structure. So Theorem 2.2.7 is used to derive conditions that ensure diagonal stability of the closed-loop parameter-dependent matrix that appears on the right hand side of (6.59). A necessary condition for the latter is that diagonal elements be negative and this implies: $k_2 < 3$ and $k_1 < 4$. Furthermore, the additional condition

$$\det A_{22} := \det \begin{bmatrix} -2 + k_1 & -1 + k_1 \\ -1 + k_1 & -4 + k_1 \end{bmatrix} > 0 \qquad (6.60)$$

implies $k_1 < 1.75$, and the corresponding quadratic diagonal simultaneous Liapunov function for system (6.59) is given by

$$V(z) = \gamma z_1^2 + z_2^2 + z_3^2, \qquad (6.61)$$

where $\gamma > \dfrac{(k_2 - 1)^2}{4(k_1 - 4)(k_2 - 3)} \geq 0$.

With this Liapunov function one gets the exact robust Hurwitz stability bounds which are $k_1 < 1.75$ and $k_2 < 3$. This is explained by the following fact. A given symmetric matrix is Hurwitz if and only if it is negative definite, in which case, the simplest diagonal Liapunov function $(z_2^2 + z_3^2)$ works for the block A_{22} in (6.59). The coefficient γ in (6.61) appears as a consequence of the matrix in (6.59) being block triangular (decomposable), so that the overall diagonal Liapunov function is given by (6.61).

In comparison, in [Šil89] using the vector Liapunov function approach, i.e., using diagonal-type Liapunov functions associated to M-matrices (that are in class $\mathcal{D}_c$), the bounds $-0.875 < k_1 < 1.75$ and $k_2 < 3$ were obtained, whilst the use of quadratic nondiagonal Liapunov functions led to the bounds $k_2 < 3$ and $|k_1| < 1.65247$.

For the same problem, some other bounds for robust stability that have been obtained are as follows: ($|k_i| < 0.52070$, [PT81]); ($|k_i| < 0.81577$, [YL86]); ((1) $k_1^2 + k_2^2 < 2.72768$; (2) $0.60521|k_1| + 0.351205|k_2| < 1$; (3) $|k_i| < 1.55328$, [ZK87]), and all of these are clearly more conservative.

Example 6.4.7 *Consider the following modification of the previous example, also analyzed in* [Šil89].

$$\dot{x} = \begin{bmatrix} -2 & 0 & 0 \\ 0 & -1 & 0 \\ 0 & 0 & -1 \end{bmatrix} x + \begin{bmatrix} 1 & 0 \\ 0 & 1 \\ 1 & \alpha \end{bmatrix} u$$

$$y = \begin{bmatrix} 1 & 0 & 0 \\ 0 & 1 & -\alpha^{-1} \end{bmatrix} x . \qquad (6.62)$$

With the decentralized output feedback

$$u = - \begin{bmatrix} k_2 & 0 \\ 0 & k_1 \end{bmatrix} y , \qquad (6.63)$$

it follows that the closed-loop system becomes

$$\dot{w} = \begin{bmatrix} -2 + k_2 & 0 & 0 \\ 0 & -1 - k_1 & \alpha k_1 \\ -k_2 & -\alpha^{-1} k_1 & -1 + k_1 \end{bmatrix} w. \tag{6.64}$$

It can be seen that the modified system retains the acyclic structure of (6.59) and although the symmetry of block A_{22} has been altered, one can identify a structural condition in the new block A_{22} since the product of the off-diagonal elements is nonpositive; and the condition for diagonal stability of this block is given by $|k_1| < 1$ together with the condition $k_2 < 2$, independently of the parameter α. The corresponding parameter-dependent diagonal Liapunov function provided by Theorem 2.2.7 is

$$V(w) = \theta w_1^2 + \text{sgn}\alpha(\alpha^{-1} w_2^2 + \alpha w_3^2), \tag{6.65}$$

where $\alpha \neq 0$ and $\alpha \not\to \infty$ and $\theta > \dfrac{k_2^2}{4(k_1 - 1)(k_2 - 2)} \geq 0$.

If $\alpha = 0$ or $\alpha \to \infty$, stability can also be ensured, in this case, with the diagonal Liapunov function

$$V(w) = \theta_1 w_1^2 + \theta_2 w_2^2 + w_3^2,$$

where $\theta_1 > \dfrac{k_2^2}{4(k_1 - 1)(k_2 - 2)} \geq 0$ as in (6.65) and $\theta_2 > \dfrac{(\alpha k_1)^2}{4(1 - k_1^2)} \geq 0$.

For comparison, the best bound obtained in [Šil89] for this problem was $|k_1| < 0.5$ and $k_2 < 2$, using the vector Liapunov function approach, which involves the moduli of the interconnection matrices and the additional condition that a comparison matrix be in $\mathcal{D}_c$ (see Section 6.1). In this example, the characteristic polynomial of the block A_{22} is Hurwitz, independently of the parameters α and k_1 so that the above conditions are still restrictive.

Once again it should be observed that the careful identification of the system structure in these examples, together with an adequate choice of Liapunov function, leads to less conservative stability results.

Some references that propose other systematic ways of solving the type of controller design problem discussed in this section are very briefly described below.

Output feedback diagonal stabilization of a parameter-dependent system

Consider the following single input, multiple output, parameter-dependent, discrete-time system

$$\begin{aligned} x(k+1) &= A(w)x(k) + Bu(k), \\ u(k) &= Ky(k), \\ y(k) &= Cx(k), \end{aligned} \tag{6.66}$$

where the system matrix $(A(w))$ is in companion form, $B = [1 \ 0 \ \cdots \ 0]^T$, $C = (c_{ij}) \in \mathbb{R}^{m \times n}$ and w belongs to a set $\mathcal{W}$ of possible parameters. The

problem here is to find a constant output feedback matrix K, such that
the closed-loop system described by (6.66) is diagonally stable. As in the
previous section, results are restricted to the class of single input, multiple
output systems. In addition, the input and output coupling matrices B and
C are constant and have no uncertainty; the only parameter dependence
(uncertainty) occurs in the system matrix A, which, moreover, unlike the
previous section, is assumed to be in companion form. First consider the
nominal system obtained by setting $w = 0$ in (6.66). The following theorem
holds.

Theorem 6.4.8 [Dus96] *The following propositions are equivalent.*

(i) *There exists an output feedback matrix* $K = [k_1 \, k_2 \cdots k_m]$ *such that
the nominal closed loop system matrix* $(A + BKC) \in \mathcal{D}_d(P)$.

(ii) *Positive real numbers* $z_i \in \mathbb{R}^+$ *and real numbers* $k_j \in \mathbb{R}$ *can be found
such that*

$$\sum_{i=1}^{n} z_i \leq 1, \tag{6.67}$$

$$\begin{bmatrix} z_i & -a_i + \sum_{j=1}^{m} c_{ji}k_j \\ -a_i + \sum_{j=1}^{m} c_{ji}k_j & z_i \end{bmatrix} \geq 0, \tag{6.68}$$

for $i = 1, \cdots, n$.

*Moreover, the diagonal elements of the matrix P must satisfy $p_1 \geq p_2 \geq
\cdots \geq p_n > 0$.* □

Now consider the problem of finding a static controller that works for
all $w \in \mathcal{W}$. Again, LMI-based conditions can be given, based on linear
fractional representations; see [Dus97] for a detailed discussion.

   Another stabilization problem is related to linear systems subject to
nonlinear perturbations of the state such that only the perturbed state
is available for feedback. Under the assumption that the nonlinearity is
diagonal and in the positive infinite sector (i.e., of class $\mathcal{S}_c^n$), the Persidskii
theorem is applicable and again the problem is to find a constant feedback
matrix K such that the closed-loop system matrix $A + BK$ is in class $\mathcal{D}_c$
or $\mathcal{D}_d$. This problem (for class $\mathcal{D}_c$) is referred to as *Persidskii design* in
[CGL97, p.289ff.], and is also discussed in [BGFB94, p.144] in a different
context (positive orthant stabilizability).

6.5 Decentralized Power-Frequency Control of Power Systems

This section assumes some familiarity with the terminology of power
systems. Thus the interested reader who is not familiar with it should

refer to [Elg71, Kun94], which contain definitions of all the terminology used here. In power systems stability analysis, and more specifically in power-frequency control, where a multimachine system is considered, the Liapunov direct method has traditionally been used by different authors (see [Pai81, Kun94] and references therein). It should be observed that these analyses, carried out for mathematical models of power systems, frequently utilize energy functions that are actually diagonal-type Liapunov functions. In this section it is shown that, for the multimachine power system model, such Liapunov functions are naturally obtained from Theorem 3.2.3, since the system model is identified to be of the Persidskii-type (3.5). Consequently the asymptotic stability of this power system model can be determined on the basis of diagonal stability of the overall system matrix.

In addition some nonintuitive results are presented in the sequel: Namely, that adequate local stabilization of each area (or machine) implies the stabilization of the overall interconnected system. This result takes into account the governor–turbine speed regulators and the nonlinearities of the network interconnections and sufficient stability conditions for the composite system can be presented explicitly in a decentralized manner, i.e., as a set of groups of noninteracting conditions, each group concerning only the parameters of one single area. Since these stability conditions are given explicitly in terms of the local feedback gains, they may be used in the design of local area regulators as well.

The stabilization referred to above is absolute stability in the positive infinite sector and, in fact, a Lur'e-type Liapunov function for the overall interconnected system is also found in a rather straightforward manner.

Power-frequency control

In order to maintain the nominal frequency of the system as well as the prescribed tie-line power schedules in spite of load variations, an appropriate control strategy must be designed; for example, stepwise load variations can be compensated by an integrator driven by the area control error (ACE_i).

In this section the mathematical models are subject to the following assumptions: Synchronous machines are represented as constant voltages behind their transient reactances; transmission line resistances are considered to be negligible; dynamical models for governor–turbine speed regulators are linear; and the nonlinearities due to network interconnections are included. Without loss of generality, only one integrator is used for each machine. The control strategy adopted is decentralized since only local information is needed for the input signal u_i in the model considered hereafter. A block diagram of the system described above is shown in Figure 6.2 below. With the assumptions above, the following system of ordinary differential equations represents the interconnected power system [Elg71, HSK81]:

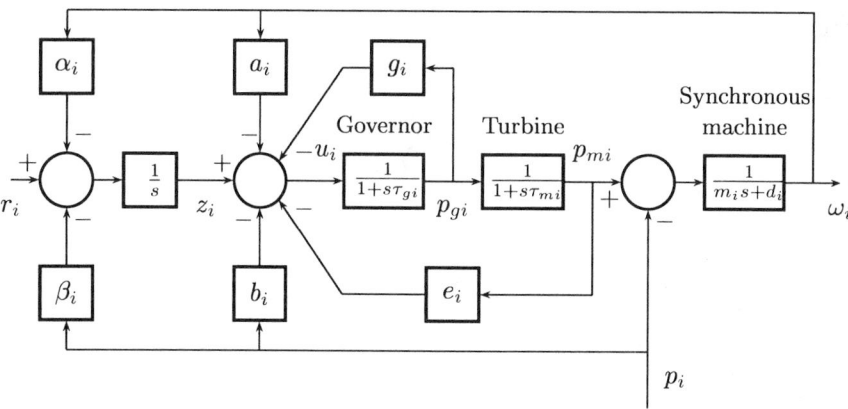

Figure 6.2. Block diagram of the local area power-frequency control. The Laplace variable is denoted s.

$$
\begin{aligned}
S_i \dot{x}_i &= A_i x_i + K_i \phi(\sigma) + B_i u_i + G_i r_i \\
\dot{\sigma} &= H x,
\end{aligned}
\tag{6.69}
$$

where, $i = 1, \ldots, n$ and

$$x_i^T := (w_i, p_{mi}, p_{gi}, z_i); \quad S_i := \text{diag}\,(m_i, \tau_{mi}, \tau_{gi}, 1); \quad x^T = (x_1^T, x_2^T, \ldots, x_n^T);$$

$$K_i^T := -[c_i, 0, 0, \beta_i c_i],$$

and where β_i is an arbitrary scalar gain, the vectors c_i are defined below, and

$$
A_i = \begin{bmatrix} -d_i & 1 & 0 & 0 \\ 0 & -1 & 0 & 0 \\ 0 & 0 & -1 & 0 \\ -\alpha_i & 0 & 0 & 0 \end{bmatrix} ; \quad
B_i = \begin{bmatrix} 0 \\ 0 \\ 1 \\ 0 \end{bmatrix} ; \quad
G_i = \begin{bmatrix} 0 \\ 0 \\ 0 \\ 1 \end{bmatrix} ;
$$

$$H = [h_1, h_2, \cdots, h_n], \quad h_i = [c_i, 0, 0, 0], \quad \sigma^T = (\sigma_1, \sigma_2, \cdots, \sigma_m),$$

where $m = (n-1)n/2$. The component σ_k is defined from the power angle difference between the ith and the jth machines, i.e., $\theta_{ij} := (\overline{\delta}_i - \overline{\delta}_j)$; therefore $\sigma_1 = \theta_{12}; \sigma_2 = \theta_{13}; \cdots; \sigma_{n-1} = \theta_{1n}; \sigma_n = \theta_{23}; \sigma_{n+1} = \theta_{24}$, etc., and $\phi(\sigma)^T = (\phi_1(\sigma_1), \phi_2(\sigma_2), \cdots, \phi_m(\sigma_m))$.

For round-rotor machines $\phi_k(\sigma_k) = \varrho_{ij} \sin(\sigma_k)$, where ϱ_{ij} is the synchronizing coefficient between the ith and jth machines. From the definition of σ_i, one obtains the vectors c_i that constitute the columns of the matrix C_n, i.e.,

$$C_n = [c_1, c_2, \cdots, c_n] \tag{6.70}$$

and $C_j (j = 2, 3, \ldots)$ are formed as follows:

$$C_2 = [1, -1]; \quad C_3 = \begin{bmatrix} 1 & -1 & 0 \\ 1 & 0 & -1 \\ 0 & 1 & -1 \end{bmatrix} =: [c_1, c_2, c_3]$$

$$C_4 = \begin{bmatrix} 1 & -1 & 0 & 0 \\ 1 & 0 & -1 & 0 \\ 1 & 0 & 0 & -1 \\ 0 & 1 & -1 & 0 \\ 0 & 1 & 0 & -1 \\ 0 & 0 & 1 & -1 \end{bmatrix} =: [c_1, c_2, c_3, c_4] ; \quad \cdots .$$

The matrix C_j has $\frac{j(j-1)}{2}$ rows and j columns, and the reader should notice the abuse of notation by which the dimensions of the vectors c_i depend on the matrix C_j they belong to. In addition the following relation holds:

$$(\sigma_n, \sigma_{n+1}, \cdots, \sigma_m)^T = -C_{n-1}(\sigma_1, \cdots, \sigma_{n-1})^T, \tag{6.71}$$

which expresses the linear dependency of the angular variables σ_k.

Without loss of generality the feedback gains corresponding to the variables p_{gi} and p_{mi} can be set to zero in this analysis. Consequently the control signals u_i, $i = 1, \cdots, n$ are given by

$$u_i = z_i - a_i w_i - b_i p_i, \tag{6.72}$$

and the operating equilibrium state of the system satisfies the following equations for the constant vector $r^T = (r_1, r_2, \cdots, r_n)$:

$$-\beta C_n^T \phi(\sigma^*) = r ; \quad w^* = 0, \tag{6.73}$$

where $w^* = (w_1^*, \cdots, w_n^*)^T$ and $\sigma^* = (\sigma_1^*, \cdots, \sigma_m^*)^T$.

If the operating point is asymptotically stable, then, after a stepwise load disturbance, the system trajectories will tend asymptotically to the equilibrium state determined by (6.73) after some transient behavior. Consequently, the problem of stability is henceforth reduced to the stability of the origin with respect to the incremental system (the post-disturbance operating point is denoted (w^*, σ^*)), which, after the feedback variables are included, has the form:

$$\begin{aligned} S_i \dot{\epsilon}_i &= \tilde{A}_i \epsilon_i + \tilde{K}_i \psi(\delta); \quad i = 1, 2, \ldots, n \\ \dot{\delta} &= H \epsilon \end{aligned} \tag{6.74}$$

where,

$$\epsilon_i = x_i - x_i^*; \ \delta = \sigma - \sigma^*; \ \psi^T = (\psi_1(\delta_1), \psi_2(\delta_2), \ldots, \psi_m(\delta_m));$$

$$\epsilon^T = (\epsilon_1^T, \epsilon_2^T, \ldots, \epsilon_n^T)$$

$$\psi_k(\delta_k) = \phi_k(\delta_k + \sigma_k^*) - \phi_k(\sigma_k^*), \ (\psi_k(0) = 0) \ ; \ k = 1, 2, \ldots, m \text{ and}$$

$$\tilde{A}_i = \begin{bmatrix} -d_i & 1 & 0 & 0 \\ 0 & -1 & 1 & 0 \\ -a_i & 0 & -1 & 1 \\ -\alpha_i & 0 & 0 & 0 \end{bmatrix} \ ; \ \tilde{K}_i = - \begin{bmatrix} c_i^T \\ 0 \\ b_i c_i^T \\ \beta_i c_i^T \end{bmatrix}.$$

The structure of the multimachine system is better understood when the regulator dynamics are neglected. In this case, the incremental system retains the form of equation (6.74), with $\epsilon_i = w_i$, $\tilde{A}_i = -d_i$, $\tilde{K}_i = -c_i^T$, $S_i = m_i$, $H = C_n$. Defining $M = \text{diag}(m_1, \cdots, m_n)$; $D = \text{diag}(d_1, \cdots, d_n)$, both positive diagonal matrices; $w^T = (w_1, \ldots, w_n)$, and the system can be represented in the form below:

$$\begin{pmatrix} \dot{w} \\ \dot{\delta} \end{pmatrix} = \tilde{A} \begin{pmatrix} w \\ \psi(\delta) \end{pmatrix}; \ \tilde{A} = \begin{bmatrix} M^{-1} & 0 \\ 0 & I_m \end{bmatrix} \begin{bmatrix} -D & -C_n^T \\ C_n & 0 \end{bmatrix}. \quad (6.75)$$

Note that the equation (6.75) has the structure of a Persidskii-type system (3.5) where the matrix $\tilde{A}$ is D-skew-symmetric (class $\mathcal{D}_{c2} \cap \mathcal{D}_{co}$). Thus the corresponding diagonal-type Liapunov function (of Theorem 3.2.3) is directly given as follows:

$$V(w, \delta) = (1/2) \sum_{i=1}^{n} m_i w_i^2 + \sum_{k=1}^{m} \int_0^{\delta_k} \psi_k(s) \, ds. \quad (6.76)$$

This function is well known in the power systems literature as an *energy function*. The novelty here is that it was derived directly by identifying the system as being of the Persidskii class. The time derivative $\dot{V}(w, \delta)$ along the trajectories of (6.75) is given by

$$\dot{V}(w, \delta) = - \sum_{i=1}^{n} d_i w_i^2$$

and together with the equality

$$(\delta_n, \delta_{n+1}, \cdots, \delta_m)^T = -C_{n-1}(\delta_1, \delta_2, \cdots, \delta_{n-1})^T \quad (6.77)$$

it can be proved that (6.75) is absolutely stable with respect to all functions ϕ_k. The reader is referred to [HSK81] for the proof of this fact that requires some additional technicalities. The main purpose here is to point out the diagonal-type structure present in these power system models which naturally entails the use of diagonal-type Liapunov functions.

Power system with governor–turbine regulators

The incremental system model in the case when governor–turbine regulators are included in the model is defined by (6.74), and can be considered as an interconnected system of the n isolated subsystems

$$\dot{\epsilon}_i = S_i^{-1}\tilde{A}_i\epsilon_i; \quad i = 1, 2, \cdots, n. \tag{6.78}$$

With this decomposition, the subsystems are interconnected through a linear combination of integrals of the scalar variables δ_i as described below.

Representing (6.74) in the form of equation (3.5) one has

$$\begin{pmatrix} \dot{\epsilon} \\ \dot{\delta} \end{pmatrix} = S^{-1}A\begin{pmatrix} \epsilon \\ \psi(\delta) \end{pmatrix}; \quad A = \begin{bmatrix} \tilde{A} & \tilde{K} \\ H & 0 \end{bmatrix}, \tag{6.79}$$

where the matrices above are defined as follows $\tilde{A} = \text{diag}(\tilde{A}_1, \cdots, \tilde{A}_n)$; $S = \text{diag}(S_1, \cdots, S_n)$, $\tilde{K} = [K_1^T, K_2^T, \cdots, K_n^T]^T$. Since the matrix A does not have any of the known properties of class $\mathcal{D}_c$, a convenient change of variables is sought in order to represent the system in a new state space description with a diagonally stable structure. This is achieved by means of the realization procedure described in Section 3.3.

In order to do this, consider the ith single machine against an infinite bus which is described by the following differential equations

$$\begin{aligned} \dot{\delta}_i &= w_i \\ \dot{\epsilon}_i &= S_i^{-1}\tilde{A}_i\epsilon_i + S_i^{-1}\tilde{b}\psi_i(\delta_i), \end{aligned} \tag{6.80}$$

where $\tilde{b}^T = -[1, 0, b_i, \beta_i]$.

Using the realization procedure of Section 3.3, the following Lur'e form state space representation can be obtained for each of the individual subsystems (6.80), i.e.,

$$\begin{pmatrix} \dot{\delta}_i \\ \dot{\zeta}_i \end{pmatrix} = \begin{bmatrix} 0 & 1 & 0 & 0 & 0 \\ 0 & -d_i/m_i & t_{23}/m_i & 0 & 0 \\ 0 & -t_{23} & -t_{33} & t_{34} & 0 \\ 0 & 0 & -t_{34} & -t_{44} & t_{45} \\ 0 & 0 & 0 & -t_{45} & -t_{55} \end{bmatrix}_i \begin{pmatrix} \delta_i \\ \zeta_i \end{pmatrix} - \begin{bmatrix} 0 \\ \frac{1}{m_i} \\ 0 \\ 0 \\ 0 \end{bmatrix}\psi_i(\delta_i). \tag{6.81}$$

Defining the parameters $q_1 = (1 + b_i)/(\tau_{mi}\tau_{gi})$; $q_o = \beta_i/(\tau_{mi}\tau_{gi})$; $p_1 = d_i/m_i - (a_i/m_i - \beta_i)$; $p_o = (d_i\beta_i/m_i - \alpha_i/m_i)/b_i$, then the entries of the tridiagonal matrices denoted $T_i := (t_{jk})_i$; $j, k = 2, 3, 4, 5$ and $i = 1, 2, \cdots, n$ are obtained through the relations below, which depend on the physical parameters of the corresponding subsystem:

$$t_{23} = -b_i m_i/\tau_{mi}\tau_{gi}; \quad t_{33} = (1/\tau_{mi}) + (1/\tau_{gi}) - p_1$$

$$t_{34} = q_1 - t_{33}p_1 - p_o; \quad t_{55} = (q_o - t_{33}p_o)/(t_{34})^2 \tag{6.82}$$

$$t_{45} = p_o - t_{44} - t_{55}; \quad t_{44} = p_1 - t_{55}.$$

Once a set of machine design parameters that ensure that all t_{ij}s are positive is determined, it follows that the system (6.81) presents a sign-stable structure and the new state vectors ζ_i that correspond to this tridiagonal realization are related to the original vectors ϵ_i by linear transformations of the type:

$$\zeta_i = \begin{bmatrix} 1 & 0 \\ L_i & R_i \end{bmatrix} \epsilon_i. \tag{6.83}$$

It should be noted that, under this transformation, the angular variables δ_i are not changed in passing from (6.80) to (6.81); thus, when the transformation $\epsilon \to \zeta$ is applied, the system (6.79) takes the form below:

$$\begin{pmatrix} \dot{\zeta} \\ \dot{\delta} \end{pmatrix} = \overline{A} \begin{pmatrix} \zeta \\ \psi(\delta) \end{pmatrix}; \quad \overline{A} = D_m^{-1} \begin{bmatrix} T & -H^T \\ H & 0 \end{bmatrix}, \tag{6.84}$$

where $D_m = \text{diag}\,(m_1, I_3, m_2, I_3, \cdots, m_n, I_3, I_m)$, I_3 (respectively I_m) is the 3×3 (respectively $m \times m$) identity matrix, $T = \text{diag}\,(T_1, T_2, \cdots, T_n)$, where the matrices T_i are defined in (6.81) and (6.82).

Recall that Theorem 6.3.7 ensures the absolute stability of an interconnected system in which all subsystems admit a sign-stable Lur'e-type realization and the interconnection matrix is in class $\mathcal{D}_c$. In a similar spirit, under the condition that each tridiagonal T_i be a sign-stable matrix, it follows that the system matrix $\overline{A}$ is D-skew-symmetric (class $\mathcal{D}_{c2}$) and since the overall system (6.84) is of the Persidskii class (3.5), Theorem 3.2.3 can be applied with the corresponding diagonal-type Liapunov function:

$$V(\zeta, \delta) = \sum_{i=1}^{n} (1/2)\, \zeta_i^T P_i \zeta_i + \sum_{k=1}^{m} \int_{\text{o}}^{\delta_k} \psi_k(\tau)d\tau, \tag{6.85}$$

where $P_i = \text{diag}\,(m_i, I_3)$. In this case too, $\dot{V}$ happens to be only negative semidefinite, and LaSalle's theorem is used to prove asymptotic stability of the equilibrium of (6.84), as in the preceding case (see [HSK81]). In order to write the Liapunov function in terms of the original variables (ϵ), the inverse of transformation (6.83) has to be used.

The discussion in this section can be summarized by the following facts.

Fact 6.5.1 *If the local feedback control parameters of each machine are chosen in such a manner that a sign-stable representation (6.81) is possible (implying asymptotic stability for each machine against infinite bus), then the overall system is absolutely stable in the positive infinite sector with respect to the network nonlinearities $\psi(\delta)$.*   □

This fact provides sufficient conditions for the absolute stability of (6.79), which have the advantage of being decentralized, i.e., a set of conditions (namely that all t_{ij}s in (6.82) are positive) is associated to the local parameters of each subsystem or machine.

Fact 6.5.2 *If the conditions of Fact 6.5.1 hold, then the composite interconnected system displays the property of absolute stability which is invariant under deletion (or addition) of transmission lines.*   □

This fact is true because the properties of the system that ensure a diagonally stable structure, which is a hereditary property (see Section 2.1.1), are preserved under the deletion or addition of transmission lines.

Note that the results above are interesting not only from the viewpoint of stability analysis, but also for the design of the multiarea decentralized robust power–frequency control of power systems, via a diagonal-type Liapunov function. By applying the above procedure, conditions for absolute stability of the overall system are obtained simply and are given in terms of the local feedback gains. This means that an appropriate local stabilization of each machine (or area) guarantees the absolute stability of the multimachine system. Consequently, an explicit nondiagonal Liapunov function of the Lur'e type is derived for the global system in a simple manner and can be used to estimate stability domains [Wil70].

Thus the power-frequency control described above is another application in which the concept of a diagonally stable structure plays an important role, first in the understanding of the system's mathematical structure and second in the possibility of using this structure for decentralized control design.

To summarize the results and examples discussed in this chapter, the authors take the liberty of paraphrasing a comment made in [Šil89]:
"It is clear that the ultimate success or failure in stability analysis and design using diagonal-type Liapunov functions depends critically on our ability to take advantage of the explicit or implicit structure of the system with respect to this type of Liapunov function. Once again, we are reminded that Liapunov's direct method is not really a method, but rather a powerful principle which requires intuition and experience of the analyst and the designer in order to achieve the best qualitative results."

6.6   Notes and References

Linearly interconnected Lur'e systems
The results in Section 6.3 for interconnected Lur'e type systems are fundamental in the derivation of stability results for interconnected power systems in Section 6.5. The former results are also closely related to those derived in [MH78], where passivity concepts were used to derive stability criteria for interconnected systems which also display diagonally stable interconnection structure.

Power-frequency control
The application of Liapunov's method to the power-frequency control problem has been the subject of numerous papers [Wil70, PMR70, DG78, SKU$^+$75] and books (see, for example, [Kun94] and the references therein). In most of these references the speed regulator dynamics is neglected, or else only simplified models are considered. In others, regulator dynamics are included and sufficient conditions for stability are found using Lur'e-type Liapunov functions [Wil70, DG78].

However, these conditions are usually given implicitly in terms of the solvability of rather complex nonlinear matrix equations, in contrast with the diagonal stability results.

6.7 Appendix: Summary of the Large Scale Systems Approach

In order to illustrate typical uses of the scalar and vector Liapunov function approaches to the analysis of large scale systems, we quote almost verbatim from the excellent survey [Vid86] only making small changes to ensure brevity and compatibility with the rest of the book.

It is assumed that the systems under study are described by a set of state-space differential equations of the form

$$\dot{x}_i(t) = f_i(x_i(t), t) + g_i(x(t), t), \quad i = 1, \ldots, \ell, \tag{6.86}$$

where $x_i(t) \in \mathbb{R}^n$ is the state vector of the ith subsystem, the vector $x(t) = (x_1(t)^T, \cdots, x_\ell(t)^T)^T$ is in $\mathbb{R}^n$, and $f_i : \mathbb{R}^{n_i} \times \mathbb{R}_+ \to \mathbb{R}^{n_i}, g_i : \mathbb{R}^n \times \mathbb{R}_+ \to \mathbb{R}^{n_i}$ are continuous. As is customary in Liapunov theory, it is assumed that the set of differential equations (6.86) has a unique solution for each initial condition, which depends continuously on the initial condition. The system

$$\dot{x}_i(t) = f_i(x_i(t), t) \tag{6.87}$$

is called the ith *isolated subsystem*, and the term $g_i(x(t), t)$ is called the ith *interaction term*.

Assume that each of the isolated subsystems is globally asymptotically stable, and that it is desired to establish the same property for the overall system. Since the origin is a globally asymptotically stable equilibrium of each system (6.87), it is reasonable to assume the existence of a continuously differentiable function $V_i : \mathbb{R}^{n_i} \times \mathbb{R}_+ \to R_+$ and class K functions $\alpha_i, \beta_i, \gamma_i$, such that

$$\begin{aligned} \alpha_i(\|x_i\|) &\leq V_i(x_i, t) \leq \beta_i(\|x_i\|), && \forall t \geq 0, \ \forall x_i \in \mathbb{R}^{n_i} \\ \alpha_i(\|x_i\|) &\to \infty, && \text{as } \|x_i\| \to \infty \\ \dot{V}_{i(I)}(x_i, t) &\leq -\gamma_i(\|x_i\|), && \forall t \geq 0, \ \forall x_i \in \mathbb{R}^{n_i}, \end{aligned} \tag{6.88}$$

where the subscript (I) on $\dot{V}_i$ denotes that the derivative is evaluated along the trajectories of the isolated subsystem (6.87).

The scalar Liapunov function approach

Theorem 6.7.1 [MM77, p. 26]: *The origin in $\mathbb{R}^n$ is a globally asymptotically stable equilibrium of the system (6.86) if the following conditions are satisfied:*

(i) There exist real constants a_{ij} such that

$$\nabla V_i(x_i, t) g_i(x, t) \leq [\gamma_i(\|x_i\|)]^{1/2} \sum_{j=1}^{\ell} a_{ij} [\gamma_j(\|x_j\|)]^{1/2} \tag{6.89}$$

for all $t \geq 0$ and for all $x \in \mathbb{R}^n$.

(ii) *There exist positive constants $r_1, \ldots, r_\ell$, such that the test matrix $T \in \mathbb{R}^{\ell \times \ell}$ defined by*

$$t_{ij} = -[r_i a_{ij} + r_j a_{ji} - \delta_{ij}]/2, \tag{6.90}$$

where δ_{ij} denotes the Kronecker delta, is positive definite.

Proof. Define the function

$$V(x,t) = \sum_{i=1}^{\ell} r_i V_i(x_i, t). \tag{6.91}$$

Then it is easy to see, since all of the constants r_i are positive, that V is positive definite, decrescent, and radially unbounded. Now, by evaluating the derivatives of V along the trajectories of (6.86), one gets

$$\dot{V}_{(C)}(x,t) = \sum_{i=1}^{\ell} r_i \dot{V}_{i(C)}(x_i, t) \tag{6.92}$$

where the subscript (C) on the various derivatives indicates that they are evaluated along the trajectories of the interconnected system (6.86). Now

$$
\begin{aligned}
\dot{V}_{i(C)}(x_i, t) &= \frac{\partial V_i(x_i, t)}{\partial t} + \nabla V_i(x_i, t)\{f_i(x_i, t) + g_i(x, t)\} \\
&= \dot{V}_{i(I)} + \nabla V_i(x_i, t) g_i(x, t) \\
&\leq -\gamma_i(\|x_i\|) + \sum_{i=1}^{\ell} a_{ij} [\gamma_i(\|x_i\|)]^{1/2} [\gamma_j(\|x_j\|)]^{1/2} .
\end{aligned}
\tag{6.93}
$$

The quantity on the right side of the last inequality can be written as a quadratic form $w^T(x) T w(x)$, where the matrix T is as defined in (6.90), and

$$w(x) = (\gamma_1(\|x_1\|)^{1/2} \ldots \gamma_\ell(\|x_\ell\|)^{1/2})^T. \tag{6.94}$$

By hypothesis, the matrix T is negative definite. As a consequence, the quadratic form is negative definite and this implies that the system (6.86) is globally asymptotically stable.

A byproduct of the above proof is that the scalar diagonal Liapunov function V defined in (6.91) has been demonstrated to be a Liapunov function for the overall system (6.86). In fact, this gives rise to the name "scalar Liapunov function approach" which, in turn, is related to diagonal stability, as discussed at the end of Section 6.1.

The vector Liapunov function approach

Now consider the so-called vector Liapunov function approach. The result stated next is couched in terms of Liapunov functions that are only directionally differentiable, and not necessarily differentiable. If $V : \mathbb{R}^n \times \mathbb{R}_+$ is directionally differentiable everywhere, and

$$\dot{x}(t) = f(x(t), t) \tag{6.95}$$

is a given differential equation, the *Dini derivative* $\dot{V}_+(x, t)$ of the function v along the trajectories of (6.95) is defined as

$$\dot{V}_+(x, t) = \lim_{\delta \to 0^+} \frac{V(x + \delta f(x, t), t) - V(x, t)}{\delta} . \tag{6.96}$$

The next result is fundamental to this approach to the stability of large scale systems (see [Šil78, p. 94] and [MM77, p. 49] and the references therein).

Theorem 6.7.2 *Consider the system (6.86) and suppose there exist Lipschitz-continuous functions $V_i : \mathbb{R}^{n_i} \times \mathbb{R}_+ \to \mathbb{R}_+$ and functions $\alpha_i, \beta_i, \gamma_i$ of class K satisfying the conditions*

$$\alpha_i(\|x_i\|) \leq V_i(x_i, t) \leq \beta_i(\|x_i\|) \tag{6.97}$$

$$\dot{V}_{i(I)^+}(x_i, t) \leq -\gamma_i(\|x_i\|) \tag{6.98}$$

$$\alpha_i(\|x\|) \to \infty, \text{ as } \|x\| \to \infty, \forall x \in \mathbb{R}^n, t \in \mathbb{R}_+. \tag{6.99}$$

Suppose the interconnection terms g_i satisfy the conditions

$$\|g_i(x, t)\| \leq \sum_{j=1}^{\ell} b_{ij} \gamma_j(\|x_j\|), \quad \forall t \in \mathbb{R}_+, \forall x \in \mathbb{R}^n, \tag{6.100}$$

where the coefficients b_{ij} are appropriate nonnegative constants. Define a test matrix W as follows:

$$w_{ij} = -\delta_{ij} + \eta_i b_{ij}, \tag{6.101}$$

where δ_{ij} is the Kronecker delta and η_i is the Lipschitz constant of V_i. Under these conditions, the system (6.86) is globally asymptotically stable if the matrix W is diagonally stable.

See Section 6.1 for a discussion of the large scale systems approach from the perspective of matrix diagonal stability.

Proof. Evaluate the Dini derivative $\dot{V}_{(C)^+}$ of the function V_i along the trajectories of the interconnected system to get

$$
\begin{aligned}
\dot{V}_{i(C)^+}(x_i, t) &= \lim_{\delta \to 0^+} \left\{ \frac{V_i(x_i + \delta(f_i(x_i, t) + g_i(x, t)), t) - V_i(x_i, t)}{\delta} \right\} \\
&= \lim_{\delta \to 0^+} \left\{ \frac{1}{\delta} [V_i(x_i + \delta(f_i(x_i, t) + g_i(x, t)), t) \right. \\
&\quad - V_i(x_i + \delta(f_i(x_i, t)), t)] \\
&\quad \left. + \frac{1}{\delta} [V_i(x_i + \delta f_i(x_i, t), t) - V_i(x_i, t)] \right\} \\
&\leq \eta_i \|g_i(x, t)\| + \dot{V}_{(I)^+}(x_i, t)
\end{aligned}
\tag{6.102}
$$

using, respectively, the Lipschitz continuity of V_i and the definition of the Dini derivative for the isolated subsystem. Now, by (6.98) and (6.100), we get

$$
\begin{aligned}
\dot{V}_{i(C)^+} &\leq \sum_{j=1}^{\ell} w_{ij} \gamma_j(\|x_j\|) \\
&= [W(q(v(x, t)))]_i
\end{aligned}
\tag{6.103}
$$

where $V(x, t)$ is the vector $(V_1(x_1, t) \ldots V_\ell(x_\ell, t))^T$, and the function $q : \mathbb{R}_+^\ell \to \mathbb{R}_+^\ell$ is defined by

$$
q(V_1, \ldots, V_\ell) = \begin{pmatrix} \gamma_1(\alpha_1^{-1}(V_1)) \\ \vdots \\ \gamma_\ell(\alpha_\ell^{-1}(V_\ell)) \end{pmatrix}.
\tag{6.104}
$$

Now the condition that W is diagonally stable implies that all eigenvalues of the comparison matrix W are negative. This, in turn, implies that,

$$\dot{V}(t) \leq Wq(V(t)), \quad t = 0, \tag{6.105}$$

and it follows that $V(t) \to 0$ as $t \to \infty$, whenever the initial condition $V(0)$ lies in the nonnegative orthant $\mathbb{R}_+^\ell$. It now follows from (6.97) that $V_i(x_i, t) \to 0$ as $t \to \infty$ for all i, which in turn establishes the global asymptotic stability of the system (6.86).

Epilogue

Much important information is contained in the structure of a system, and this book takes some steps in showing how to take advantage of a so-called diagonally stable structure. This perspective can be seen as a useful qualitative complement to other methods such as the large scale method and, in particular, to the powerful linear matrix inequality computational approach that has recently become very popular. The main concern in this book is not so much with numerical issues as with the analysis and the understanding of how the formulation of a problem in terms of matrix diagonal stability, which corresponds to a particular type of linear matrix inequality (one with a diagonal solution) has several payoffs in terms of desirable qualitative system properties such as robust and absolute stability.

Inspection of the many results and examples that use diagonal-type Liapunov functions shows that, starting from nondiagonal Liapunov functions, it would be either difficult, restrictive, or impractical to perform the kind of manipulations necessary to conclude the negative definiteness of the corresponding $\dot{V}$s and ΔVs. This suggests that matrix diagonal stability and the associated diagonal-type Liapunov functions are necessary to achieve the best results for certain classes of systems, in the sense that one cannot do better with nondiagonal ones. However, although a large amount of favorable evidence has been accumulated, this unifying "necessity conjecture" has not been proved and is mentioned here in closing as a challenge to the skillful reader.

References

[AAI⁺93]   D. Amitai, A. Averbuch, M. Israeli, S. Itzikowitz, and E. Turkel. A survey of asynchronous finite-difference methods for parabolic PDEs on multiprocessors. *Applied Numerical Mathematics*, 12:27–45, 1993.

[Abe86]   E. H. Abed. Strong D-stability. *Systems and Control Letters*, 7(3):207–212, 1986.

[ABFK98]   H. L. S. Almeida, A. Bhaya, D. M. Falcão, and E. Kaszkurewicz. A team algorithm for robust stability analysis and control design of uncertain time-varying linear systems using piecewise quadratic Liapunov functions. In *Proceedings of the 37th IEEE Conference on Decision and Control*, pages 4410–4415, Tampa, FL, 1998.

[AD96]   F. Albertini and D. D'Alessandro. Asymptotic stability of continuous-time systems with saturation nonlinearities. *Systems and Control Letters*, 29:175–180, 1996.

[AFMM91]   G. Avitabile, M. Forti, S. Manetti, and M. Marini. On a class of nonsymmetrical neural networks with application to ADC. *IEEE Transactions on Circuits and Systems*, CAS-38:202–209, 1991.

[Aga92]   R. P. Agarwal. *Difference Equations and Inequalities*. Marcel Dekker, New York, 1992.

[AJM76]   B. D. O. Anderson, E. I. Jury, and M. Mansour. Schwartz matrix properties for continuous and discrete time systems. *International Journal of Control*, 23(1):1–16, 1976.

[AK72]   M. Araki and B. Kondo. Stability and transient behavior of composite nonlinear systems. *IEEE Transactions on Automatic Control*, 17:537–541, 1972.

[AKKK92]   E. A. Asarin, V. S. Kozyakin, M. A. Krasnosel'skii, and N. A. Kuznetsov. *Stability Analysis of Desynchronized Discrete-Event Systems,* (in Russian). Nauka, Moscow, 1992.

[AM58]   K. J. Arrow and M. McManus. A note on dynamic stability. *Econometrica,* 26:448–454, 1958.

[AM81]   B. D. O. Anderson and J. B. Moore. Detectability and stabilizability of time-varying discrete-time linear systems. *SIAM Journal on Control and Optimization,* 1(19):20–32, 1981.

[Ara75]   M. Araki. Application of M-matrices to the stability problems of composite dynamical systems. *Journal of Mathematical Analysis and Applications,* 52:309–321, 1975.

[AT86]   E. H. Abed and A. L. Tits. On the stability of multiparameter time-scale systems. *International Journal of Control,* 44(1):211–218, 1986.

[AT98]   S. Arik and V. Tavsanoglu. A comment on "Comments on 'Necessary and sufficient condition for absolute stability of neural networks'". *IEEE Transactions on Circuits and Systems–I: Fundamental Theory and Applications,* 45(5):595–596, 1998.

[Bar71]   S. Barnett. *Matrices in Control Theory.* Van Nostrand, New York, 1971.

[Bar88a]   N. E. Barabanov. Liapunov indicator of discrete inclusions–Parts I,II,III. *Automation and Remote Control,* 49(2,3,5):152–157,283–287,558–565, 1988.

[Bar88b]   N. E. Barabanov. On the Kalman problem. *Siberian Mathematical Journal,* 29(3):333–341, 1988.

[Bau63]   F. L. Bauer. Optimally scaled matrices. *Numerische Mathematik,* 5:73–87, 1963.

[Bau78]   G. M. Baudet. Asynchronous iterative methods for multiprocessors. *Journal of the Association of Computing Machinery,* 15:226–244, 1978.

[BBP78]   G. P. Barker, A. Berman, and R.J. Plemmons. Positive diagonal solutions to the Lyapunov equations. *Linear and Multilinear Algebra,* 5:249–256, 1978.

[BC73]   R. K. Brayton and C. C. Conley. Some results on the stability and instability of the backward differentiation methods with non-uniform time steps. In John J. H. Miller, editor, *Topics in Numerical Analysis,* Proceedings of the Royal Irish Academy Conference on Numerical Analysis, pages 13–33. Academic Press, New York, 1973.

[BCK95]   S. P. Bhattacharyya, H. Chapellat, and L. H. Keel. *Robust Control: The Parametric Approach.* Prentice-Hall, Upper Saddle River, New Jersey, 1995.

[BCZ95]   A. Betser, N. Cohen, and E. Zeheb. On solving the Lyapunov and Stein equations for a companion matrix. *Systems and Control Letters,* 25:211–218, 1995.

[BEN88]   R. Bru, L. Elsner, and M. Neumann. Models of parallel chaotic iteration methods. *Linear Algebra and its Applications*, 103:175–192, 1988.

[Ber82]   C. S. Berger. Proof of a certain conjecture on the stability of linear discrete systems. *International Journal of Control*, 36(3):545–546, 1982.

[Ber83]   D. P. Bertsekas. Distributed asynchronous computation of fixed points. *Mathematical Programming*, 27:107–120, 1983.

[BGFB94]   S. Boyd, L. El Ghaoui, E. Feron, and V. Balakrishnan. *Linear Matrix Inequalities in System and Control Theory*. SIAM, Philadelphia, 1994.

[BH83]   A. Berman and D. Hershkowitz. Matrix diagonal stability and its implications. *SIAM Journal on Algebraic and Discrete Methods*, 4:377–382, 1983.

[BJ90]   P. H. Bauer and E. I. Jury. A stability analysis of two-dimensional nonlinear digital state-space filters. *IEEE Transactions on Acoustics, Speech, and Signal Processing*, 38:1578–1586, 1990.

[BK91]   A. Bhaya and E. Kaszkurewicz. Robust, diagonal and D-stability via QLF's: The discrete-time case. In *Proceedings of the 30th IEEE Conference on Decision and Control*, volume 3, pages 2624–2629, Brighton, 1991. Journal version published as [BK93].

[BK93]   A. Bhaya and E. Kaszkurewicz. On discrete-time diagonal and D-stability. *Linear Algebra and its Applications*, 187:87–104, 1993.

[BK95]   A. Bhaya and E. Kaszkurewicz. Global stability results for Hopfield neural networks. In *Proceedings of the 2nd Latin American Seminar on Advanced Control (LASAC'95)*, pages 137–143, Santiago, Chile, 1995.

[BK96]   A. Bhaya and E. Kaszkurewicz. Comments regarding "On the stability of interval matrices". *IEEE Transactions on Automatic Control*, 41(5):762–766, 1996.

[BK99]   A. Bhaya and E. Kaszkurewicz. A note on a robust stability problem. *International Journal of Control*, 1999. to appear.

[BKB96]   B. Barán, E. Kaszkurewicz, and A. Bhaya. Parallel asynchronous team algorithms: convergence and performance analysis. *IEEE Transactions on Parallel and Distributed Systems*, 7(7):677–688, 1996.

[BKK96]   A. Bhaya, E. Kaszkurewicz, and V. S. Kozyakin. Existence and stability of a unique equilibrium in continuous-valued discrete-time asynchronous Hopfield neural networks. *IEEE Transactions on Neural Networks*, 7(3):620–628, 1996.

[BKM91]   A. Bhaya, E. Kaszkurewicz, and F. C. Mota. Asynchronous block-iterative methods for almost-linear equations. *Linear Algebra and its Applications*, 154:487–508, 1991.

[BL78]   I. M. Bakhilina and D. M. Lerner. Properties of Lyapunov equations with nonnegative free term matrices. *Automation and Remote Control*, 39(5):772–774, 1978.

[BL91]   P. H. Bauer and L.-J. Leclerc. A computer-aided test for the absence of limit cycles in fixed-point digital filters. *IEEE Transactions on Signal Processing*, 39(11):2400–2410, 1991.

[BM93]   M. Bahi and J. C. Miellou. Contractive mappings with maximum norms: Comparison of constants of contraction and application to asynchronous iterations. *Parallel Computing*, 19:511–523, 1993.

[BM94]   A. Bhaya and F. C. Mota. Equivalence of stability concepts for discrete time-varying systems. *International Journal of Robust and Nonlinear Control*, 4:725–740, 1994.

[BNS89]   A. Berman, M. Neumann, and R. J. Stern. *Nonnegative Matrices in Dynamic Systems*. John Wiley, New York, 1989.

[BP79]   A. Berman and R. J. Plemmons. *Nonnegative Matrices in the Mathematical Sciences*. Academic Press, New York, 1979.

[BP90]   P. H. Bauer and K. Premaratne. Robust stability of time-variant interval matrices. In *Proceedings of the 29th IEEE Conference on Decision and Control*, volume 2, pages 434–435, Honolulu, 1990.

[BP94]   A. Berman and B. Plemmons. *Nonnegative Matrices in the Mathematical Sciences*. Classics in Applied Mathematics. SIAM, Philadelphia, 1994.

[BR80]   G. V. Baliga and M. V. C. Rao. On symmetric and unity interconnections between three nonlinear subsystems. *Automatica*, 16:711–713, 1980.

[Bro92]   D. P. Brown. Stability of n order discrete systems with overflow arithmetic. *Journal of the Franklin Institute*, 329(1):15–23, 1992.

[BS70]   S. Barnett and C. Storey. *Matrix Methods in Stability Theory*. Thomas Nelson, London, 1970.

[BS91]   A. Berman and D. Shasha. Inertia preserving matrices. *SIAM Journal on Matrix Analysis and Applications*, 12(2):209–219, 1991.

[BSS93]   M. S. Bazaraa, H. D. Sherali, and C. M. Shetty. *Nonlinear Programming: Theory and Algorithms*. John Wiley, New York, 1993.

[BT79]   R. K. Brayton and C. H. Tong. Stability of dynamical systems: a constructive approach. *IEEE Transactions on Circuits and Systems*, CAS-26(4):224–234, 1979.

[BT80]   R. K. Brayton and C. H. Tong. Constructive stability and asymptotic stability of dynamical systems. *IEEE Transactions on Circuits and Systems*, CAS-27(11):1121–1130, 1980.

[BT82]   J. Bernussou and A. Titli. *Interconnected Dynamical Systems: Stability, Decomposition and Decentralisation*. North-Holland, Amsterdam, 1982.

[BT89]   D. P. Bertsekas and J. N. Tsitsiklis. *Parallel and Distributed Computation: Numerical Methods*. Prentice-Hall, Englewood Cliffs, New Jersey, 1989.

[BT92]   T. Bose and D. A. Trautman. Two's complement quantization in two-dimensional state-space digital filters. *IEEE Transactions on Signal Processing*, 40(10):2589–2592, 1992.

[BW78]   A. Berman and R. C. Ward. $\mathcal{ALPS}$: classes of stable and semi-positive matrices. *Linear Algebra and its Applications*, 21:163–174, 1978.

[BW92]   M. A. Berger and Y. Wang. Bounded semigroups of matrices. *Linear Algebra and its Applications*, 166:21–27, 1992.

[BY89]   S. Boyd and Q. Yang. Structured and simultaneous Lyapunov functions for system stability problems. *International Journal of Control*, 49(6):2215–2240, 1989.

[Cai76]   B. E. Cain. Real, 3×3, D-stable matrices. *Journal of Research of the National Bureau of Standards–B. Mathematical Sciences*, 80B:75–77, 1976.

[Cai84]   B. E. Cain. Inside the D-stable matrices. *Linear Algebra and its Applications*, 56:237–243, 1984.

[Cai90]   B. E. Cain. Item #39010. *Mathematical Reviews*, 90c:1551, 1990.

[Car84]   D. Carlson. Controllability, inertia, and stability for tridiagonal matrices. *Linear Algebra and its Applications*, 56:207–220, 1984.

[CC96]   R. S. Cantrell and C. Cosner. Models for predator-prey systems at multiple scales. *SIAM Review*, 38(2):256–286, 1996.

[CDHJ98]   B. Cain, L. M. DeAlba, L. Hogben, and C. R. Johnson. Multiplicative perturbations of stable and convergent operators. *Linear Algebra and its Applications*, 268:151–169, 1998.

[CDJ82]   D. Carlson, B. N. Datta, and C. R. Johnson. A semi-definite Lyapunov theorem and the characterization of tridiagonal *D*-stable matrices. *SIAM Journal on Algebraic and Discrete Methods*, 3(3):293–304, 1982.

[CFY95]   J. Chen, M. K. H. Fan, and C. C. Yu. On D-stability and structured singular values. *Systems and Control Letters*, 24(1):19–24, 1995.

[CG99a]   Y.-C. Chu and K. Glover. Bounds of the induced norm and model reduction errors for systems with repeated scalar nonlinearities. *IEEE Transactions on Automatic Control*, 44(3):471–483, 1999.

[CG99b]   Y.-C. Chu and K. Glover. Stabilization and performance synthesis for systems with repeated scalar nonlinearities. *IEEE Transactions on Automatic Control*, 44(3):484–496, 1999.

[CGL97]   P. Colaneri, J. C. Geromel, and A. Locatelli. *Control Theory and Design*. Academic Press, San Diego, 1997.

[CHS92]   D. Carlson, D. Hershkowitz, and D. Shasha. Block diagonal semistability factors and Lyapunov semistability of block triangular matrices. *Linear Algebra and its Applications*, 172:1–25, 1992.

[CHS97]   B. Cain, D. Hershkowitz, and H. Schneider. Theorems of the alternative for cones and Lyapunov regularity of matrices. *Czechoslovak Mathematical Journal*, 47(122):487–499, 1997.

[CL93]   N. Cohen and I. Lewkowicz. A necessary and sufficient criterion for the stability of a convex set of matrices. *IEEE Transactions on Automatic Control*, 38(4):611–615, 1993.

[CL97a]   N. Cohen and I. Lewkowicz. Convex invertible cones and the Lyapunov equation. *Linear Algebra and its Applications*, 250:105–131, 1997.

[CL97b]   N. Cohen and I. Lewkowicz. Convex invertible cones of state space systems. *Mathematics of Control, Signals and Systems*, 10:265–285, 1997.

[CLR97]   N. Cohen, I. Lewkowicz, and L. Rodman. Exponential stability of triangular differential inclusion systems. *Systems and Control Letters*, 30:159–164, 1997.

[CM69]   D. Chazan and W. L. Miranker. Chaotic relaxation. *Linear Algebra and its Applications*, 2:190–222, 1969.

[CM94]   P. J. Campo and M. Morari. Achievable closed-loop properties of systems under decentralized control: conditions involving the steady state gain. *International Journal of Control*, 39(5):932–943, 1994.

[Cox94]   G. E. Coxson. The P-matrix problem is co-NP-complete. *Mathematical Programming*, 64:173–178, 1994.

[Cro78]   G. W. Cross. Three types of matrix stability. *Linear Algebra and its Applications*, 20:253–263, 1978.

[CT89]   J. L. Calvet and A. Titli. Overlapping vs partitioning in block-iteration methods: Application in large-scale system. *Automatica*, 25(1):137–145, 1989.

[CU93]   A. Cichocki and R. Unbehauen. *Neural Networks for Optimization and Signal Processing*. John Wiley, Chichester, 1993.

[CvdEG+97]   A. Cima, A. van den Essen, A. Gasull, E. Hubbers, and F. Mañosas. A polynomial counterexample to the Markus-Yamabe conjecture. *Advances in Mathematics*, 131(2):453–457, 1997.

[Dat74]   B. N. Datta. A constructive method for finding the Schwarz form of a Hessenberg matrix. *IEEE Transactions on Automatic Control*, AC-20(2):616–617, 1974.

[Dat78]   B. N. Datta. Stability and D-stability. *Linear Algebra and its Applications*, 21:135–141, 1978.

[DBR98]   A. Djebbari, M. F. Belbachir, and J. M. Rouvaen. A fast exhaustive search algorithm for checking limit cycles in fixed-point digital filters. *Signal Processing*, 69:199–205, 1998.

[DDB95]   M. Dahleh and I. J. Diaz-Bobillo. *Control of Uncertain Systems: A Linear Programming Approach*. Prentice-Hall, Upper Saddle River, New Jersey, 1995.

[Del88]   D. F. Delchamps. *State Space and Input-Output Linear Systems*. Springer-Verlag, New York, 1988.

[DFK92]   I. C. Decker, D. M. Falcão, and E. Kaszkurewicz. Parallel implementation of a power system simulation methodology using the conjugate gradient method. *IEEE Transactions on Power Systems*, 7(1):458–465, 1992.

[DG78]   R. Doraiswami and U. C. M. Gondar. Design and stability of a multiarea load-frequency control under varying loads. *IEEE Transactions on Power Apparatus and Systems*, 97:1005, 1978.

[Djo83]   M. Djordjevic. Stability analysis of large-scale systems whose subsystems may be unstable. *Large Scale Systems*, 5:252–262, 1983.

[DK74]   J. L. Daleckii and M. G. Krein. *Stability of Solutions of Differential Equations in Banach Space.* American Mathematical Society, Providence, Rhode Island, 1974.

[DL92]   I. Daubechies and J. C. Lagarias. Sets of matrices all infinite products of which converge. *Linear Algebra and its Applications*, 161:227–263, 1992.

[Dor87]   P. Dorato, editor. *Robust Control.* IEEE Press, New York, 1987.

[DTS71]   Y. P. Dusonchet, S. N. Talukdar, and H. E. Sinnot. Load flows using a combination of point Jacobi and Newton's methods. *IEEE Transactions on Power Apparatus and Systems*, PAS-90:941–949, 1971.

[Dus96]   S. Dussy. Robust stabilization of discrete-time parameter-dependent systems: the finite precision problem. In *Proceedings of the 35th IEEE Conference on Decision and Control*, pages 3976–3981, Kobe, Japan, 1996.

[Dus97]   S. Dussy. On the robust control design of quantized systems. In *Proceedings of the 36th IEEE Conference on Decision and Control*, pages 1297–1298, San Diego, CA, 1997.

[DY90]   P. Dorato and R. K. Yedavalli, editors. *Recent Advances in Robust Control.* IEEE Press, New York, 1990.

[DZ98]   P. V. D. Driessche and X. Zou. Global attractivity in delayed Hopfield neural network models. *SIAM Journal on Applied Mathematics*, 58(6):1878–1890, 1998.

[EAF79]   N. G. El-Agizi and M. Fahmy. Two-dimensional digital filters with no overflow oscillations. *IEEE Transactions on Acoustics, Speech, and Signal Processing*, ASSP-27(5):465–469, 1979.

[EF97]   L. Elsner and S. Friedland. Norm conditions for convergence of infinite products. *Linear Algebra and its Applications*, 250:133–142, 1997.

[Ela96]   S. N. Elaydi. *An Introduction to Difference Equations.* Springer-Verlag, New York, 1996.

[Elg71]   O. I. Elgerd. *Electric Energy Systems Theory.* McGraw-Hill, New York, 1971.

[Els95]   L. Elsner. The generalized spectral-radius theorem: an analytic-geometric proof. *Linear Algebra and its Applications*, 220:151–159, 1995.

[EM85]   K. T. Erickson and A. N. Michel. Stability analysis of fixed-point digital filters using computer generated Lyapunov functions–part I:

direct form and coupled form filters. *IEEE Transactions on Circuits and Systems*, CAS-32(2):113–132, 1985.

[Eng90]   J. C. Engwerda. Stabilizability and detectability of discrete-time time-varying systems. *IEEE Transactions on Automatic Control*, 35(4):425–429, 1990.

[ET82]   M. N. El Tarazi. Some convergence results for asynchronous algorithms. *Numerische Mathematik*, 39:325–340, 1982.

[FGL$^+$98] R. Fleming, G. Grossman, T. Lenker, S. Narayan, and S. C. Ong. On Schur *D*-stable matrices. *Linear Algebra and its Applications*, 279:39–50, 1998.

[Fil64]   A. F. Filippov. Differential equations with discontinuous right-hand side. *American Mathematical Society Translations*, 42(2):199–231, 1964.

[FK98]   Y. Fang and T. G. Kincaid. Global properties for a class of dynamical neural circuits. *Journal of the Franklin Institute*, 335B(1):163–177, 1998.

[FM76]   E. Fornasini and G. Marchesini. State-space realization theory of two-dimensional filters. *IEEE Transactions on Automatic Control*, AC-21(4):484–491, 1976.

[FM78]   E. Fornasini and G. Marchesini. Doubly-indexed dynamical systems: state-space models and structural models. *Mathematical Systems Theory*, 12:59–72, 1978.

[FMM92]   M. Forti, S. Manetti, and M. Marini. A condition for global convergence of a class of symmetric neural circuits. *IEEE Transactions on Circuits and Systems–I: Fundamental Theory and Applications*, 39(6):480–483, 1992.

[FMM94]   M. Forti, S. Manetti, and M. Marini. Necessary and sufficient condition for absolute stability of neural networks. *IEEE Transactions on Circuits and Systems–I: Fundamental Theory and Applications*, 41(7):491–494, 1994.

[For94]   M. Forti. On global asymptotic stability of a class of nonlinear systems arising in neural network theory. *Journal of Differential Equations*, 113:246–264, 1994.

[FP62]   M. Fiedler and V. Pták. On matrices with non-positive off-diagonal elements and positive principal minors. *Czechoslovak Mathematical Journal*, 12:382–400, 1962.

[FP67]   M. Fiedler and V. Pták. Diagonally dominant matrices. *Czechoslovak Mathematical Journal*, 92(17):420–433, 1967.

[FT95]   M. Forti and A. Tesi. New conditions for global stability of neural networks with application to linear and quadratic programming problems. *IEEE Transactions on Circuits and Systems–I: Fundamental Theory and Applications*, 42(7):354–366, 1995.

[Gan59]   F. R. Gantmacher. *The Theory of Matrices, vols. I & II*. Chelsea, New York, 1959.

[GC87]   J. C. Geromel and J. J. Da Cruz. On the robustness of optimal regulators for nonlinear discrete-time systems. *IEEE Transactions on Automatic Control*, AC-32(8):703–710, 1987.

[Ger85]   J. C. Geromel. On the determination of a diagonal solution of the Lyapunov equation. *IEEE Transactions on Automatic Control*, 30(4):404–486, 1985.

[GM81]   H. J. Greenberg and J. S. Maybee. *Computer-assisted Analysis and Model Simplification*. Academic Press, New York, 1981.

[GMM71]   N. S. Goel, S. C. Maitra, and E. W. Montroll. *Nonlinear Models of Interacting Populations*. Academic Press, New York, 1971. Originally printed in *Reviews of Modern Physics*, 43(2): 231-276, 1971.

[GMW81]   P. E. Gill, W. Murray, and M. H. Wright. *Practical Optimization*. Academic Press, San Diego, 1981.

[Goh76]   B. S. Goh. Nonvulnerability of ecosystems in unpredictable environments. *Theoretical Population Biology*, 10:83–95, 1976.

[Goh77]   B. S. Goh. Global stability in many-species systems. *The American Naturalist*, 111:135–143, 1977.

[Goh78]   B. S. Goh. Sector stability of a complex ecosystem model. *Mathematical Biosciences*, 40:157–166, 1978.

[GOH98]   J. C. Geromel, M. C. De Oliveira, and L. Hsu. LMI characterization of structural and robust stability. *Linear Algebra and its Applications*, 285(1-3):69–80, 1998.

[Gol96]   R. M. Golden. *Mathematical Methods for Neural Network Analysis and Design*. MIT Press, Cambridge, Massachusetts, 1996.

[GPS95]   J. C. Geromel, P. L. D. Peres, and S. R. Souza. A convex approach to the mixed $\mathcal{H}_2/\mathcal{H}_\infty$ control problem for discrete-time uncertain systems. *SIAM Journal on Control and Optimization*, 33(6):1816–1833, 1995.

[GR72]   D. D. Givone and R. P. Roesser. Multidimensional iterative circuits–general properties. *IEEE Transactions on Computers*, C-21:1067–1073, 1972.

[GŠ73]   L. T. Grujić and D. D. Šiljak. On stability of discrete composite systems. *IEEE Transactions on Automatic Control*, AC-18(5):522–524, 1973.

[GS86]   J. C. Geromel and A. O. E. Santo. On the robustness of linear continuous time dynamic systems. *IEEE Transactions on Automatic Control*, AC-31(12):1136–1138, 1986.

[Gur95]   L. Gurvits. Stability of discrete linear inclusion. *Linear Algebra and its Applications*, 231:47–85, 1995.

[Gut95]   G. Gutierrez. A solution to the bidimensional global asymptotic stability conjecture. *Annales de L'Institut Henri Poincaré–Analyse Non Lineaire*, 12(6):45–74, 1995.

[Hah67]   W. Hahn. *Stability of Motion*. Springer-Verlag, New York, 1967.

[Har80]   D. J. Hartfiel. Concerning the interior of the D-stable matrices. *Linear Algebra and its Applications*, 30:201–207, 1980.

[Hay99]   S. Haykin. *Neural Networks: A Comprehensive Foundation*. Prentice-Hall, Upper Saddle River, New Jersey, 1999.

[HC96]   L. Hsu and R. R. Costa. Adaptive control with sliding modes: Theory and applications. In *Minicourse Proceedings of the XI Congresso Brasileiro de Automática*, pages 39–60, São Paulo, Brazil, 1996.

[Heb94]   K. J. Hebel. Structural properties of a class of state space digital filters that are free of overflow oscillations. *IEEE Transactions on Signal Processing*, 42(11):3263–3264, 1994.

[Her88]   D. Hershkowitz. Lyapunov diagonal semistability of acyclic matrices. *Linear and Multilinear Algebra*, 22:267–283, 1988.

[Her92]   D. Hershkowitz. Recent directions in matrix stability. *Linear Algebra and its Applications*, 171:161–186, 1992.

[Her98]   D. Hershkowitz. On cones and stability. *Linear Algebra and its Applications*, 276:249–259, 1998.

[Hin93]   T. Hinamoto. 2-D Lyapunov equation and filter design based on the Fornasini-Marchesini second model. *IEEE Transactions on Circuits and Systems–I: Fundamental Theory and Applications*, 40:102–110, 1993.

[HJ85]   R. A. Horn and C. R. Johnson. *Matrix Analysis*. Cambridge University Press, Cambridge, U.K., 1985.

[HJ91]   R. A. Horn and C. R. Johnson. *Topics in Matrix Analysis*. Cambridge University Press, Cambridge, U.K., 1991.

[HKB99]   L. Hsu, E. Kaszkurewicz, and A. Bhaya. Matrix-theoretic conditions for the realizability of sliding manifolds. In *Preprints of the 14th IFAC World Congress*, Beijing, P.R. China, 1999.

[HM93]   A. V. M. Herz and C. M. Marcus. Distributed dynamics in neural networks. *Physical Review E*, 47(3):2155–2161, 1993.

[HM98]   L. Hou and A. N. Michel. Asymptotic stability of systems with saturation constraints. *IEEE Transactions on Automatic Control*, 43(8):1148–1154, 1998.

[Hma94]   A. Hmamed. Comments on "Vector norms as Lyapunov functions for linear systems". *IEEE Transactions on Automatic Control*, 39(12):2522–2522, 1994.

[HNC65]   F. Harary, R. Z. Norman, and D. Cartwright. *Structural Models: An Introduction to the Theory of Directed Graphs*. John Wiley, New York, 1965.

[Hol69]   J. G. Holbrook. The recurrent continuant method of transfer function synthesis. *The Radio and Electronic Engineer*, 38(2):73–79, 1969.

[Hop82]   J. J. Hopfield. Neuronal networks and physical systems with emergent collective computational abilities. *Proceedings of the National Academy of Sciences*, 79:2554–2558, 1982.

[Hop84]   J. J. Hopfield. Neurons with graded response have collective computational properties like those of two-state neurons. *Proceedings of the National Academy of Sciences*, 81:3088–3092, 1984.

[Hor95]   B. G. Horne. Lower bounds for the spectral radius of a matrix. Technical Report NECI 95-14, NEC Research Institute, 1995. ftp://ftp.nj.nec.com/pub/horne/spectral.ps.Z.

[HQ98]   C.-H. Hou and J.-X. Qian. Stability analysis for neural dynamics with time-varying delays. *IEEE Transactions on Neural Networks*, 9(1):221–223, 1998.

[HS49]   D. Hawkins and H. Simon. Note: Some conditions of macroeconomic stability. *Econometrica*, 17:53–56, 1949.

[HS74]   M. W. Hirsch and S. Smale. *Differential Equations, Dynamical Systems and Linear Algebra*. Academic Press, New York, 1974.

[HS85a]   D. Hershkowitz and H. Schneider. Lyapunov diagonal semistability of real H-matrices. *Linear Algebra and its Applications*, 71:119–149, 1985.

[HS85b]   D. Hershkowitz and H. Schneider. Scalings of vector spaces and the uniqueness of Lyapunov scaling factors. *Linear and Multilinear Algebra*, 17:203–226, 1985.

[HS88]   D. Hershkowitz and D. Shasha. Cones of real positive semidefinite matrices associated with matrix stability. *Linear and Multilinear Algebra*, 23:165–181, 1988.

[HS98]   J. Hofbauer and K. Sigmund. *Evolutionary Games and Population Dynamics*. Cambridge Univ. Press, Cambridge, U.K., 1998.

[HSK81]   L. Hsu, L. A. Salgado, and E. Kaszkurewicz. Structural properties in the stability problem of interconnected systems. In *Proceedings of the 2nd IFAC Symposium on Large Scale Systems—Theory and Applications*, pages 67–77, Toulouse, 1981. Pergamon Press Oxford.

[HT85]   J. J. Hopfield and D. W. Tank. Neural computation of decisions in optimization problems. *Biological Cybernetics*, 52:141–152, 1985.

[HT86]   J. J. Hopfield and D. W. Tank. Computing with neural circuits: A model. *Science*, 233:625–633, 1986.

[Hu87]   H. Hu. An algorithm for rescaling a matrix positive definite. *Linear Algebra and its Applications*, 96:131–147, 1987.

[Hu92]   H. Hu. A projective method for rescaling a diagonally stable matrix to be positive definite. *SIAM Journal on Matrix Analysis and Applications*, 13(4):1255–1263, 1992.

[Hur67]   J. Hurt. Some stability theorems for ordinary difference equations. *SIAM Journal on Numerical Analysis*, 4(4):582–596, 1967.

[IOŠ90]   M. Ikeda, Y. Ohta, and D. D. Šiljak. Parametric stability. In G. Conte, A. M. Perdon, and B. Wyman, editors, *New Trends in Systems Theory – Proc. of the Univ. Genova - The Ohio State University Joint Conference*, 1990.

[Jac96]   L. B. Jackson. *Digital Filters and Signal Processing*. Kluwer Academic, Boston, 3rd edition, 1996.

[JE87]   C. R. Johnson and L. Elsner. The relationship between Hadamard and conventional multiplication for positive definite matrices. *Linear Algebra and its Applications*, 92:231–240, 1987.

[Jef74]   C. Jeffries. Qualitative stability and digraphs in model ecosystems. *Ecology*, 55(6):1415–1419, 1974.

[JGN95]   L. Jin, M. M. Gupta, and P. N. Nikiforuk. Stability criteria of discrete-time analog neural networks. In *Proceedings of the 34th IEEE Conference on Decision and Control*, pages 3040–3041, New Orleans, LA, 1995. Paper FA09.

[JKD77]   C. Jeffries, V. Klee, and P. V. D. Driessche. When is a matrix sign stable? *Canadian Journal of Mathematics*, XXIX(2):315–326, 1977.

[JKD87]   C. Jeffries, V. Klee, and P. V. D. Driessche. Qualitative stability of linear systems. *Linear Algebra and its Applications*, 87:1–48, 1987.

[Joh74a]   C. R. Johnson. Second, third, and fourth order D-stability. *Journal of Research of the National Bureau of Standards*, 78B(1):11–13, 1974.

[Joh74b]   C. R. Johnson. Sufficient conditions for D-stability. *Journal of Economic Theory*, 9:53–62, 1974.

[Joh75]   C. R. Johnson. A characterization of the nonlinearity of D-stability. *Journal of Mathematical Economics*, 2:87–91, 1975.

[Joh77]   C. R. Johnson. Price stability in union of markets. In John R. Graef, editor, *Stability of Dynamical Systems: Theory and Applications*, pages 127–135. Marcel Dekker, 1977.

[Joh98]   C. R. Johnson. Olga, matrix theory and the Taussky unification problem. *Linear Algebra and its Applications*, 280:39–49, 1998.

[Jon49]   J. R. E. Jones. A further ecological study of a calcareous stream in the "Black Mountain" district of South Wales. *Journal of Animal Ecology*, 18(2):142–159, 1949.

[Kac85]   T. Kaczorek. *Two-dimensional Linear Systems*. Springer-Verlag, Berlin, 1985.

[Kai80]   T. Kailath. *Linear Systems*. Prentice-Hall, Englewood Cliffs, New Jersey, 1980.

[Kal63]   R. E. Kalman. On a new characterization of linear passive systems. In *Proceedings of the 1st Allerton Conference on Circuit and System Theory*, pages 456–470, University of Illinois, 1963.

[KAS92]   H. Kiendl, J. Adamy, and P. Stelzner. Vector norm Lyapunov functions for linear systems. *IEEE Transactions on Automatic Control*, 37(6):839–842, 1992.

[KB59]   R. E. Kalman and J. Bertram. A unified approach to the theory of sampling systems. *Journal of the Franklin Institute*, 267:405–436, 1959.

[KB89]   E. Kaszkurewicz and A. Bhaya. Qualitative stability of discrete-time systems. *Linear Algebra and its Applications*, 17:65–71, 1989.

[KB92]   E. Kaszkurewicz and A. Bhaya. Comments on "Overflow oscillations in state-space digital filters". *IEEE Transactions on*

Circuits and Systems–II: Analog and Digital Signal Processing, 39(9):675–676, 1992.

[KB93a]   E. Kaszkurewicz and A. Bhaya. A delay-independent robust stability condition for linear discrete-time systems. In *Proceedings of the 32nd IEEE Conference on Decision and Control*, volume 4, pages 2624–2629, San Antonio, TX, 1993.

[KB93b]   E. Kaszkurewicz and A. Bhaya. Robust stability and diagonal Liapunov functions. *SIAM Journal on Matrix Analysis and Applications*, 14(2):508–520, 1993.

[KB94]   E. Kaszkurewicz and A. Bhaya. On a class of globally stable neural circuits. *IEEE Transactions on Circuits and Systems–I: Fundamental Theory and Applications*, 41(2):171–174, 1994.

[KB95]   E. Kaszkurewicz and A. Bhaya. Comments on "Necessary and sufficient condition for absolute stability of neural networks". *IEEE Transactions on Circuits and Systems–I: Fundamental Theory and Applications*, 42(8):497–499, 1995.

[KB99a]   E. Kaszkurewicz and A. Bhaya. Comments on "New sufficient conditions for absolute stability of neural networks". *IEEE Transactions on Circuits and Systems–I: Fundamental Theory and Applications*, 1999. To appear.

[KB99b]   E. Kaszkurewicz and A. Bhaya. The large scale system approach from the perspective of diagonal stability. *Nonlinear Analysis*, 35(1):143–152, 1999.

[KBŠ90]   E. Kaszkurewicz, A. Bhaya, and D. D. Šiljak. On the convergence of parallel asynchronous block-iterative computations. *Linear Algebra and its Applications*, 131:139–160, 1990.

[KH79]   E. Kaszkurewicz and L. Hsu. Stability of nonlinear systems: A structural approach. *Automatica*, 15:609–614, 1979.

[KH80]   E. Kaszkurewicz and L. Hsu. Sign stable structures and passive networks: Stability and realizability. In *Proceedings of the 3rd Brazilian Congress on Automatic Control*, pages 255–266, Rio de Janeiro, 1980.

[KH84a]   E. Kaszkurewicz and L. Hsu. A note on the absolute stability of nonlinear discrete-time systems. *International Journal of Control*, 40(4):867–869, 1984.

[KH84b]   E. Kaszkurewicz and L. Hsu. On two classes of matrices with positive diagonal solutions to the Lyapunov equation. *Linear Algebra and its Applications*, 59:19–27, 1984.

[Kha78]   V. L. Kharitonov. Asymptotic stability of an equilibrium position of a family of systems of linear differential equations. *Differentsial'nye Uravneniya*, 14:2086–2088, 1978.

[Kha82]   H. K. Khalil. On the existence of positive diagonal P such that $PA + A'P < 0$. *IEEE Transactions on Automatic Control*, 27:181–184, 1982.

[Kha92]   H. K. Khalil. *Nonlinear Systems*. Macmillan, New York, 1992.

[KHH84]   E. Kaszkurewicz, A. Hermeto, and L. Hsu. A result on stability of nonlinear discrete time systems and its application to recursive digital filters. In *Proceedings of the 23rd IEEE Conference on Decision and Control*, pages 100–102, Las Vegas, NV, 1984.

[KKKK83]   A. F. Kleptsyn, V. S. Kozyakin, M. A. Krasnosel'skii, and N. A. Kuznetsov. Effect of small synchronization errors on stability of complex systems. I. *Automation and Remote Control*, 44(7):861–867, 1983.

[KKKK84a]   A. F. Kleptsyn, V. S. Kozyakin, M. A. Krasnosel'skii, and N. A. Kuznetsov. Effect of small synchronization errors on stability of complex systems. II. *Automation and Remote Control*, 45(3):309–314, 1984.

[KKKK84b]   A. F. Kleptsyn, V. S. Kozyakin, M. A. Krasnosel'skii, and N. A. Kuznetsov. Effect of small synchronization errors on stability of complex systems. III. *Automation and Remote Control*, 45(8):1014–1018, 1984.

[KLMK77]   S.-Y. Kung, B. C. Lévy, M. Morf, and T. Kailath. New results in 2-D systems theory, part II: 2-D state-space models–realization and the notions of controllability, observability and minimality. *Proceedings of the IEEE*, 65(6):945–961, 1977.

[KLS89]   M. A. Krasnosel'skii, J. A. Lifshits, and A. V. Sobolev. *Positive Linear Systems: The Method of Positive Operators*. Heldermann Verlag, Berlin, 1989.

[Koi94]   P. Koiran. Dynamics of discrete-time, continuous-state Hopfield networks. *Neural Computation*, 6:459–468, 1994.

[KR83]   E. Kaszkurewicz and P. Resende. On the absolute stability of linearly interconnected non-linear systems. In *Proceedings of the 3rd IFAC/IFORS Symposium on Large Scale Systems: Theory and Applications*, pages 233–238, Warsaw, Poland, 1983.

[Kra57]   G. M. Kranc. Compensation of an error sampled system by a multirate controller–part ii. *AIEE Transactions*, 76:149–158, 1957.

[Kra91]   J. F. B. M. Kraaijevanger. A characterization of Lyapunov diagonal stability using Hadamard products. *Linear Algebra and its Applications*, 151:245–254, 1991.

[Kua93]   Y. Kuang. *Delay Differential Equations: With Applications in Population Dynamics*. Academic Press, San Diego, 1993.

[Kun93]   S. Y. Kung. *Digital Neural Networks*. Prentice-Hall, Englewood Cliffs, New Jersey, 1993.

[Kun94]   P. Kundur. *Power System Stability and Control*. McGraw-Hill, New York, 1994.

[KYF89]   S. R. Kolla, R. K. Yedavalli, and J. B. Farison. Robust stability bounds on time-varying perturbations for state-space models of linear discrete-time systems. *International Journal of Control*, 50(1):151–159, 1989.

[LA80]   R. M. Lewis and B. D. O. Anderson. Necessary and sufficient conditions for delay-independent stability of linear autonomous systems. *IEEE Transactions on Automatic Control*, AC-20(4):735–739, 1980.

[Lad76]   G. S. Ladde. Cellular systems II, stability of compartmental systems. *Mathematical Biosciences*, 30:1–21, 1976.

[LaS76]   J. P. LaSalle. *The Stability of Dynamical Systems*. SIAM, Philadelphia, 1976.

[LaS86]   J. P. LaSalle. *The Stability and Control of Discrete Processes*. Springer-Verlag, New York, 1986.

[LdS97]   X. Li and C. E. de Souza. Delay-dependent robust stability and stabilization of uncertain linear delay systems: A linear matrix inequality approach. *IEEE Transactions on Automatic Control*, 42(8):1144–1148, 1997.

[Lew99]   I. Lewkowicz. Convex invertible cones of matrices: a unified framework for the equations of Sylvester, Lyapunov and Riccati. *Linear Algebra and its Applications*, 286:107–133, 1999.

[Lia49]   A. M. Liapunov. *Problème Général de la Stabilité du Mouvement*. Princeton University Press, Princeton, 1949. *Annals of Mathematics Studies no. 7*. French translation of 1892 paper in Russian. Also published as Lyapunov Centenary Issue of the *International Journal of Control*, 55(3), 1992.

[LL61]   J. P. LaSalle and S. Lefschetz. *Stability by Lyapunov's Direct Method with Applications*. Academic Press, New York, 1961.

[LL69]   V. Lakshmikantham and S. Leela. *Differential and Integral Inequalities, vol. I*. Academic Press, New York, 1969.

[LM86]   B. Lubachevsky and D. Mitra. A chaotic asynchronous algorithm for computing the fixed point of a nonnegative matrix of unit spectral radius. *Journal of the Association of Computing Machinery*, 33(1):130–150, 1986.

[LM94]   D. Liu and A. N. Michel. *Dynamical Systems with Saturation Nonlinearities*. Springer-Verlag, London, 1994.

[LNF91]   D. K. Le, O. D. I. Nwokah, and E. Frazho. Multivariable decentralized integral controllability. *International Journal of Control*, 54(2):481–496, 1991.

[Log87]   D. O. Logofet. On the hierarchy of subsets of stable matrices. *Soviet Mathematics Doklady*, 34(2):247–250, 1987.

[Log89]   D. O. Logofet. Do there exist diagonally stable matrices without dominating diagonal? *Soviet Mathematics Doklady*, 38(1):113–115, 1989.

[LPR98]   K. Loskot, A. Polański, and R. Rudnicki. Further comments on "Vector norms as Lyapunov functions for linear systems". *IEEE Transactions on Automatic Control*, 43(2):289–291, 1998.

[LT85]   P. Lancaster and M. Tismenetsky. *The Theory of Matrices*. Academic Press, San Diego, 1985.

[Lu94]   W.-S. Lu. On a Lyapunov approach to stability analysis of 2-D digital filters. *IEEE Transactions on Circuits and Systems–I: Fundamental Theory and Applications*, 41(10):665–669, 1994.

[LW95]   J. C. Lagarias and Y. Wang. The finiteness conjecture for the generalized spectral radius of a set of matrices. *Linear Algebra and its Applications*, 214:17–42, 1995.

[LW98a]   X.-B. Liang and L.-D. Wu. A comment on "Comments on 'Necessary and sufficient condition for absolute stability of neural networks'". *IEEE Transactions on Circuits and Systems–I: Fundamental Theory and Applications*, 45(5):594, 1998.

[LW98b]   X.-B. Liang and L.-D. Wu. New sufficient conditions for absolute stability of neural networks. *IEEE Transactions on Circuits and Systems–I: Fundamental Theory and Applications*, 45(5):584–586, 1998.

[LW99]   X.-B. Liang and J. Wang. A proof of Kaszkurewicz and Bhaya's conjecture on absolute stability of neural networks in the two-neuron case. *IEEE Transactions on Circuits and Systems–I: Fundamental Theory and Applications*, 1999. to appear.

[MADF98]   M. W. McConley, B. D. Appleby, M. A. Dahleh, and E. Feron. Computational complexity of Lyapunov stability analysis problems for a class of nonlinear systems. *SIAM Journal on Control and Optimization*, 36(6):2176–2193, 1998.

[Man88]   M. Mansour. Sufficient conditions for asymptotic stability of interval matrices. *International Journal of Control*, 47(6):1973, 1988.

[Mar66]   M. Marden. *Geometry of Polynomials*. American Mathematical Society, Providence, Rhode Island, 1966.

[Mat92]   K. Matsuoka. Stability conditions for nonlinear continuous neural networks with asymmetric connection weights. *Neural Networks*, 5:495–500, 1992.

[May66]   J. S. Maybee. New generalizations of Jacobi matrices. *SIAM Journal on Applied Mathematics*, 14(5):1033–1039, 1966.

[May74]   J. S. Maybee. Combinatorially symmetric matrices. *Linear Algebra and its Applications*, 8:529–537, 1974.

[MB96]   F. C. Mota and A. Bhaya. On D-stable and D-semistable matrices and the structured singular value. In *Proceedings of the 35th IEEE Conference on Decision and Control*, volume 2, pages 1284–1286, Kobe, Japan, 1996.

[Mer94]   R. M. Mersereau. *Digital Filtering*. John Wiley, New York, 1994.

[Met45]   L. A. Metzler. Stability of multiple markets: the Hicks conditions. *Econometrica*, 13:227–292, 1945.

[MFP89]   A. N. Michel, J. A. Farrell, and W. Porod. Qualitative analysis of neural networks. *IEEE Transactions on Circuits and Systems*, 36(2):229–243, 1989.

[MH78]   P. J. Moylan and D. J. Hill. Stability criteria for large-scale systems. *IEEE Transactions on Automatic Control*, AC-23(2):143–149, 1978.

[MI75]   N. H. McClamroch and G. D. Ianculescu. Global stability of two
            linearly interconnected nonlinear subsystems. *IEEE Transactions
            on Automatic Control*, 20:678–682, 1975.

[Mie74]   J. C. Miellou. Itérations chaotiques à retards. *Comptes Rendus
            Hebdomadaires des Seances de L'Academie Des Sciences Serie A*,
            278(14):957–960, 1974.

[Min88]   H. Minc. *Nonnegative matrices*. John Wiley, New York, 1988.

[MK86]   T. Mori and H. Kokame. A necessary and sufficient condition
            for stability of linear discrete systems with parameter variations.
            Journal of the Franklin Institute, 321(3):135–138, 1986.

[MK87]   T. Mori and H. Kokame. Convergence property of interval ma-
            trices and interval polynomials. *International Journal of Control*,
            45(2):481–484, 1987.

[MKB92]   F. Mota, E. Kaszkurewicz, and A. Bhaya. Robust stabilization of
            time-varying discrete interval systems. In *Proceedings of the 31st
            IEEE Conference on Decision and Control*, volume 1, pages 341–
            346, Tucson, AZ, 1992.

[MM64]   M. Marcus and H. Minc. *A Survey of Matrix Theory and Matrix
            Inequalities*. Prindle, Weber and Schmidt, Boston, MA, 1964.

[MM77]   A. N. Michel and R. K. Miller. *Qualitative Analysis of Large Scale
            Dynamical Systems*. Academic Press, New York, 1977.

[MM85]   A. N. Michel and R. K. Miller. Stability analysis of discrete-time
            interconnected systems via computer generated Lyapunov functions
            with applications to digital filters. *IEEE Transactions on Circuits
            and Systems*, CAS-32(8):737–753, 1985.

[MM97]   A. P. Molchanov and M. V. Morozov. Sufficient conditions for ro-
            bust stability of linear nonstationary control systems with periodic
            interval constraints. *Automation and Remote Control*, 58(1, part
            2):82–87, 1997.

[MMK98]   Y. Mori, T. Mori, and Y. Kuroe. Classes of discrete linear systems
            having common quadratic Lyapunov functions. In *Proceedings of
            the 1995 American Control Conference*, pages 3364–3365, 1998.

[MMR78]   W. L. Mills, C. T. Mullis, and R. A. Roberts. Digital filter re-
            alizations without overflow oscillations. *IEEE Transactions on
            Acoustics, Speech, and Signal Processing*, ASSP-26(4):334–338,
            1978.

[Moy77]   P. J. Moylan. Matrices with positive principal minors. *Linear
            Algebra and its Applications*, 17:53–58, 1977.

[MP72]   A. N. Michel and D. W. Porter. Stability analysis of composite sys-
            tems. *IEEE Transactions on Automatic Control*, AC-17:222–226,
            1972.

[MP89]   A. P. Molchanov and Y. S. Pyatnitskii. Criteria of asymptotic sta-
            bility of differential and difference inclusions encountered in control
            theory. *Systems and Control Letters*, 13:59–64, 1989.

[MQ69]   J. S. Maybee and J. P. Quirk. Qualitative problems in matrix
            theory. *SIAM Review*, 11:30–51, 1969.

[MSV96]   R. Mantri, A. Saberi, and V. Venkatasubramanian. Stability analysis of continuous time planar systems with state saturation nonlinearity. In *Proceedings of the 1996 IEEE International Symposium on Circuits and Systems*, pages 60–63, Atlanta, GA, 1996.

[Mur93]   J. D. Murray. *Mathematical Biology*. Springer-Verlag, Berlin, 1993.

[MW89]   C. M. Marcus and R. M. Westervelt. Dynamics of iterated-map neural networks. *Physical Review A*, 40(1):501–504, 1989.

[MY60]   L. Markus and H. Yamabe. Global stability criteria for differential systems. *Osaka Journal of Mathematics*, 12:305–317, 1960.

[MZ89]   M. Morari and E. Zafiriou. *Robust Process Control*. Prentice-Hall, Englewood Cliffs, New Jersey, 1989.

[NB94]   K. S. Narendra and J. Balakrishnan. A common Lyapunov function for stable LTI systems with commuting A-matrices. *IEEE Transactions on Automatic Control*, 39:2469–2471, 1994.

[Nik68]   H. Nikaido. *Convex Structures and Economic Theory*. Academic Press, New York, 1968.

[NT73]   K. Narendra and J. Taylor. *Frequency Domain Criteria for Absolute Stability*. Academic Press, New York, 1973.

[OF97]   T. Ooba and Y. Funahashi. Stability robustness for linear state space models. *Systems and Control Letters*, 29:191–196, 1997.

[OGH99]   M. C. De Oliveira, J. C. Geromel, and L. Hsu. LMI characterization of structural and robust stability: the discrete-time case. *Linear Algebra and its Applications*, 296(1-3):27–38, 1999.

[Oku78]   K. Okuguchi. Matrices with dominant diagonal blocks and economic theory. *Journal of Mathematical Economics*, 5:43–52, 1978.

[OR70]   J. M. Ortega and W. C. Rheinboldt. *Iterative Solutions of Non-linear Equations in Several Variables*. Academic Press, New York, 1970.

[Ort72]   J. M. Ortega. *Numerical Analysis—A Second Course*. Academic Press, New York, 1972.

[Ort73]   J. M. Ortega. Stability of difference equations and convergence of iterative processes. *SIAM Journal on Numerical Analysis*, 10(2):268–282, 1973.

[OS62]   A. Ostrowski and H. Schneider. Some theorems on the inertia of general matrices. *Journal of Mathematical Analysis and Applications*, 4:72–84, 1962.

[OŠ85]   Y. Ohta and D. D. Šiljak. Overlapping block diagonal dominance and existence of Liapunov functions. *Journal of Mathematical Analysis and Applications*, 112:396–410, 1985.

[Ost55]   A. Ostrowski. Determinanten mit überwiegender Hauptdiagonal und die absolute Konvergenz von linearen Iterationsprozessen. *Commentarii Mathematici Helvetici*, 30:175–210, 1955.

[Ost61]   A. Ostrowski. On some metrical properties of operator matrices and matrices partitioned into blocks. *Journal of Mathematical Analysis and Applications*, 2:161–209, 1961.

[OT51]   A. Ostrowski and O. Taussky. On the variation of the determinant of a positive definite matrix. *Neder. Akad. Wet. Proc.*, A54:383–386, 1951.

[Pai81]   M. A. Pai. *Power System Stability*. Elsevier North-Holland, Amsterdam, The Netherlands, 1981.

[Par62]   P. C. Parks. A new proof of the Routh-Hurwitz stability criterion using the second method of Lyapunov. *Proceedings of the Cambridge Philosophical Society*, 58(4):694–702, 1962.

[Per69]   S. K. Persidskii. Problem of absolute stability. *Automation and Remote Control*, 12:1889–1895, 1969.

[PMR70]   M. A. Pai, M. A. Mohan, and J. G. Rao. Power system transient stability regions usings Popov's method. *IEEE Transactions on Power Apparatus and Systems*, PAS-89(5):788–801, 1970.

[Pol71]   E. Polak. *Computational Methods in Optimization: A Unified Approach*. Academic Press, New York, 1971.

[Pol82]   B. J. Poljak. Convergence and convergence rate of iterative stochastic algorithms. *Automation and Remote Control*, 42(12):83–94, 1982.

[Pol97]   A. Polański. Lyapunov function construction by linear programming. *IEEE Transactions on Automatic Control*, 42(7):1013–1016, 1997.

[PR67]   A. A. Piontkovskii and L. D. Rutkovskaya. Investigation of certain stability theory problems by the vector Lyapunov function method. *Automation and Remote Control*, 28(10):1422–1429, 1967.

[PR93]   S. Poljak and J. Rohn. Checking robust nonsingularity is NP-hard. *Mathematics of Control, Signals and Systems*, 6:1–9, 1993.

[PS93]   T. Parthasarathy and M. Sabatini. Some new results on the global asymptotic stability Jacobian conjecture. *Bulletin of the Polish Academy of Sciences-Mathematics*, 41(3):221–228, 1993.

[PT81]   R. V. Patel and M. Toda. Quantitative measures of robustness for linear multivariable systems. In *Proceedings of the Joint Automatic Control Conference*, San Francisco, 1981. paper TP8.

[PTS77]   R. V. Patel, M. Toda, and S. Sridhar. Robustness of linear quadratic state feedback designs in the presence of uncertainty. *IEEE Transactions on Automatic Control*, AC-22(6):945–949, 1977.

[QD86]   L. Qiu and E. J. Davison. New perturbation bounds for the robust stability of linear state space models. In *Proceedings of the 25th IEEE Conference on Decision and Control*, volume TA-1, pages 751–755, Athens, Greece, 1986.

[QR65]   J. P. Quirk and R. Ruppert. Qualitative economics and the stability of equilibrium. *Review of Economic Studies*, 32:311–326, 1965.

[RCDP93]   M. A. Rotea, M. Corless, D. Da, and I. R. Petersen. Systems with structured uncertainty: relations between quadratic and robust stability. *IEEE Transactions on Automatic Control*, 35(5):799–803, 1993.

[Red85a]   R. Redheffer. Volterra multipliers - I. *SIAM Journal on Algebraic and Discrete Methods*, 6(4):592–611, 1985.

[Red85b]   R. Redheffer. Volterra multipliers - II. *SIAM Journal on Algebraic and Discrete Methods*, 6(4):612–623, 1985.

[Reg92]   P. A. Regalia. On finite precision Lyapunov functions for comparison matrices. *IEEE Transactions on Automatic Control*, 37(10):1640–1644, 1992.

[RHK82]   J. I. Rego, L. Hsu, and E. Kaszkurewicz. Synthesis of active RC filters via state space realization in arrow form. In *Proceedings of the IEEE International Symposium on Circuits and Systems*, pages 1078–1081, Rome, 1982.

[RHL77]   N. Rouche, P. Habets, and M. Laloy. *Stability Theory by Liapunov's Direct Method*. Springer-Verlag, New York, 1977.

[Rit89]   J. H. F. Ritzerfeld. A condition for the overflow stability of second-order digital filters that is satisfied by all scaled state-space structures using saturation. *IEEE Transactions on Circuits and Systems*, CAS-36(8):1049–1057, 1989.

[RKH86]   P. Resende, E. Kaszkurewicz, and L. Hsu. On the absolute stability of multivariable nonlinear systems using a block tridiagonal state-space realization. *International Journal of Control*, 43(2):635–643, 1986.

[RM87]   R. A. Roberts and C. T. Mullis. *Digital Signal Processing*. Addison-Wesley, Reading, Massachusetts, 1987.

[Rob69]   F. Robert. Bloc *H*-matrices et convergence des methods iteratives classiques par blocs. *Linear Algebra and its Applications*, 2:223–265, 1969.

[Rob76]   F. Robert. Contraction en norme vectorielle: convergence d'itérations chaotiques pour des equations non linéaires de point fixe à plusieurs variables. *Linear Algebra and its Applications*, 13:19–35, 1976.

[Roe75]   R. P. Roesser. A discrete state-space model for linear image processing. *IEEE Transactions on Automatic Control*, AC-20:1–10, 1975.

[Ros63]   H. H. Rosenbrock. A Lyapunov function with applications to some nonlinear physical systems. *Automatica*, 1:31–53, 1963.

[Ros88]   T. Roska. Some qualitative aspects of neural computing systems. In *Proceedings of the 1988 IEEE International Symposium on Circuits and Systems*, pages 751–754, 1988.

[Roy68]   H. L. Royden. *Real Analysis*. Macmillan, New York, 2nd edition, 1968.

[RW84]   R. Redheffer and W. Walter. Solution of the stability problem for a class of generalized Volterra prey-predator systems. *Journal of Differential Equations*, 52:245–263, 1984.

[RZ81]   R. Redheffer and Z. Zhiming. Global asymptotic stability for a class of many-variable Volterra prey-predator systems. *Nonlinear Analysis: Theory, Methods and Applications*, 5:1309–1329, 1981.

[SB82]   I. H. Suh and Z. Bien. On stabilization by local state feedback for for discrete-time large-scale systems with delays in interconnections. *IEEE Transactions on Automatic Control*, AC-27(3):744–746, 1982.

[SB98]   Y. Su and A. Bhaya. On the Bose-Trautman condition for the stability of 2-D linear models. *IEEE Transactions on Signal Processing*, 46(7):2069–2070, 1998.

[SBK98]   Y. Su, A. Bhaya, and E. Kaszkurewicz. A general asynchronous block iterative model with related convergence conditions. *Journal of Computational and Applied Mathematics*, 91(2):261–273, 1998.

[SBKK98] Y. Su, A. Bhaya, E. Kaszkurewicz, and V. S. Kozyakin. Further results on the stability of asynchronous discrete-time linear systems. *Linear Algebra and its Applications*, 281(1-3):11–24, 1998.

[Sch56]   H. R. Schwarz. Ein Verfahren zur Stabilitatsfrage bei Matrizen Eigenwerteproblem. *Zeitschrift für Angewandte Mathematik und Physik*, 7:473–500, 1956.

[Sej86]   T. J. Sejnowski. Open questions about computation in cerebral cortex. In *Parallel Distributed Processing: Explorations in the Microstructure of Cognition*, pages 372–389. M.I.T. Press, Cambridge, MA, 1986.

[Sen72]   E. Seneta. *Nonnegative Matrices*. John Wiley, New York, 1972.

[SH91]   B. Shafai and C. V. Hollot. Robust nonnegative stabilization of interval discrete systems. In *Proceedings of the 30th IEEE Conference on Decision and Control*, volume 1, pages 49–51, Brighton, 1991.

[Shi99]   M.-H. Shih. Simultaneous Schur stability. *Linear Algebra and its Applications*, 287(1-3):323–336, 1999.

[Šil78]   D. D. Šiljak. *Large Scale Dynamic Systems: Stability and Structure*. Elsevier North-Holland, Amsterdam, The Netherlands, 1978.

[Šil89]   D. D. Šiljak. Parameter space methods for robust control design: A guided tour. *IEEE Transactions on Automatic Control*, 34(7):674–688, 1989.

[Šil91]   D. D. Šiljak. *Decentralized Control of Complex Systems*. Academic Press, San Diego, 1991.

[SKU$^+$75] O. Saito, K. Koizumi, M. Udo, M. Sato, H. Mukae, and T. Tsuji. Security monitoring systems including fast transient stability studies. *IEEE Transactions on Power Apparatus and Systems*, PAS-94(5):1789–1805, 1975.

[SL83]   Y. M. Svirezhev and D. O. Logofet. *Stability of Biological Communities*. Mir Publishers, Moscow, 1983.

[Smi88]   H. L. Smith. Systems of ordinary differential equations which generate an order preserving flow. A survey of results. *SIAM Review*, 30(1):87–113, 1988.

[Smi95]   H. L. Smith. *Monotone Dynamical Systems: An Introduction to the Theory of Competitive and Cooperative Systems*. American Mathematical Society, Providence, Rhode Island, 1995.

[Son98]   E. Sontag. *Mathematical Control Theory: Deterministic Finite Dimensional Systems*. Springer-Verlag, New York, 1998.

[SP94]   D. Shevitz and B. Paden. Lyapunov stability theory of nonsmooth systems. *IEEE Transactions on Automatic Control*, 39(9):1910–1914, 1994.

[SŠ88]   M. E. Sezer and D. D. Šiljak. Robust stability of discrete systems. *International Journal of Control*, 48(5):2055–2063, 1988.

[SŠ91]   M. E. Sezer and D. D. Šiljak. Nested epsilon decompositions of linear systems: Weakly coupled and overlapping blocks. *SIAM Journal on Matrix Analysis and Applications*, 12:521–533, 1991.

[SŠ94]   M. E. Sezer and D. D. Šiljak. On stability of interval matrices. *IEEE Transactions on Automatic Control*, 39(2):368–371, 1994.

[SŠ97]   D. M. Stipanović and D. D. Šiljak. Convex M-matrices and polytopic dynamic systems. In *Proceedings of the 36th IEEE Conference on Decision and Control*, volume 5, pages 4366–4368, San Diego, CA, 1997.

[ST91]   P.S. Souza and S.N. Talukdar. Genetic algorithms in asynchronous teams. In *Proceedings of the Fourth International Conference on Genetic Algorithms*, pages 392–397, California, 1991.

[Str88]   G. Strang. *Linear Algebra and its Applications*. Harcourt Brace Jovanovich, San Diego, 3rd edition, 1988.

[SVD97]   J. A. K. Suykens, J. Vandewalle, and B. L. R. DeMoor. NL_q theory: Checking and imposing stability of recurrent neural networks for nonlinear modeling. *IEEE Transactions on Signal Processing*, 45(11):2682–2691, 1997.

[SW62]   J. Stoer and C. Witzgall. Transformations by diagonal matrices in a normed space. *Numerische Mathematik*, 4:158–171, 1962.

[SW69]   I. W. Sandberg and A. N. Willson, Jr. Some network-theoretic properties of nonlinear dc transistor networks. *Bell System Technical Journal*, 48:1293–1311, 1969.

[SW98]   M.-H. Shih and J.-W. Wu. On a discrete version of the Jacobian conjecture of dynamical systems. *Nonlinear Analysis: Theory, Methods and Applications*, 34(5):779–789, 1998.

[SWC91]   F. Salam, Y. Wang, and M. Choi. On the analysis of dynamic feedback neural nets. *IEEE Transactions on Circuits and Systems*, 38(2):196–201, 1991.

[Szy98]   D. B. Szyld. Partially asynchronous overlapping iterations. *Computational and Applied Mathematics*, 17(1):101–115, 1998.

[TAT78]   Y. Takeuchi, N. L. Adachi, and H. Tokumaru. The stability of generalized Volterra equations. *Journal of Mathematical Analysis and Applications*, 62:453–473, 1978.

[Tau49]   O. Taussky. A recurring theorem on determinants. *American Mathematical Monthly*, 56:672–676, 1949.

[Tau61]   O. Taussky. A generalization of a theorem by Lyapunov. *Journal of the Society for Industrial and Applied Mathematics*, 9:640–643, 1961.

[Tau64]   O. Taussky. Matrices C with $C^n \to 0$. *Journal of Algebra*, 1:5–10, 1964.

[TB97a]   J. N. Tsitsiklis and V. D. Blondel. The Lyapunov exponent and joint spectral radius of pairs of matrices are hard–when not impossible–to compute and to approximate. *Mathematics of Control, Signals and Systems*, 10(1):31–40, 1997.

[TB97b]   J. N. Tsitsiklis and V. D. Blondel. Lyapunov exponents of pairs of matrices, a correction (vol 10, pg 31, 1997). *Mathematics of Control, Signals and Systems*, 10(4):381, 1997.

[TBT90]   P. Tseng, D. P. Bertsekas, and J. N. Tsitsiklis. Partially asynchronous, parallel algorithms for network flow and other problems. *SIAM Journal on Control and Optimization*, 28(3):678–710, 1990.

[TG86]   M. Takeda and J. W. Goodman. Neural networks for computation: number representations and programming complexity. *Applied Optics*, 25(18):3033–3046, 1986.

[TH86]   D. W. Tank and J. J. Hopfield. Simple neural optimization networks: An A/D converter, signal decision circuit, and a linear programming network. *IEEE Transactions on Circuits and Systems*, CAS-33:533–541, 1986.

[TPM83]   S. N. Talukdar, S. S. Pyo, and R. Mehrotra. Designing algorithms and assignments for distributed processing. Research Project 1764-3 Report EPRI EL-3317, Electric Power Research Institute, 1983.

[TV90]   A. Tesi and A. Vicino. Robust stability of state-space models with structured uncertainties. *IEEE Transactions on Automatic Control*, 35(2):191–195, 1990.

[ÜD92]   A. Üresin and M. Dubois. Asynchronous iterative algorithms: Models and convergence. In L. Kronsjö and D. Shumsheruddin, editors, *Advances in Parallel Algorithms*, chapter 10, pages 302–342. Oxford, London, 1992.

[Utk78]   V. I. Utkin. *Sliding Modes and Their Applications in Variable Structure Systems*. Mir, Moscow, 1978.

[Utk92]   V. I. Utkin. *Sliding Modes in Control and Optimization*. Springer-Verlag, Berlin, 1992.

[Van83]   A. Vanelli. *Solution Techniques for 0-1 Indefinite Quadratic Programming Problems with Applications to Decomposition*. PhD thesis, University of Waterloo, 1983.

[Var62]   R. S. Varga. *Matrix Iterative Analysis*. Prentice-Hall, Englewood Cliffs, New Jersey, 1962.

[VFK92]   M. H. M. Vale, D. M. Falcão, and E. Kaszkurewicz. Electrical power network decomposition for parallel computations. In *Proceedings of the 1992 IEEE International Symposium on Circuits and Systems*, volume 6, pages 2761–4, San Diego, CA, 1992.

[Vid78] M. Vidyasagar. On matrix measures and convex Liapunov functions. *Journal of Mathematical Analysis and Applications*, 62(1):90–103, 1978.

[Vid86] M. Vidyasagar. New directions of research in nonlinear system theory. *Proceedings of the IEEE*, 74(8):1060–1091, 1986.

[Vid92] M. Vidyasagar. Improved neural networks for analog-to-digital conversion. *Circuits, Systems and Signal Processing*, 11(3):387–398, 1992.

[Vid93] M. Vidyasagar. *Nonlinear Systems Analysis*. Prentice-Hall, Englewood Cliffs, New Jersey, 2nd edition, 1993.

[VL87]   P. P. Vaidyanathan and V. Liu. An improved sufficient condition for absence of limit cycles in digital filters. *IEEE Transactions on Circuits and Systems*, CAS-34(3):319–322, 1987.

[Vol31] V. Volterra. *Leçons sur la théorie mathématique de la lutte pour la vie*. Gauthier-Villars, Paris, 1931.

[WB78]   A. Wörz-Busekros. Global stability in ecological systems with continuous time delay. *SIAM Journal on Applied Mathematics*, 35:123–134, 1978.

[WH94]   J. W. Wu and K. S. Hong. Delay-independent exponential stability criteria for time-varying discrete delay systems. *IEEE Transactions on Automatic Control*, 39:811–814, 1994.

[Whi86] R. E. White. Parallel algorithms for nonlinear problems. *SIAM Journal on Algebraic and Discrete Methods*, 7(1):137–149, 1986.

[Wie82] G. Wiener. Some structural results concerning certain classes of qualitative matrices and their inverses. *Linear Algebra and its Applications*, 48:161–175, 1982.

[Wil70] J. L. Willems. *Stability Theory of Dynamical Systems*. Thomas Nelson, London, 1970.

[Wil76] J. C. Willems. Liapunov functions for diagonally dominant systems. *Automatica*, 12:519–523, 1976.

[Wil86] J. L. Willems. Structural controllability and observability. *Systems and Control Letters*, 8:5–12, 1986.

[Wim98] H. K. Wimmer. Diagonal matrix solutions of a discrete-time Lyapunov inequality. *IEEE Transactions on Automatic Control*, 43(3):442–445, 1998.

[WML94] K. Wang, A. N. Michel, and D. Liu. Necessary and sufficient condition for the Hurwitz and Schur stability of interval matrices. *IEEE Transactions on Automatic Control*, 39(6):1251–1255, 1994.

[Xin87] L. X. Xin. Stability of interval matrices. *International Journal of Control*, 45:203–210, 1987.

[YA88]   J. Yong and A. Arapostathis. Stabilization of discrete-time linear systems with a time-delay in the feedback loop. *International Journal of Control*, 48(4):1475–1485, 1988.

[Yed85]   R. K. Yedavalli. Perturbation bounds for robust stability in linear state space models. *International Journal of Control*, 42:1507–1517, 1985.

[YF90]   C.-C. Yu and M. K. H. Fan. Decentralized integral controllability and D-stability. *Chemical Engineering Science*, 45(11):3299–3309, 1990.

[YL86]   R. K. Yedavalli and Z. Liang. Reduced conservatism in stability robustness bounds by state transformation. *IEEE Transactions on Automatic Control*, AC-31:863–866, 1986.

[YN89]   E. Yaz and X. Niu. Stability robustness of linear discrete-time systems in the presence of uncertainties. *International Journal of Control*, 50(1):173–182, 1989.

[You71]   D. M. Young. *Iterative Solutions of Large Linear Systems*. Academic Press, London, 1971.

[ZD89]   C. S. Zhou and T. L. Deng. Stability analysis of gray discrete-time system. *IEEE Transactions on Automatic Control*, 34(2):173–175, 1989.

[ZDG96]   K. Zhou, J. C. Doyle, and K. Glover. *Robust and Optimal Control*. Prentice-Hall, Upper Saddle River, New Jersey, 1996.

[Zie68]   H. Ziegler. *Principles of Structural Stability*. Blaisdell, Waltham, Massachusetts, 1968.

[Zió90]   M. Ziółko. Application of Lyapunov functionals to studying stability of linear hyperbolic systems. *IEEE Transactions on Automatic Control*, 35(10):1173–1176, 1990.

[ZK87]   K. Zhou and P. P. Khargonekar. Stability robustness bounds for linear state-pace models with structured uncertainty. *IEEE Transactions on Automatic Control*, AC-32:621–623, 1987.

[ZŠ94]   A. I. Zečević and D. D. Šiljak. A block-parallel Newton method via overlapping epsilon decompositions. *SIAM Journal on Matrix Analysis and Applications*, 15(3):824–844, 1994.

Index